‘With Our Backs to the Ocean’

ENVIRONMENTAL HISTORIES OF THE NORTH ATLANTIC WORLD

Previously published volumes in this series are listed at the back of the book.

Volume 5

‘With Our Backs to the Ocean’

Land, Lordship, Climate Change, and Environment in the North-West European Past

Essays in Memory of Alasdair Ross

Edited by

Richard Oram

BREPOLS

British Library Cataloguing in Publication Data

A catalogue record for this book is available from the British Library.

D/2023/0095/140
ISBN: 978-2-503-59699-0
e-ISBN: 978-2-503-59701-0
DOI: 10.1484/M.EHNAW-EB.5.126283
ISSN: 2565-8131
e-ISSN: 2565-9502

Printed in the EU on acid-free paper

Ally Ross at Kilwinning Abbey
(Photo: Michael Penman).

Contents

List of Illustrations

Simon Taylor

Philip Slavin

Richard C. Hoffmann

Abbreviations

AASS – *Acta Sanctorum*

AC – 'Annals of Connacht'

AI – 'Annals of Inishfallen'

ALC – 'Annals of Loch Cé'

ALI – *Acts of the Lords of the Isles 1336–1493*

AU – 'Annala Uladh'

BPNR Database – *The Berwickshire Place-Name Resource*

DMLBS – *Dictionary of Medieval Latin from British Sources*

DOST – *A Dictionary of the Older Scottish Tongue (up to 1700)*

eDIL – Electronic Dictionary of the Irish Language

MGH – Monumenta Germaniae Historica

PoMS – *People of Medieval Scotland*

RMS – *Registrum Magni Sigilli Regum Scottorum/*
Register of the Great Seal of Scotland

RPS – *The Records of the Parliaments of Scotland to 1707*

SDBM – *Schoenberg Database of Manuscripts*

WSA – Wiltshire and Swindon Archives

Figure 1.1. Loch A'an from Carn Etchachan (photo by author).

Introduction: Alasdair Ross and the Environmental History of Medieval and Early Modern Scotland

Richard Oram
History, Heritage, and Politics, University of Stirling

Loch A'an, Loch A'an, hoo deep ye lie!
Tell nane yer depth and nane shall I.
Bricht though yer deepmaist pit may be,
Ye'll haunt me till the day I dee.
Bricht, an' bricht, an' bricht as air,
Ye'll haunt me noo for evermair.[1]

It is rare testimony to the academic achievement of any scholar for their contribution to the production of useful knowledge to be recognized in a volume of essays written by colleagues from both sides of the Atlantic in Canada, the USA, Great Britain and Scandinavia. It is even greater testimony to the reach and impact of that achievement to win recognition from a group of scholars whose work explores a wide geographical range along the eastern shores of the North Atlantic and which spans nearly two millennia of human/environmental interaction in that region. Perhaps the greatest marker which it presents is in its acknowledgement of the scale of the contribution made, which is itself all the more impressive in that it was the output of little more than ten years of academic endeavour. These essays, written by friends and colleagues, former students and research collaborators, are celebratory of the career of Dr Alasdair

[1] These lines are from Alasdair's own 1934 copy of 'Loch Avon' in Nan Shepherd's anthology, *In the Cairngorms*, and are imbued with the enduring love that he felt for those mountains and the loch at their heart, above whose head rear the cliffs where he spent many days in his teens and twenties climbing with his friends. The poem was read at his funeral.

'With Our Backs to the Ocean': Land, Lordship, Climate Change, and Environment in the North-West European Past. Essays in Memory of Alasdair Ross, ed. by Richard Oram (Turnhout: Brepols, 2023)
EHNAW 5, pp. 1–14 BREPOLS PUBLISHERS 10.1484/M.EHNAW-EB.5.131268

'Ally' Ross, Reader in Medieval Environmental History and former Director of the Centre for Environment, Heritage and Policy at the University of Stirling, whose death in August 2017 deprived Scottish and wider European Medieval Environmental History of an outstanding scholar who achieved much in his short academic career, but who was poised to deliver so much more. The papers presented here are no random gathering of contributions from colleagues and friends but are a considered collection that illustrates the dazzling breadth of Alasdair's interests, nonetheless underscoring the coherence and cohesion within both his research and that interplay of disciplines and methodologies which constitutes 'environmental history'. Spanning a range of historical themes that were central to his work — climate and weather, environmental resources and their management, disease and its impacts, critical analysis of documentary sources and place-name records, and the intimate connections between people, environment and place — they are intended to be both a tribute to his memory and a call-to-arms for others to pick up the multiple strands of interdisciplinary research upon which Ally was engaged before his untimely death. Principally environmental historical in their thrust, this gathering contains new research whose value transcends purely documents-based history, illuminating themes central to the work of archaeologists, palaeoecologists and environmental scientists, disciplines with which Ally engaged and collaborated throughout his career. There is much here to inspire interest and debate in undergraduate and taught prostgraduate classes and yet more to stimulate researchers across the environmentally focused historical disciplines. Collectively, these papers are a signal of new directions, new methods, and new opportunities to strengthen the voice of environmental history in the discourse and debate surrounding climate emergency and humanity's future.

It would be an understatement to say that Alasdair Ross exploded onto the academic scene in the mid-1990s when he signed up for the University of Aberdeen's Summer School for Access programme. As the then course convenor for the Scottish History module within that programme, I encountered a man in his thirties who, after years of running his own business as a fish-merchant in Aberdeen and having recently gone through substantial upheaval in his personal life, had decided that he was finally ready to go to University, intending to study Archaeology. By his own confession, he had been switched off from formal education in his mid-teens, being uninspired by the traditional curriculum of his secondary school on Deeside and preferring to spend every hour he could with his schoolmates climbing on the conglomerate cliffs of the Mearns coast and, especially, on the granite crags of the Cairngorm Mountains. There was his inspiration, communing with the rock and developing his talents as one of the most gifted Scottish mountaineers of his generation. There, also, he became increasingly aware of the delicate poise present in the varied landscapes and seascapes around him between the seemingly changeless face of the land and the cicatrices of ancient and still-unfolding human influences on its appearance. Here, too, he found the lyric writing of Nan Shepherd (1893–1981), the Aberdeenshire Scottish Modernist poet whose *In the Cairngorms* (first published in 1934) and short non-fiction book *The Living Mountain* (written in the 1940s but not published until 1977)[2] chronicled her experiences as a hill-walker and climber, and her own deep awareness of the interplay of nature and humanity in her beloved mountains. These remained an inspiration to him in manifold subtly changing ways for the remainder of his life. Ally, like Nan Shepherd, *understood* the land and loved both it and its slowly revealed secrets and surprises with deep and enduring joy and passion.

University life was a revelation and a re-birth for Ally; he excelled academically and was inspired to progress to higher degree programmes. It was not all about academic endeavour; it was there that he met and courted his wife Sonja, who was first his academic tutor, then his research collaborator, and finally his soulmate and partner. From the first, he excelled in History and was persuaded by his course tutors to keep his focus there and to embrace the contributions of landscapes studies, archaeology and architectural history to an interdisciplinary approach to historical studies rather than focus exclusively on Archaeology. His focus was drawn swiftly to Scottish History, his enthusiasm triggered by that first Summer School programme on 'The Scottish Wars of Independence 1296–1357', but combined with Celtic to give him a deeper awareness of the rich cultural heritage of his north-east Scottish homeland. Securing an outstanding First Class degree, he progressed to doctoral studies and kept alive his strong interdisciplinary focus although he was by then based principally within the History department at Aberdeen. By the early 2000s, with his Ph.D. secured and various options for post-doctoral research open to him, he took the first steps into Environmental History, a field that would dominate — but not monopolize — his work from that time on.

Many people unfamiliar with Alasdair's background are surprised at the *prima facie* eclectic nature of its subject matter. But his published research simply underscores both his original degree studies background and the consequent breadth and depth of interest that was sparked by that History and Celtic combination. There is, throughout, a logic in the diversity of his work,

[2] Shepherd, *The Living Mountain*; Shepherd, *In the Cairngorms*.

which resolves itself into connections between people, place and environment. His publications range from an essay on the Early Medieval Pictish sculptural evidence for harps,[3] through family politics and landholding in twelfth- to fourteenth-century Scotland,[4] to failed late eighteenth- and nineteenth-century estate-led Improvement programmes in central Highland Scotland.[5] In all his work he demonstrated remarkable versatility, with an unerring ability to find subject-redefining evidence embedded within even the best-known source materials. The bulk of Ally's energies, however, was focused on two main themes that became his driving passion: early and high medieval northern Scottish economy and society, and the environmental history of central and northern Highland Scotland. At their root lay a fascination with the sinews of political power and a drive to understand better the mechanisms for social control and resource exploitation that gave substance to that power. He was not inspired by 'great men's' history but by the social and economic structures upon which they founded their positions and, critically, the systems that they constructed to manage the natural resources of the land to sustain the populations who gave substance to their claims of lordship. In a decade of prodigious productivity, he delivered a succession of monographs, research papers and commissioned reports, edited collections and popular magazine editions that collectively have placed the tools for better understanding of Scotland's past environment into the hands of his peers, students and the wider public. We can only wonder what might have lain ahead as he began to apply his theories beyond Scotland to the wider North Atlantic regions of Europe.

It is surprising even to those of us who knew Ally well that his post-doctoral career spanned barely twelve years and yet he delivered such a profound impact in his research field and widely beyond. The origins of much of his work lay in his doctoral thesis and the mass of data that he had gathered which proved too great for inclusion in that tightly prescribed exercise. His first thoughts had been to research the history of the Earldom of Mar, the medieval province which encompassed the southern part of his beloved Cairngorms, but he was persuaded to focus instead on the previously under-researched north-

[3] Ross, '"Harps of their Owne Sorte"?'.

[4] Cameron and Ross, 'The Treaty of Edinburgh and the Disinherited (1328–1332)'; Ross, 'Men for All Seasons? The Strathbogie Earls of Atholl and the Wars of Independence, c. 1290–c. 1335: Part 1'; Ross, 'Men for All Seasons? The Strathbogie Earls of Atholl and the Wars of Independence, c. 1290–c. 1335: Part 2'; Ross, 'The Lords and Lordship of Glencarnie'; Ross, 'The Identity of the "Prisoner of Roxburgh"'; Ross, 'Moray, Ulster and the MacWilliams'.

[5] Ross, 'Improvement on the Grant Estates in Strathspey in the Eighteenth Century'.

ern Scottish region of Moray, for which there was a richer surviving medieval record.[6] Moray — not to be confused with the much smaller modern Scottish local government administrative unit of the same name — was the heartland of the early medieval Pictish kingdom of Fortriu. That kingdom was the homeland of a rival lineage to that which came to hold the kingship of the Scots from the mid eleventh century and later became the nucleus of the great earldom which King Robert I created for his nephew, Thomas Randolph, to assert Bruce lordship over the central Highlands of Scotland. Stretching from the Cairngorms to the sea and encompassing the central Highland watershed of the second greatest river catchment in Scotland, the Spey, Moray presented an opportunity to explore the entire range of Scotland's exploitable environmental niches from the high, sub-arctic montane zone of the Cairngorm plateau to the shifting sand-dunes and tidal lagoons of the Laich of Moray. Although he did not realize it yet, it provided the perfect laboratory in which to explore the fundamental relationships of people, land and lordship, and the influences of changing climatic and environmental conditions on those relationships, across the span of the Medieval Climate Anomaly and the 'little ice age'.

So it was that Ally embarked on what he intended would be a socio-political analysis of Moray and its eleventh- to thirteenth-century lordship structures. Very quickly, however, he came to realize that the evidence he was amassing contained a detailed record of something much older and, in truth, far more interesting. This was the *dabhach* (plural *dabhaichean*), the base unit of a system of land and resource organization whose origins and function had been the subject of debate since the mid-nineteenth century. Ally's research overturned older theories and established that the *dabhach* was an entity that dated from at least *c.* 900 and probably had its origins much earlier still in north-eastern Scotland's Pictish culture. More importantly, from the perspectives of socio-economic and environmental history, he demonstrated that it remained in some form of use as an identifiable social unit in parts of central Highland Scotland into the 1930s, enabling long-term analysis of land-use patterns, availability of key resources, and of greatest value, environmental change across a millennium. For historians familiar with the long historiographies of the Carolingian *mansus* or the Anglo-Saxon *hide*, it may seem unremarkable that one of Ally's key findings was that these *dabhaichean* represented an apparently planned or systematized division of the land. Their purpose, he determined, was to give settled communities access to a suite of key resources: arable land, summer and

[6] Ross, 'The Province of Moray c. 1000–1230'.

winter grazing, water, fuel and building materials. From a Scottish historical perspective, that discovery adds strength to revisionist historiography of the Picto-Scottish state of the late ninth and tenth centuries which has argued for a greater level of political, social, and, importantly, fiscal sophistication in Pictish Fortriu and its successor state, Alba.[7] His analysis provided this revision with the evidence for an over-arching authority exercising effective political and legal domination over populations spread from the fertile and heavily-settled lowlands to the thinly-peopled remote upland districts of central and north-west Highland Scotland. Far from being a primitive, tribal society lacking in institutions of government and administration, the social, economic/fiscal and political sophistication of that kingdom was exposed through Ally's evidence of a systematized pattern of *dabhaichean*.

The full significance of these findings was not at first obvious even to Ally when he submitted his doctoral thesis in 2003. Over the next two years, however, while he was engaged in his first post-doctoral research fellowship at the University of Stirling (an interdisciplinary environmental history project examining the documentary and palaeoenvironmental records for post-medieval agricultural improvement around Loch Tay),[8] he began to examine the suites of resources intrinsic to the *dabhach*. Through this work, he began to identify commonalities in how each *dabhach* was structured to ensure provision of adequate access to the different environmental niches that supplied those resources. A first published exploration of the *dabhach* and his new hypothesis for their origin, structure and function appeared in 2006 as 'The *Dabhach* in Moray: A New Look at an Old Tub'.[9] The title of that paper alludes to late eighteenth- and nineteenth-century understanding of the name, which saw it as derived from the Gaelic term for a grain tub or seed-vat. From that identification, most nineteenth-century scholarship saw the term as linked to a physical object; a seed-vat. What it represented in daily experience, however, had remained open to debate, with views varying from it being an expression of the extent of arable sown with the vat's contents, the yield of a measured area gathered in such a container, or the volume of cereal rendered to a social superior from a terri-

[7] See, for example, recent work on the northern Pictish kingdoms by historians, archaeologists and environmental historians: Noble and Evans, *The King in the North*; Woolf, *From Pictland to Alba, 789–1070*. The beginnings of the revision of thinking surrounding the Pictish kingdom can be seen in syntheses like Smyth's *Warlords and Holy Men*.

[8] *Written in the Hills: Human Impact on Upland Diversity, Past, Present and Future* [University of Stirling/Leverhulme Trust research project].

[9] Ross, 'The *Dabhach* in Moray'.

tory of defined extent. As a 'measuring tub', the *dabhach* had come to be seen as an originally arable unit that had progressively acquired a fiscal identity as a unit of assessment. Most discussion of the *dabhach* down to the early 2000s thus concentrated on the arable character that the linkage with seed sown or gathered implied, focusing on medieval references to ploughed land and grain yields from such units. What this discussion overlooked were references to 'a competent quantity of mountain and grazing ground' also attached to them. It was this which made Ally reconsider what was actually being expressed through the use of the *dabhach* term.

Guided by the evidence from his *Written in the Hills* Loch Tay research that revealed the longevity of farming communities whose land spanned a transect from lowland loch-side arable to high montane summer grazing, Ally undertook a closer examination of the *dabhaichean* which he had identified in his thesis. From that reassessment, he concluded that this traditional prescription was wrong in every element; it had no link to grain — whether as seed, yield or render — and arguments which focused on measured extents were insupportable. It became clear that the generally accepted interpretation was founded on the devolving premise of a yield of a specific volume being the produce from cultivated ground of a fixed area, an interpretation rooted in the grain-based hypothesis. To test his new thesis, Ally undertook the first systematic mapping of *dabhaichean* in northern Scotland. That exercise confirmed that they were of widely varying extent, partly determined by topography, altitude, and predominant landscape character, that they frequently comprised of non-contiguous parts to ensure access to necessary resources to sustain a community expressed as a number of households, and that the component elements had roughly equivalent carrying capacities for livestock, expressed (questionably, as he was to argue elsewhere) in some sources as *soums*. This broad suite of equivalent capacities convinced him that the *dabhaichean* originated as the planned base-unit in a common assessment system that served both fiscal and military purposes. That fiscal role was a key determinant in their later bundling into larger blocks that gained religious-economic significance as parishes.

During his research, Ally had begun to examine how access to, and exploitation of key resources was allocated and managed within the *dabhaichean*. Soums had long been postulated as a possible expression of carrying capacity of the grazing land and, working on the basis of that understanding, he began a systematic analysis of evidence for grassland management, meadows and hay production, key but often contested elements in the historiography of medieval Scottish rural society. In this, he engaged in close collaboration with colleagues from the environmental sciences, whose paleoecological and palaeoenvironmental research evidenced the impact of past exploitation regimes in ways that the generalized legal prose of the parchment record could not. Collectively, the interdisciplinary evidence pointed to a sophisticated mechanism that provided adequate resources for both summer and winter grazing for agreed numbers of livestock and, contrary to a deeply-entrenched historiographical convention, access also to meadow-land for winning of a hay-crop for supplementary winter fodder.[10] That this was a system established and overseen by an over-arching authority was evident in arrangements made to provide summer grazing for lowland, primarily coastal *dabhaichean* in distant upland areas. For example, the people of the *dabhaichean* of the Moray coastal district — the so-called Laich — rich in arable, fishing-rights, and winter grazing but poor in summer grass resources — took their livestock over thirty miles inland to the Forest of Stratha'an in the northern valleys of the Cairngorm Mountains. While recognising that soums were not the systematized mechanism that previous research had suggested but were instead subject to manifold local variations and regularly negotiated settlements between landowners and land users, he nevertheless saw that souming levels were regulated and enforceable by legal sanction, with individual or communal breaches being subject to punitive fines. Such responses demonstrated the eagerness of both lords and tenants to maximize potential profits from the land but also their recognition of the threat and consequences of over-grazing and the effectiveness of the policing of resource use by a superior authority.[11]

Over this same period, his work began increasingly to intersect with that of other historians of Scotland's medieval environment, principally Fiona Watson, Mairi Stewart, Richard Tipping, and me, who were exploring exploitation of other common resources, principally wood (for wood pasture, fuelwood, building timber and other materials) and peat. Applying that research to his analysis of soums, where he had found that the ratios by which carrying capacities were measured changed over time, sometimes with great frequency and rapidity, he proposed that this reflected at first a combination of local responses to market conditions (principally through the drive to maximize income from cattle hides and wool) and population pressure. He interpreted this as evidence for settlements having expanded to the limits of the resources available to them

[10] This work culminated in Ross, *Literature Review of the History of Grassland Management*.

[11] This research was a further riposte to the lingering 'Tragedy of the Commons' mentality that continued to pervade some publications on the regulation of access to common resources.

and placing an insupportable burden on grazing. Several high-profile disputes between landowners and between communities with shared rights in the contested resource document such conditions. Increased evidence for litigation over access to grazing or to regulate grazing levels do support this analysis and reveal a system coming under great stress before the end of the thirteenth century. The picture, however, becomes increasingly complex for the fourteenth and fifteenth centuries.[12] In those centuries, a growing body of evidence for wider ecosystemic stress and associated human ecodynamic change presents itself across the spectrum of basic natural resources exploited by Scots.

With a broad suite of environmental niches available for exploitation, dependency traps and other path-dependent theories that were then being applied to research into, for example, the failure of the Norse Greenland colonies, were for him inadequate explanations for the socio-economic changes that occurred in later medieval Scotland. Rejecting the unconscious environmental determinism towards which this evidence could have drawn him, he looked instead at the interplay of human and environmental agency, examining the responses to changing conditions recorded with mounting frequency in rentals and baronial court-books in the fifteenth and earlier sixteenth centuries. Focusing on the surviving and largely unpublished archives of three of the greatest central Highland land-owning families — the Gordons of Huntly, the Grants of Freuchie and the Campbells of Breadalbane — he began to identify patterns of commonality in responses by local communities to both environmental and economic factors. Despite the probable — but for Scotland unquantifiable — release of human population pressure consequent on plague mortality, greater levels of regulation, enforcement of new soums, restrictions on access to certain resources, or abandonment of arable land and switches from arable to pastoral regimes, all pointed to a protracted environmental crisis that affected rural communities in upland and lowland Scotland.

We can see Ally wrestling with this data and trying to reconcile it with the climate proxy data that was newly available in the later 2000s in his analysis — published in 2013 — of two documents from the Gordon archive relating to litigation over management of timber resources in Stratha'an, a major valley system that penetrates the Cairngorms from the north-east.[13] Dating from the

[12] Ross, *Assessing the Impact of Past Grazing Regimes*; Ross, 'Scottish Environmental History and the (Mis)use of Soums'.

[13] Ross, 'Two 1585 × 1612 Surveys of Vernacular Buildings and Tree Usage in the Lordship of Strathavon, Banffshire'.

late 1580s and early 1610s respectively, the documents contain lists of buildings in the various settlements within a cluster of *dabhaichean* which made up the central district of the Gordon family's lordship of Stratha'an, noting the number of couples — the paired roof-support timbers — in each and calculating the volume of timber expressed as 'trees' required to renew and maintain the buildings; the seven-year rolling total required, 781,865 trees, is staggering, even when it is understood the Scots term *tree* = major tree-limb or branch in modern English usage, and most probably coppice shoot in late sixteenth-century cases. It is a remarkable record, not only of the voracious consumption of timber by peasant communities and the systematized coppicing of the birch, hazel, willow, alder tree species which it must have entailed, but also of the continued use of the *dabhach* as the unit for apportionment and regulation of access to these resources. Furthermore, it revealed the overarching authority of Stratha'an's lords as regulators of felling, managers of coppicing and replanting, and active pursuers of transgressors. All of this, moreover, could be set against a backdrop of rapid climate change as the 'little ice age' entered one of its coldest phases and delivered a new set of challenges to montane communities like the tounships of upper Stratha'an, with increased snowfall, extended periods of intense cold — affecting grass-growing season length — cool summers and wet autumns. All terrestrial resources were affected, with even the availability of fuel affected by the wet summer and autumn impact on dried peat provision. The situation in the valley, however, was complicated further by local political conflict, competing resource-management agendas at the level of peasant tenants, local and superior lordship, and a net inflow of population at the direction of one of the rival parties in the lawsuit. All of this was occurring at a time of increasing environmental stress where the available resources could scarcely meet the demands of the pre-existing communities. What Ally's analysis of the data yielded was a richly textured record of the consequences of human choices made in the context of a supply crisis and how human agency interacting with environmental change delivered unforeseen outcomes, both negative and positive.

In parallel with his research into grassland and timber management, Ally also initiated a programme of research into Scotland's medieval salmon fisheries, which formed the basis of a highly productive research collaboration with Richard Hoffmann, designed as the springboard for an in-depth re-evaluation of medieval Scotland's riverine fisheries.[14] Some of the fruits of that collabora-

[14] See Hoffmann, this volume, and Hoffmann, '*Salmo salar* in Late Medieval Scotland'; Hoffmann and Ross, 'This Belongs to Us!'.

tion are presented in Richard's contribution to this present volume. But it was in terrestrial resources — grass, timber, and fuel — and the structures within which their exploitation and management were regulated that his primary interest lay. Drawing together his original doctoral research, he first presented his developed thinking on the interplay of socio-economic, political and environmental factors that underpinned the evolution of the medieval Scottish state in his 2011 monograph, *The Kings of Alba*.[15] Ostensibly a discussion of the political processes that led to the demise of the northern Scottish kingship based on Moray and the forcible integration of their realm into the central Scottish kingdom of Alba in the late eleventh and early twelfth centuries, this masterly re-imagining of the origins of the medieval Scottish state also presented in outline his new model for the locally operative structures upon which those polities were founded. By 2014, his continued research into those structures and the key resources they controlled had provided abundant materials, while fuller engagement with archaeological evidence and climate proxy data gave a broader and more academically mature perspective from which to embark on the expansion and substantial revision of his initial work on *dabhaichean*. The outcome, published in 2015, was his ground-breaking monograph, *Land Assessment and Lordship in Medieval Northern Scotland*.[16] This was the capstone of his now fully-developed thesis for the significance of the *dabhach* as the fundamental socio-economic building-block of the early medieval Picto-Scottish state and the primary source for evidence of long-term environmental change in upland Scotland. But his book also set out a manifesto for future research which would have extended his model out of his core area to examine the other land-units recorded in culturally different parts of the medieval kingdom — the *arachor*, *ceathramh* and *pennyland* — and into a fuller comparison with English and mainland European systems of assessment. Ally had started that work, collaborating with colleagues at Stirling and Glasgow particularly on research into parish boundaries as a preliminary stage in the identification of the socio-economic building-blocks beneath them. Much of this took the form of annotated maps, where he could set out in two dimensions what his documentary research and physical engagement with the land through ground-truthing had told him. It was a project that had advanced to the stage of preliminary discussion and in early 2017 he was starting to lay out the basics of an application for funding to complete the detailed research into

[15] Ross, *The Kings of Alba c. 1000–c. 1130*.

[16] Ross, *Land Assessment and Lordship in Medieval Northern Scotland*.

a series of sample districts. His death has left that as an unfulfilled ambition for others to complete.

While much of Alasdair's early research output was avowedly Scottish in its focus, in respect of his environmental history work he was at pains to stress its international applicability. The circumstances that he explored and the models he constructed hold relevance for the experience of the whole north-eastern sector of the lands bordering the eastern shores of the North Atlantic, from Cork and Kerry in south-west Ireland to Vatnsfjörður in north-western Iceland. It was continuities and continua that he was researching, working across a millennial scale and within an oceanic region which marked the interface between the core zones of Celtic and Scandinavian cultures. Few scholars currently engaged in research into the environment of the medieval and later North Atlantic region will not encounter the work of Ally Ross on resource exploitation, the socio-economic structures which controlled and regulated access to those resources, and how the structures evolved in response to human political and economic drivers and the stimuli of long-term climate change. An insight into the continuing development of his thinking on such long-term socio-economic processes and environmental influences affecting human decision-making is to be seen in the paper included within this collection[17] which, along with my own contribution below, was conceived within the framework of a research project on 'Mortality Crises Before the Black Death' instituted by Tim Newfield during his time as a post-doctoral Research Fellow in the Centre for Environment, Heritage and Policy at the University of Stirling. In it, he was drawing together historical record evidence and climate proxy data to offer a radical revision of traditional interpretations of Scottish economic performance across the transition from the Medieval Climate Anomaly into the 'little ice age', stripped clean of the rose-tinted perspectives of fourteenth- and early fifteenth-century Scottish chroniclers and the gushing enthusiasm and positivity of more recent political historians. His cool reappraisal of Scotland's experience in that era of successive environmental, political and economic shocks has applicability far beyond that small medieval kingdom, opening compelling avenues for exploration of contemporary medieval experiences from the declining Norgesveldet to the crises of late Plantagenet England, but, equally, they also set out a roadmap for explorations of how our own culture and society might respond to the current shocks of climate emergency.

[17] See Ross, 'A Post-Plague Golden Age?', this volume.

In all his interactions with his peers, from his membership of the European Society for Environmental History and the European and World Congresses, to the postgraduate seminars in the Centre for Environment, Heritage and Policy, Ally stressed the importance of disciplinary consilience, of interrogation and critique of the culturally-framed perceptions through which the increasingly irrelevant duality of the human/environment opposition is negotiated, and the construction of synthetic records across the longue durée using data derived from multiple sources and as many disciplines as possible. He always searched for the links that led from past experience to current circumstance, but without a deterministic teleology that created an artificial straitjacket of path dependence. Unconsciously, even within his earliest environmental history research, he was adopting a whole systems approach to his analysis that placed the organism he was researching — from individuals to *dabhaichean* to states — within the context of the networks from the local to the global that affected its evolution, and were in turn themselves affected by its responses, across the centuries.

It is in that deep awareness of the interconnectedness of human cultures and global environment that the significance of Alasdair's research for the next generation of scholars lies. He was only beginning to chafe at the limits which traditional methodologies were placing on the active engagement of many of his peers in contemporary climate change debates. Ally encouraged the interdisciplinary syntheses of environmental data from historical record, archaeological, and palaeoecological sources that are now transforming our understanding of political, social, economic and cultural change across the pre-Modern North Atlantic and the European Atlantic regions generally, but was frustrated by how much of it was still focused on the micro rather than macro scales or on 'failed' or 'completed experiments' such as tenth- to fifteenth-century Greenland. Certainly, such sources are supplying increasingly high-resolution environmental data, in terms of both temporal and geographic precision, and creating a much more nuanced and multifaceted understanding of the intersection of environments, mentalities, culture, economics and politics in the historic past. But the disjunction between the historic communities who functioned within their limits and the modern communities that now occupy the same zones constitutes a telling weakness in arguments about how environmental history and archaeology can deliver valuable insights for application in current climate-change modelling. No matter how robust the interdisciplinary research which draws together such data, or how transformative the multivariable modelling of past trends and their ecological and human impacts has been, the unescapable bottom line is that the society under examination failed. Elsewhere around the North Atlantic periphery, and, as Alasdair Ross's work demonstrated, especially in countries such as Scotland where unbroken sequences of land-use records, palaeoenvironmental profiles and climate proxy data link the earlier medieval cultures directly to their modern successors, we have 'continuing experiments' that do permit us to investigate the circumstances that established resilience and delivered sustainable practices across successive waves of profound climate change. Even here, however, the focus remains on what in modern core-periphery terms are outlying areas, albeit ones that in the Middle Ages were hubs of international trade within networks that extended from the Baltic and Mediterranean to the Greenland colonies. While their experience perhaps comes closer to providing models for societies that benefited from a broader suite of environmental resources and which had, consequently, greater ranges of choice in how to respond to climate change, they were still non-urban and pre-industrial in their socio-economic and political structures. This is where Ally's legacy can have its greatest impact, for it is one that presents environmental historians with a series of models through which to establish their relevance in debates over our continuing, fumbling responses to the climate emergency. It was scalable and replicable experiments within the experience of a still functioning system that he sought out. He chose to focus on a still-living experiment and present evidence that contributes an answer to questions around cultural limits to adaptation and social limits to change in a complex and globally-connected society. It is up to us to apply the lessons of the experiences he uncovered as we face the most far-reaching set of choices yet to confront humanity.

Saints on the Shore: Coastal Encounters in the Early Medieval West

Ellen Arnold*
University of Stavanger

Introduction

As anyone who has spent time along the seacoasts of Northern Europe knows, the sea and the shore are compelling, alluring, and dangerous. Coasts offer glimpses of the unknown and the eternal, and waves both draw our gaze out to the infinite and bring us crashing back into the present moment. The changeless and yet ever-changing interaction of sea and shore lulls, soothes, and frightens, and calls the attention of many an observer to human fragility and the power and beauty of nature. How could this not be a shared experience across time?

We have to be careful here. As Alain Corbin notes in his foreword to his beautiful history, *The Lure of the Sea: The Discovery of the Seaside in the Western World, 1750–1840*, it is dangerous to suppose that our experiences match those of past people. 'Psychological anachronism', he warns, can lead to fundamental misunderstandings of the past, and 'there is no other means for understanding people from the past than attempting to see through their eyes and live with their feelings'.[1] Corbin traces the European experience of the sea coast and explores the development of European 'beach culture' and the 'desire for

* I owe a giant debt of friendship to Ally Ross. He supported me when I was a junior scholar, and was generous with his time, energy, and heart. He encouraged me to keep pursuing the quirkier and more subjective side of environmental history. He also modeled how to be engaged in broader scholarly circles, how to love places, and how to show and share that love with others. He is sorely missed, and I hope that he would have taken delight in some of these strange little stories.

[1] Corbin, *The Lure of the Sea*, p. xvi.

the shore'. He assesses what Europeans knew about the oceans and coasts, how these marginal, boundary areas fit into their understanding of the world, and how they interpreted and dealt with the duality of coasts as havens and dangers. It is an erudite and emotive work that takes both space and culture seriously. And yet.

As is often the case with cultural histories of nature appreciation, Corbin begins his work by quickly dismissing the world before 1750 as largely irrelevant to a study of the history of the seashore and the coast. He writes that 'with few exceptions, the classical period knew nothing of the attraction of seaside beaches, the emotion of a bather plunging into the waves, or the pleasures of a stay at the seaside. A veil of repulsive images prevented the seaside from exercising its appeal'.[2] He follows up this comment with a brief assay into the Biblical flood and the impact of Christian cosmology in creating a view of the ocean abyss as chaotic, tumultuous, and of spiritual value because the roaring and crashing of the waves were 'reminders of the sins of the first humans, doomed to be engulfed'. He does not engage with medieval voices and ideas — the classical world, dismissed in a sentence, is immediately followed by examples from early modern and modern writers. The Middle Ages are reduced here to Biblical cosmogony, St Brendan, and Beowulf.

This problem is not Corbin's alone; it seeps through even into works that have been written directly about medieval ideas of nature, most notably Lynn White's famous essay on Christian attitudes to nature,[3] in which he reduced medieval Christian thought to a handful of non-representative sources, leaving out the many, mellifluent voices of medieval people who wrote about nature. We are attuned to thinking about such comments as being a part of a modern approach to landscape appreciation. In *Wilderness and the American Mind*, Roderick Nash argues that the appreciation for wild beauty only became possible in Europe with the Enlightenment: 'The ideal focus for any Christian in the Middle Ages was the attainment of heavenly beatitudes, not enjoyment of his present situation. Such a point of view tended to check any appreciation of natural beauty. Thus, during the Renaissance, Christianity offered considerable resistance to the development of joy in perceiving wild landscapes'.[4]

But what if we take Corbin's fascinating questions, and direct them towards the early medieval world? What were the experiences of early medieval people

[2] Corbin, *Lure of the Sea*, p. 1.

[3] White, 'The Historical Roots of Our Ecologic Crisis'.

[4] Nash, *Wilderness and the American Mind*, p. 19.

on the ocean's edge? Rivers were essential to their economy and experiences, and, to use a well-worn cliché, rivers run into the sea — how did medieval people interpret the fluid and mutable boundaries not only between sea and shore, but also between river and sea? How was the shore as liminal space experienced and interpreted? And if they did not see the seashore in quite the same way that modern people have, how did they see it?

The Poetry of Venantius Fortunatus

I would like to begin where so many of my recent historical thoughts have begun — with the voice of the sixth-century bishop and poet Venantius Fortunatus. In a poem written to Bishop Felix of Nantes, Fortunatus gives us evidence of a medieval person enjoying leisure time along the sea, examples of the medieval ability to evoke the coastal environment, and the power of the coast to be a complex metaphor in the hands of skilled writers. Felix had written to Fortunatus, and he describes the moment he read his friend's letter:

> I was yawning away by the seacoast, enticed by the seduction of my natural sloth, and lounging over-long on the edge of the shore, when suddenly I had the experience of being soaked by the rolling breakers of your eloquence as if by salt spray sent up by a barrier of rocks.[5]

This was not the only time the poet used the tangible characteristics of the seashore as metaphors for writing. In a poem to another bishop, who had sent him some poems (perhaps for his approval?), he wrote: 'Navigating your poems that foam with your verses swell, I believed I was launching my sails on to the surging sea. Your page, though flat, threw up stormy breakers, and from your spring poured forth ocean-like waves'. And, in a letter to a friend about their long-distance friendship, Fortunatus imagined himself as emotionally and physically bound by the coast:

> The heaving surface of the ocean washes round me, while Paris holds you, dear friend, in its keep. The Seine is your dwelling place, mine circled by the Breton tide: one love binds us together, though we are geographically apart. The fury of the deep does not steal from me your countenance here, nor the North wind carry off your name, my friend. To my heart you return as often, my dear one, as the sea-wave to the shore in stormy weather. As the sea tosses when the east wind blows, so

[5] Venantius Fortunatus, poem 3.4, in his *Poems*, pp. 131–32.

> my mind, dear friend, is agitated without you. A pleasurable storm seethes in my tranquil heart ...[6]

In another instance, Fortunatus appears to describe a trip where, momentarily stuck on what may be an estuarial island of the Gironde because of bad weather, he whiled away his time hunting for seashells, one of which he sent as a present to one of his correspondents and the wife of Bishop Leontius, Placidina. He has a tradition of sending little poems to friends along with tangible gifts; in this case, he is sending the present plucked 'from the midst of the waves, the ocean suppressed its heaving and murmuring waters'. The modern translator, Michael Roberts, points out in his notes that 'it is suggested that the object in question is an especially attractive seashell, which Fortunatus found on the island ... but the circumstances of the poem are somewhat obscure'.[7] Fortunatus continues, further describing his 'eagerness to explore this stretch of the sea', recalling in his letter how 'the frenzy of the breakers rolling in from the North drove me back'.

Here we have a fifth-century writer spending free time (planned and not) wandering the coasts; he has experience of estuarial environments and coastal landscapes, and is able to convey aspects of the unique characteristics of those places where water and waves meet land, where the solid and known meet the unknown and unstable, and where secrets and beauty and danger all meet. What is this, if not a medieval acknowledgement of the lure of the sea?

Early Frankish, Carolingian and Northumbrian Writers

The sea coasts attracted the attention of early medieval writers, who observed and commented on the liminality of the spaces; the ways in which land and sea merge and blur and become differentiated. Of course, one of the most visible and powerful ways in which this is observed is with the tides. Famously, Bede wrote a treatise on tides, which is often connected by scholars to his interest in dating Easter — however, this ecclesiastical issue should not overshadow an equally important interest in observing and understanding the way that God works in the natural world. As Sarah Foot points out in her discussion of the religious interpretations of natural signs, scholars like Bede 'showed an impressive capacity to observe and report natural phenomena such as the phases of

[6] Venantius Fortunatus, poem 3.26, *Poems*, pp. 202–03.

[7] Venantius Fortunatus, poem 1.17, *Poems*, pp. 52–55.

the moon, movements of planets and the sequence of tides'. But, she points out immediately, 'in the holistic worldview prevalent in early medieval Christian Europe, the hand of God was always implicitly manifest in guiding the forces of nature'.[8]

Bede, though perhaps the best known, was not the only early medieval writer to spend time pondering the mysteries of the tide, and to use them to meditate on how God shaped the world in tangible, real, and immediate ways. Gregory of Tours, bishop and historian, wrote about what he considered to be the wonders (*miracula*) of nature as part of his work on the stars. Of all the wondrous things in God's creation (which also included seeds, the phoenix, Mt. Etna, a burning fountain near Grenoble, the sun, and the moon), he listed the tides as the first.[9] He explained that 'The first [wonder] of all, therefore, is the tidal movement of the sea and the ocean, in which [the sea] every single day expands so much that, reaching the edge of the land, it covers the beach, and, receding to where it came from, leaves [the beach] dry again'. Though he is primarily interested in the motion of the tide, he also explains the ways in which people benefit from this, and how it can be interpreted as a gift from God: 'An abundant multitude of fish and various weeds is then collected by the people walking on the [now] dry ground. God provided this first wonder for the human race, [one] which is very much to be admired and which corresponds to his power'.[10]

Depictions of coastal landscapes appear in poems throughout early medieval Europe; the famous Anglo-Saxon poem 'The Seafarer' is by far the best-known example. In it, a hoary veteran describes his life at sea, his nights on duty facing the wild waves as his ship 'tossed near the rocks'. The poet depicts a multi-sensory winter coast, where the sailor 'heard nothing there but the sea booming — the ice-cold wave, at times the song of the swan'.[11] He describes the sounds of many different sea birds, both pleasant and a reminder of his isolation: 'The cry of the gannet was all my gladness, the call of the curlew, not the laughter of men, the mewing gull, not the sweetness of mead. There, storms beat the rocky cliffs; the icy-feathered tern answered them; and often the eagle, dewy-winged, screeched overhead'. Sea birds seem to have caught the

[8] Foot, 'Plenty, Portents and Plague', p. 17.

[9] Gregory of Tours, 'De cursu stellarum ratio', p. 410.

[10] Translation by Giselle de Nie, quoted from her article 'The Spring, the Seed and the Tree', p. 93.

[11] Crossley-Holland, *The Anglo-Saxon World*, p. 53.

imagination of many a poet. They also appear in 'The Wanderer', a companion poem found in the same manuscript (the Exeter book). The poem adds sights to the sounds: 'the sea-birds bathing, spreading their feathers, frost and snow falling mingled with hail'.[12] Alcuin, one of the shining lights of the Carolingian Renaissance, asked of a 'truant pupil': 'Why do you fly through the great palaces of kings like a bird, playful on the ocean's sounding main?'[13]

Alcuin also wrote a lengthy poem about the destruction by Vikings of the monastery of Lindisfarne, a cultural centre; he reminds readers of the ways that 'Unkindly chance always throws sadness and prosperity together, as the sea waves return with their ebb and flow'.[14] We like to think of things as 'regular as the tide', but coastal environments are anything but. Like a sneaker wave, bad weather could come up out of nowhere. Bede, who had a clear and compelling sense of his coastal Northumbrian home, and who carefully studied the tides, describes this changeableness of the sea powerfully. In a story in which travellers are stranded on a Northern island while travelling, he evokes the grim coastal environments, 'bleak with snow, clouds lour in the sky, there is a gale raging and the sea is a fury of waves'.[15] Reflecting on the meaning of this shifting space, he observes that 'All the ways of this world are as fickle and unstable as a sudden storm at sea'.[16]

The Exeter book, which contains 'The Wanderer' and 'The Seafarer', also holds a series of Anglo-Saxon riddle poems, one of which takes the perspective of a large storm that disturbs the coast. The storm whips up the ocean and 'the foam is rolled; the whale-sea roars, loudly rages, streams beat the shore, sometimes throw stone and sand onto steep cliffs, seaweed and wave'.[17] The dangerous coast, powerful, storm-tossed, and dangerous, is here clearly still appreciated for its power and strength. This is not the abject loathing and fear that Corbin would have us imagine.

[12] Crossley-Holland, *The Anglo-Saxon World*, p. 50.

[13] Godman, *Poetry of the Carolingian Renaissance*, p. 123.

[14] Godman, *Poetry of the Carolingian Renaissance*, p. 127.

[15] Bede, 'Life of Cuthbert', p. 59.

[16] Bede, 'Life of Cuthbert', p. 55; Arnold, 'Hagiography and Environmental History'.

[17] Dale, *The Natural World in the Exeter Book Riddles*, p. 183.

The Perils of the Sea

Many early medieval accounts emphasize the risks of too close an encounter with coastal environments. Shipwrecks at sea were an anticipated risk of long-distance travel — but the coast brought unanticipated dangers. In the life of St Cuthbert, Bede tells of an incident when monks were caught up suddenly in coastal risk. They 'used to bring the wood they needed for the house by raft, from a good distance away along the current of the river'. On one such trip, the rafts were blown to sea by a sudden gale. As the peasants watched, 'the rafts floating out to sea looked like five birds bobbing up and down on the waves'. Eventually, Cuthbert's prayers allow a safe return.[18]

In addition to this story of the sudden shift from river to sea that could be experienced at river mouths and estuaries, Bede shows us the seabirds and the waves and the wind and storms (and otters[19]) and the fogs that could all isolate someone as quickly and menacingly as any rocky hermitage. While Corbin might read this as a horror of the coast, Bede explains in the life of St Cuthbert that such isolation was, in fact, a lure. Cuthbert, he writes, would happily 'live in a tiny dwelling on a rock in the ocean, surrounded by the swelling waves, cut off from the knowledge and the sight of all'. And yet, he noted, even then he 'would still not be free from the cares of this fleeting world'.[20]

The life of Audomar contains a story of a moment when the river/sea divide unexpectedly shifted, and seemingly safe riverbanks were suddenly entrée into a dangerous sea. In this instance, a young boy who served the bishop Audomar climbed into a boat unsupervised while his patron was taking a nap. 'He wanted to take the boat out between the riverbanks, but he was caught up by the ocean current, and pulled out to sea. Then suddenly a large storm swelled up, and the little boat was tossed about on the waves. Out of control, it got swept into the English Channel. He was suddenly at a place where he could see neither coast, but through prayer he was able to fight the wind and the waves. The boy landed safely on the Saxon shore'.[21] The miracle does not end there — 'fearful of being taken captive, knowing that he was at fault because of his disobedience, he cried and called out to his patron, Audomar. Putting his trust in God, he got back into the boat. And accompanied by the rushing winds, through the wonder of

18 Bede, 'Life of Cuthbert', pp. 48–49.

19 Bede, 'Life of Cuthbert', p. 58.

20 Bede, 'Life of Cuthbert', p. 56.

21 'Vita Audomari Altera', chs 8–9.

God, he was returned to that very site from which he had left, within the hour he was speedily restored to the shore. Climbing out of the boat, he recognized the fields, and penitent, he ran back to his patron'.

Wondrous Beasts

It was not only waves and weather that could lead to surprising developments on the seashore; there were also whales. Even today, whales dominate our imagination in terms of what it means for people to go out into the ocean, and the risks that ocean dwellers experience when they encounter the shore. Whale-watching tours may be the most common way (other than perhaps cruises) that many modern people who do not live on the shore actually go out on the ocean, and whale- and dolphin-strandings or other similar events (such as the whale in the Thames in 2006) remain cultural touchstones, moments when we communally gaze on the denizens of the deep and recognize their otherness, the ways that they do not belong with and among us.[22]

Medieval people similarly recognized the strangeness of both cetacea and their appearance within the human sphere.[23] The late seventh century vita Filiberti describes a mass stranding of dolphins at the mouth of the Loire when the animals were washed up on the shore: 'a great multitude of the fish that are called *marsuppas* (harbour porpoises) found themselves in the river. When the sea receded, two hundred and thirty-seven of them remained on the dry banks'.[24] Yet, rather than reacting with horror and grief, as we might, the medieval monks rejoiced, as the dolphins represented a reprieve from a season of want and famine. Similarly, rather than celebrating the sighting of whales while at sea, medieval authors and those whose stories they recorded, who were not actively seeking the beasts out reacted with amazement and fear.

Another seventh-century author (that of the life of St Gertrude) describes an encounter with a living whale, labelling it straightaway as a time when 'we faced peril on the sea'. They were out at sea, and the waves began to roil all around them; 'And behold, a great and terrible whale appeared to us as if thrown up from the deep; we could not see all of it, but only a part of its back'. Terror led the monks aboard the ship to pray to St Gertrud for deliverance, and, 'Surely, as I heard and saw, I swear to you: as those words were repeated

[22] On medieval encounters with whales, see Szabo, *Monstrous Fishes and the Mead-Dark Sea*.

[23] Parts of this discussion are drawn from my earlier online article: Arnold, 'Washed Ashore'.

[24] 'Vita Filiberti', ch. 39.

three times, leaving, [the whale] sought out the depths, and we were able to make port safely that night'.[25]

It was not only whales and dolphins that attracted seaside attention. The ninth-century life of St Condedus contains a story about a remarkable profusion of fish and a strong tide in the Seine estuary; this story contains remarkable descriptions of the littoral landscape, and through this unusual event we can see the power of seaside marvels to attract crowds; here we glimpse the curiosity of medieval people towards the wonders of the sea. The author explains that at one point the island of Belcinnaca (which was an estuarial island of the Seine that appeared and disappeared several times before the regulation of the Seine) had a sudden 'beaching' of a host of marine fish. 'Such a great number of fish arose abundantly from the sea and the Seine' the author writes, 'that the fishers from other larger/deep rivers who came to investigate it said that they had never seen anything like it, nor heard of such an incredible thing having been seen'.[26] The mass beaching of the fish seems to have been related to an unusual tidal event combined with strange coastal weather. The Seine used to be a much more dynamic river, especially at the estuary. Before modern engineering, the mouth had what is termed a 'tidal bore' — where tides could rush deep inland.

Liminality and Transgression

This seems to have been the case at the time of the great profusion of fish. The author describes the striking sea: 'At that time the waves of the ocean were apple green and crashing against themselves throughout the day and night as anyone walking [along the coast?] could not fail to see with what vigour and strength they were moving, so that they were pushing beyond the island, so that they pushed back the waters of the Seine on the Eastern coast 60 miles, and they reached all the way to the place that was known as Pistas'. So struck was the writer by this tumultuous tide that he compared it to Charybdis, noting that it simply consumed the regular tide: the waves 'pulled back, and again pushed forth [...] the shores of the Seine were seen to recede and come back with the excessively swift flow'. The river was 'filled up suddenly by this inundation, so much that a person found to be even a little bit close to the shore was scarcely able to escape'.

[25] 'Vita Geretrudis', ch. 5.

[26] 'Vita S. Condedi', ch. 5.

In a turn of phrase very typical for descriptions of the tidal bore on the Garonne River further south, he concludes the story by noting the disruption to the river, the ways that this tide not only swallowed up the coast, but inverted and confused the river itself: 'Indeed, the river was seen to turn its course around from below and return most swiftly, and in the manner of the tallest tree and with the greatest roar it rose up, and for many miles around the clear waters of the river turned bitter/brackish. These things are said about the flow and the location of the place, and the flooding of the river'.

In this instance, the tidal surge seems to have been responsible only for a striking display of nature's power. Not all coastal incidents were so benign. Storm surges and giant tides could turn coasts into sites of mass risk and danger. On 26 December 838, according to the Annals of St-Bertin, 'a great flood far beyond the usual coastal tides covered nearly the whole of Frisia. So great was the inundation that the region became almost like the mounds of sand common in those parts which they call the dunes'. The annalist of St-Bertin, clearly not a local to the coastlands, nonetheless has some degree of knowledge about the traditional landscape and also about the extensiveness of this flood event, as he continues, 'Every single thing the sea rolled over, men as well as other living creatures and houses too it destroyed. The number of people drowned was very carefully counted: 2437 deaths were reported'.[27]

The annalist then pointed out that 'then in February an army of fiery red and other colours could often be seen in the sky, as well as shooting stars trailing fiery tails'. The links made between this devastating flood and signs in the sky is clearly one made in part because of the connection between The Flood and the rainbow, but is also a reminder of the ways in which medieval observers saw the shores and their tides as drawn into the larger movements of the celestial heavens, as part of a complex and beautiful natural order.

Some would have us think that the only religious interpretation of this could be fear and abhorrence; Dado of Rouen, the author of the vita of St Eligius, leaves us with a powerful reminder that the sea and the shore and the sky were all knitted together as parts of a breath-taking Creation: 'Heaven is high indeed, and the earth vast, and the sea immense and the stars beautiful but more immense and more beautiful by necessity is he who created them'.[28]

[27] *The Annals of St-Bertin*, p. 42.

[28] Dado of Rouen, 'Life of St Eligius of Rouen', p. 158.

'Verbalist Ingenuity' and the Evidential Basis for Virgin-Soil Smallpox Epidemics in the Sixth Century: From Iona to Ṣanʿāʾ

Timothy P. Newfield*
Departments of History and Biology, Georgetown University

with Stephanie Marciniak (*Department of Anthropology, Pennsylvania State University*) and Bryna Cameron-Steinke (*Department of History, Georgetown University*)

Introduction

The origins of most major pathogenic diseases are murky at best. Smallpox's early history is no different. It is now, and has been for centuries, in flux. When physicians wrote authoritative medical history in-house, what constituted the earliest evidence for smallpox could be a subject of serious debate. Disagreements occasionally boiled over. Few pulled their punches about Johann Hahn's proposal, advanced in his 1733 *Variola antiquitates*, that good evidence for smallpox was to be found in a multitude of ancient Greek and Roman texts (historical, medical, poetic, and theatrical).[1] Hahn may have only been following the lead of Claude Saumaise, the prolific philologist, who in an aside in his 1000-plus-page critique on numerical mysticism, published in 1648, suggested smallpox was hardly unknown to some Greco-Roman

* This paper is dedicated to the memory of Alasdair Ross. I was lucky to spend many evenings with Alasdair at Mr Singh's Indian Cottage while at Stirling in 2013–2014. Before and after then, he was a friend I very much looked forward to seeing. He was a wonderful person. His all-round good nature and dry humour are deeply missed. I will not forget that rainy November afternoon when he took my family for a tour of a Roman fort along the Antonine Wall.

The authors also wish to thank Benan Grams, Laura Goffman, David Gyllenhaal, Michael Morony, and Rachel Singer for e-conversations about alleged evidence for smallpox in the sixth century.

[1] Hahn, *Variolarum antiquitates*.

'With Our Backs to the Ocean': Land, Lordship, Climate Change, and Environment in the North-West European Past. Essays in Memory of Alasdair Ross, ed. by Richard Oram (Turnhout: Brepols, 2023)
EHNAW 5, pp. 25–82 BREPOLS PUBLISHERS 10.1484/M.EHNAW-EB.5.131270

writers.[2] Nevertheless, Hahn's assertion that smallpox was masquerading as 'anthrax', 'carbuncle', and hazy references to pitted complexions in ancient texts was lambasted. Physician-poet Paul Werlhof wasted no time disassembling the claim,[3] and many countenanced his efforts. Over time Hahn's 'scholastic prating' became the stuff of 'libellers' and was considered 'so absurd that it cannot for a moment be listened to'.[4]

[2] Saumaise, *De annis climactericis*, pp. 726–27. Hahn's thinking contrasted sharply with others at the time, such as Johannes Friend, who voiced his belief that the Greeks were unacquainted with smallpox in his history of medicine, published and translated multiple times both before and after the appearance of Hahn's treatise: Freind, *The History of Physick*, I, p. 274 and II, pp. 188–89; Friend, *Histoire de la Medecine*, trans. by Coulet, I, pp. 145–46 and II, pp. 99–100; Friend, *Histoire de la Medecine*, trans. by Senac, pp. 112, 202–04; Friend, *Historia Medicinae*, pp. 137–38. Daniel Le Clerc, in his continuation of his 1696 *Histoire de la médecine*, likewise asserted that Greeks and Romans said nothing of smallpox, believing it either took its rise after antiquity or that it had simply failed to reach the Greco-Roman world: Le Clerc, 'Essai d'un plan pour servir à la continuation de l'histoire de la medecine', I, pp. 776–77. Many earlier physicians agreed: not Greek, not Roman, but Arab medical texts furnished the first references to smallpox: de Fonseca, *Consultationes medicae singularibus remediis refertae*, p. 170; Sennert, *De Febribus Libri IV*, p. 464; Porchon, *Nouveau traitte du pourpre*, pp. 8–9. Thomas Sydenham, the 'English Hippocrates', too was sure, '[...] nec apud Hippocratem vestigium ullum nec apud Galenum invenitur [...]': *Observationes medicae*, pp. 260–61. Jean-Jacques Paulet was critical of Saumaise's ancient variola: *Histoire de la petite vérole*, I, pp. 53, 70.

[3] Werlhof published within two years of Hahn: *Disquisitio medica et philologica de variolis et anthracibus*.

[4] Shortly after Werlhof, Richard Mead remarked that those who see smallpox in Greco-Roman writings are 'widely mistaken': *De Variolis et Morbillis Liber*, p. 2; Mead, *A Discourse on the Smallpox and Measles*, p. 2. Paulet concurred: *Histoire*, I, pp. iii, 2, 4, 25, 44–45, 51, 57. For Paulet, Hahn had forced the issue: 'À travers cette vaste érudition dont son livre est rempli, on le voit toujours à la torture, forcer tous les passages où íl recontre ce mot' (*Histoire*, I, p. 37). Further, he waxed 'Ceux qui cherchent la petite vérole dans l'Antiquité, ne la voyant jamais décrite, pa trouvent partout, parce qu'elle n'y est point' (*Histoire*, I, p. 57). Woodville, *The History of Inoculation of the Smallpox*, I, pp. 4–6, was certain Greco-Romans did not know the smallpox he knew. Hahn's efforts, he noted, were 'judiciously refuted by Welhoff' and 'do not seem to have produced one convert'. James Moore, another tough critic of Hahn's work, thought Hahn's efforts 'fruitless' and 'satisfactorily refuted', 'ridiculed agreeably' and 'little worth preserving': *The History of the Smallpox*, pp. 1–4. Although many others took it as fact that smallpox was not to be found in ancient texts, like Ainslie, 'Observations Respecting the Small-Pox', p. 52, Robert Willan followed Hahn's lead, finding smallpox almost everywhere he looked in Greco-Roman and early medieval sources. Alarmingly, he seems not to have cited Hahn: 'An Inquiry into the Antiquity of the Smallpox, Measles, and Scarlet Fever'. Slightly better, in his preface to Willan's text, Ashby Smith briefly alludes to Hahn's efforts, underlining that most do not share Willan or Hahn's view: Smith, 'Preface', pp. vi–vii. Soon after Baron

This debate over smallpox's origins and early geography, which carried on well into the twentieth century,[5] pivoted on the interpretation of diverse strands of written evidence. Passages from the Hippocratic corpus and Galen were essential, as were the remarks of Rhazes, but more arcane, obscure and dubious bits of material factored significantly too. There was, for instance, the account of François d'Entrecolles, once head of the French Jesuit mission to China, about various 'livres chinois' on variolation which he leafed through in the Forbidden City. He wrote about his discoveries in a letter dated 11 May 1726 to historian Jean-Baptiste Du Halde who soon published it in the *Lettres édifiantes et curieuses* which were made available in several European languages mid-century. The letter included transcriptions of passages 'fidèlement traduit' about nasal insuffiation and allusions to recipes against smallpox surviving the alleged book burning of Qin Shi Huang (213 BCE) and supposedly predating (Golden-Age) Greece itself.[6] Then there was the speculation of the onetime

suggested Thucydides furnished 'as accurate an account of the leading symptoms of Variola as could possibly be expected from any historian not medical': *Life of Edward Jenner*, p. 177. While some more recent works, like Dixon, *Smallpox*, pp. 187, 188, have argued that smallpox was not known in the ancient Mediterranean region, the tide began to turn in favour of Hahn in the mid-nineteenth century. Notably, Haeser was confident that the Athenian plague was not smallpox but that Greco-Romans, and Mediterraneans before them, possibly knew the disease. He wagered the disease is invisible in ancient texts because of ambiguous ancient terminology and the symptoms of the disease may have changed overtime: *Lehrbuch der Geschichte der Medicin und der Volkskrankheiten*, pp. 17, 78–79, 143, 255. More in line with Hahn and Willan yet, Hirsch thought that 'andeutungen' of the disease could be found in Greco-Roman medical texts, though if Galen knew the disease he did not describe it in detail, and the reported breadth of some disease outbreaks in the first Common Era centuries itself suggested smallpox: 'Blattern', pp. 214–16. Baas, *Grundriss der geschichte der medicin und des heilenden standes*, pp. 147–48, thought Roman-era Mediterranean plagues were possibly smallpox. Now, again, there is little doubt. Zinsser expressed doubt about Greco-Roman smallpox, but then suggested it could have contributed to the Athenian and Antonine plagues: *Rats, Lice, and History*, pp. 123–24, 136–37, 139. Some scholars have pushed the idea more than others, notably Littman, 'The Athenian Plague: Smallpox', pp. 261–75; Littman, 'The Plague of Syracuse: 396 BC', pp. 110–16; Littman and Littman, 'Galen and the Antonine Plague', pp. 243–55; Littman, 'The Plague of Athens: Epidemiology and Paleopathology', pp. 456–67. Tucker, *Scourge*, p. 6, and Hopkins, *The Greatest Killer*, pp. 19–23, had little doubt smallpox spread across the ancient Mediterranaean world.

[5] The pursuit of smallpox's origins continues, of course, but now with molecular clocks and ancient viral DNA. See below. For an early account of what could be called the first failed phase of the debate, see: Hirsch, 'Blattern', pp. 214–15.

[6] Dentrecolles, 'Lettre', pp. 304–61. Another Jesuit missionary in Beijing, Pierre-Martial Cibot, wrote a smallpox treatise in the 1760s which further popularized the idea that small-

Bengalese governor and accused forger John Holwell, published in 1767, that smallpox's Indian antiquity must have been very deep for what he called the '*Aughtorrah Bhade* scriptures' referred to the very same 'Gootee ka Tagooran' or 'Goddess of Spots' appealed to amidst the smallpox season in his own day. Those scriptures, Holwell maintained, were published 3366 years before his writing.[7]

As assorted as the evidence may have been, it was not enough. This hunt for smallpox's cradle failed. It was not for a lack of effort. None of the evidence was not challenging to interpret or altogether indisputable. This was true of the physical material introduced too. Take the visible 'rash' (desiccated pustules?) on the mummified cheeks of Ramesses V (d. *c.* 1145 BCE). Shortly after the thirty-something pharaoh was unwrapped on 25 June 1905 some concluded, solely on the basis of his approximately 3000-year-worn complexion, that smallpox first showed itself in Egypt.[8] Although others were not convinced and

pox had persisted in China for millennia, 'trios mille ans' or since 1122 BCE according to his calculation: Cibot, 'De La Petite Vérole', pp. 392, 397. Not long afterwards this accuracy was considered 'specious', for example Woodville, *History*, pp. 10–11 and Wilks, *Historical Sketches*, III, p. 17. Wilks claimed reliable evidence for inoculation in India as early as the sixth century CE and thought that it was more than probable that smallpox was a new disease then in India and perhaps in China too: *Historical Sketches*, pp. 17, 21. More carefully, Nicholas argued the earliest evidence for smallpox in India predates the seventh century: 'The Goddess Śītalā and Epidemic Smallpox in Bengal', pp. 21–44; Fenner and others, *Smallpox*, pp. 211, 214, followed. For Hopkins, who drew on Nicholas too, smallpox was in India for 'for a very long time'. He likewise thought smallpox present early in China, but not as early as most had speculated. Proposing 'Huns' introduced it in 249 BCE, he picked the end date of the Chou dynasty rather than the start date (1122 BCE) which d'Entrecolles picked up: *Greatest Killer*, pp. 16–18. Nevertheless, on the basis of d'Entrecolles and Holwell's writings it was asserted commonly that smallpox was present very early in India and China, for example: Schnurrer, *Chronik der seuchen, I*, pp. 4, 16, 54, 156; Haeser, *Lehrbuch*, p. 255; Hirsch, 'Blattern', pp. 215–16; Zinsser, *Rats*, pp. 106–07, 107 n. 1; Dixon, *Smallpox*, p. 188; Tucker, *Scourge*, p. 6.

7 Holwell, *An Account of the Manner of Inoculating of Small Pox in the East Indies*, pp. 7–8. Holwell as forger or fraud victim: App, *The Birth of Orientalism*. Others, on yet weaker grounds, speculated the practice of inoculation, and thereby smallpox, had been established in the Caucasus very early: Chais, *Essai apologétique sur la méthode de communiquer la petite vérole par inoculation*, p. 6.

8 Smith, *The Royal Mummies*, pp. 90–92, plate LVI. Ramesses was one of several purportedly poxed mummies. Fergunson and Ruffer thought Ramesses' mummified rash smallpox, but faced criticism early on: Ruffer, 'An Eruption Resembling that of Variola', pp. 32–34; Ruffer, 'Pathological Notes', pp. 175. Fergunson and Ruffer seem to have dismissed ('Pathological Notes', p. 172) Smith's description of 'a large irregularly triangular deep ulcer' in the right groin suggestive of 'an open bubo' (*Royal Mummies*, p. 91). Among others, Fenner and others, *Small-*

doubt was promptly cast both on Ramesses' diagnosis and the Egyptian genesis, some continue to hold the rash proves smallpox's early presence in north Africa.[9]

This pursuit for smallpox's origins failed because it was flawed from the outset. Not only was there disagreement about what was enough for a smallpox diagnosis, or what source type could afford a diagnosis, but the supposition that smallpox would have left a decent trace about where and around when it first emerged was wishful if not delusional thinking.[10] Complicating the matter, attempts to reveal smallpox's 'remotest age' sometimes privileged hints of alleged remedies over actual descriptions of disease. So, Dio Cassius' needles, illicitly tainted, the Roman historian tells us, to spread illness in 189 CE, were interpreted as early variolization efforts and exploited to put smallpox on the Roman map.[11] More concerning, many today would argue that the hunt was on in the eighteenth and nineteenth centuries for something that could not be seen, or which was, in fact, unseen; that to locate a disease's cradle one must follow the trail of the pathogen, not the disease. Without pathogens, let alone the Variola virus (VARV), the cause of smallpox, there was, by current standards at least, no hope.[12]

pox, pp. 210–11, speculate, on the basis of mummified soft tissues, that variola occurred in Egypt earlier than elsewhere. The very roughly dated Ebers Papyrus, which was found between the legs of a mummy in Thebes and describes a disease of the skin some have thought indicative of smallpox, has been used occasionally to uphold Ramesses' diagnosis. Hopkins, *Greatest Killer*, pp. 15–16.

[9] As Ruffer noted in 1911 ('An Eruption Resembling that of Variola'). To this day the speculative diagnosis of Ramesses' rash remains integral to accounts of smallpox's history despite several failed attempts to obtain a laboratory confirmation. Failed lab attempts: Hopkins, 'Ramses V', pp. 22–26; Lewin, '"Mummy" Riddles Unravelled', pp. 3–8; McCollum and others, 'Poxvirus Viability', pp. 177, 180, 181. For tentative smallpox diagnoses of Ramesses and other Egyptian mummies given prominent place in accounts of the virus' history, see: Zinsser, *Rats*, p. 108; Dixon, *Smallpox*, pp. 188–89; Fenner and others, *Smallpox*, pp. 210–11; Tucker, *Scourge*, pp. 5–6; Hopkins, *The Greatest Killer*, pp. 14–16; Khalakdina, Costa, and Briand, 'Smallpox', p. 257. Were the diagnosis to be, like Ramesses, dethroned, more scholars might consider the possibility that smallpox is not a multi-millennia-old pathogen.

[10] Haeser, *Lehrbuch*, p. 255; Hirsch, 'Blattern', p. 216.

[11] Dio Cassius, *Roman History IX*, LXXIII.14.3–4, pp. 100–01. Many have turned this into a smallpox diagnosis, for example: Willan, 'Inquiry', p. 51.

[12] Variola was identified in the laboratory in the twentieth-century interwar period. Naturally, the study of origins before germ theory demands attention, so too non-linear histories of smallpox-like disease, but pre-germ-theory conclusions about smallpox's geography will satisfy few today interested in the origins of VARV.

There were yet more problems. As physician-cum-medical-historian Charles Creighton observed in 1891, the whole enterprise suffered from a hefty dosage of 'bigotry and intolerance'.[13] The epidemiological othering of many Europeans interested in the subject ensured that the disease originated outside of Europe. Indeed, that smallpox was born in Europe was nearly unheard of.[14] Were it present in the Greco-Roman world, as Hahn wanted, most would have undoubtedly argued it arrived from elsewhere.[15]

That smallpox's cradle was primordial was another guess, one which could likewise skew how sources were read.[16] No matter whether one looked to Arabia, China, Greece or India, there was a push to assign smallpox to deep antiquity.[17] Not to do so after Edward Jenner was in some eyes to 'belittle the advantages' of vaccination, but by Jenner's time that smallpox was ancient was already old news.[18] Even those who admitted that the disease 'originated from causes so perfectly incomprehensible, as to set at defiance all rational conjecture', as did one early advocate of Jennerian vaccination, argued the disease had undeniably persisted more than a millennium.[19] Whatever the impetus, few questioned that smallpox had been around for a long time.[20]

[13] Creighton, *History of Epidemics in Britain*, p. 440. Creighton himself, however, was confident that smallpox was known in ancient China and India, that it was not originally a disease of temperate regions and that it made its way to Europe via Arabia: *History*, pp. 441, 444–45.

[14] Willan repeatedly underscored the significance of the early evidence for smallpox in Greco-Roman texts. Seeking only to disprove the claim that the disease appears in Arabic texts before Greek or Latin ones, he paid no attention to alleged evidence for the disease elsewhere: 'Inquiry', pp. 1, 16–17, 115.

[15] Cf., for example, Schnurrer, Haeser and Hirsch in notes 5 and 6 above.

[16] So 'absurdes' and 'mal fondés', and evidently so popular, was the idea that humanity's creation and smallpox were coeval that Paulet devoted most of a chapter of his work on petite vérole attempting to refute it: *Histoire*, I, pp. ii, 3–24. It had been argued smallpox and humankind were innate, for example: Willis, 'De febribus', ch. 15, p. 171.

[17] Additional to most of the aforementioned, like those of Hahn, Paulet, Woodville and Willan, see Schnurrer, *Chronik*, I, pp. 4, 16, 54, 156; Haeser, *Lehrbuch*, pp. 17, 254–55; Hirsch, 'Blattern', pp. 214–16. The thinking that smallpox is many millennia old remains present in high-profile publications, despite the lack of good evidence, archaeological, palaeogenomic, written or otherwise, for example: Khalakdina, Costa, and Briand, 'Smallpox', p. 257; Kirby, 'WHO Celebrates', p. 174.

[18] Creighton, *History*, p. 440.

[19] Woodville, *History*, p. 2.

[20] Notably, Richard Mead, a chief advocate of the Ethiopian origin of smallpox, thought the disease modern. Like others who favoured a less than ancient origin, however, Mead advo-

Throughout all of this, the only thing that seemed certain, or nearly so, was that people in Arabia, in the vicinity of Mecca, encountered smallpox in the mid-sixth century CE. No matter whether one thought the disease existed earlier or not, for many early commentators on smallpox's past, there was no doubting that the disease was a notable element of the disease burden in Southwest Asia not only in Rhazes' lifetime, but in Mohammad's too.[21] Byzantine Alexandria, in the early seventh century, was the only other place the disease was regularly identified,[22] though some, as we will see, argued as well that smallpox diffused through parts of Western Europe in the sixth century. In other words, late antiquity was the disease's surest 'remotest age'.

The following pages rehearse, but do not revive, these old ideas about smallpox's past, now relics of late antique medical history overshadowed by the Justinianic Plague (541–544) and the first plague pandemic (541–750).[23] It may seem like debates and musings a century-or-more-old have little bearing, or should have little bearing, on any history of smallpox in late antiquity we write now, but that could not be further from the truth. Our preconceptions about smallpox's past, the very ideas we possess when we start researching and writing about it, and how, therefore, we come to our sources — whatever they may be and no matter how novel they are — have been shaped, knowingly or

cated that smallpox was 'a modern disease' because it 'was not known to the ancient Greek and Roman physicians'. The disease was modern in the sense that it postdated Galen but predated Rhazes: Mead, *De Variolis*, pp. 2–3, 7; Mead, *Discourse*, pp. 2–3, 8. Paulet might not have thought the disease born in Ethiopia (*Histoire*, I, p. 62), but he too considered smallpox 'une maladie nouvelle' because we meet it first in the sixth century: *Histoire*, I, pp. 4, 25, 77, 92. Earlier authors too thought smallpox something novel because ancient authorities wrote nothing about it, Fonseca, *Consultationes*, p. 170; Sennert, *De Febribus*, p. 464; Le Clerc, 'Essai', pp. 776–77; Freind, *History*, I, p. 274 and II, p. 188.

21 This is discussed below. For most early commentators, though, smallpox appeared in Arabia a year or three before it did in Europe. So Freind has smallpox 'first observ'd…and describ'd by the Mahometans': *History*, II, p. 188. Notably, as Paulet was quick to point out, Voltaire, 'qui aperçoit tout', blamed smallpox on Mohammad: 'La petite verole que l'on doit, dit-il, a Mahomet': *Histoire*, I, p. 2. Voltaire went further to say that smallpox, which killed more than war, originated in Arabia: *Histoire de l'empire de Russie sous Pierre le Grand*, p. 74.

22 Thanks to the writings of Aaron of Alexandria: Paulet, *Histoire*, I, pp. 95, 103–04; Willan, 'Inquiry', p. 2; Moore, *History*, pp. 56, 61–62; Haeser, *Lehrbuch*, p. 255; Fenner and others, *Smallpox*, pp. 212, 214; Hopkins, *Greatest Killer*, p. 26.

23 On late antique plague see, to start, Biraben and Le Goff, 'La peste dans le Haut Moyen Age'; Horden, 'Mediterranean Plague in the Age of Justinian'; Little, ed., *Plague and the End of Antiquity*; Keller and others, 'Ancient *Yersinia pestis*'; Mordechai and others, 'The Justinianic Plague'.

not, by the disease's previous interpreters. While histories of premodern illness often fail to acknowledge the influence of writers of earlier generations, here we try not to lose sight of the weight of previous thinking and how it shapes exisiting narratives.

This paper addresses the evidential basis for 'virgin-soil' smallpox in mid-sixth-century Europe and asks whether smallpox lurks in the writings of Gregory or Tours, Marius of Avenches, an anonymous Irish annalist, or Adomnán of Iona, writers who, at one point or another, have been drawn on in the quest for smallpox's cradle. In doing so we offer an overdue corrective to the epidemiological othering that has led many commentators on smallpox's past to assert that the disease either emerged, or took root, in Arabia before being introduced to other regions in Southwest Asia, North Africa and Europe by the armies of the Rashidun and Umayyad caliphates.

That smallpox was hardly unknown in Arabia in the mid-sixth century and that it facilitated Islamic conquests over the next two centuries were once fashionable claims, ones still commonly repeated, that require careful evaluation.[24] While physician-historians were convinced centuries ago that the 'irruption des Sarrazins' went hand-in-hand with the diffusion of smallpox, and that 'to follow the progress of the Small Pox is to proceed with the history of the Arabs',[25] they were divided on the scale of a mid-sixth-century epidemic considered pivotal in most histories of the disease. Led astray by their orientalist and Islamophobic thinking that Arab armies had to have introduced smallpox to Europe, some argued that the epidemic observed at Mecca in the sixth century was not smallpox or that it remained limited and the disease confined to Arabia and possibly East Africa until the 630s. Others, however, were confident a large outbreak spilled out of Arabia about the time of Mohammad's birth and arrived in Europe soon afterwards. Then there were those who thought both scenarios true.[26] We are convinced of neither.

There is no doubting that good evidence for those old, contemptuous, orientalist claims is lacking, no matter how often they continue to crop up. At the same time, the available written sources are problematic enough that

[24] This is particularly so considering that more recent surveys of smallpox's past have perpetuated this idea. For instance, Zinsser, *Rats*, pp. 124–25 n. 8; Dixon, *Smallpox*, pp. 189, 190; Tucker, *Scourge*, p. 6; Hopkins, *Greatest Killer*, p. 26.

[25] Paulet, *Histoire*, I, p. 96; Moore, *History*, p. 63.

[26] As discussed below, Paulet wagered that smallpox spread widely in the 570s, from Arabia to continental Europe, but he also ventured that the disease diffused anew in with Arab armies from the seventh century, albeit after they contracted the disease in Egypt.

it is unclear what should be done with them. With similar collections of late antique accounts of disease outbreaks, historians have traditionally gone one of three ways.[27] One approach would be to read the evidence as comprising a series of interconnected smallpox epidemics that spanned the Hebrides to Arabia, in effect a variola virgin-soil pandemic. A second approach would argue that the reported epidemics only faintly resemble smallpox and that grounds for stitching the individual epidemics together are weak; so, no smallpox and no pandemic. A third approach, a middle ground, would find smallpox in some of our texts, but not necessarily a pandemic.

However alluring the pandemic hypothesis may be, it demands much more evidence than is presently available. At the same time, we cannot pretend to claim conclusively that the epidemics our authors report were entirely distinct and altogether non-variolous. Some of the passages assessed below read more like smallpox than others and each presents a number of serious complications. The mid-sixth-century European evidence for a smallpox-like disease, however, does rank among the best available for smallpox before Rhazes. Not only is it much better than what d'Entrecolles or Holwell offered, but it is good enough, we argue, to tentatively put smallpox, or something like it, in mid-sixth-century Europe and thereby to unravel the reckless claim that Arab armies first introduced smallpox to the wider Mediterranean region and Europe in the seventh and eighth centuries, easing the business of conquest.[28]

While smallpox-like disease may have been introduced repeatedly to many areas of Europe before cities there were populous enough to sustain it, the written evidence for smallpox in sixth-century Europe is far superior to the evidence for it there in the seventh or eighth centuries. Further undercutting the claim that Arab conquests initially disseminated the disease, plausible evidence for smallpox in the context of those conquests is nonexistent or very nearly so.[29]

[27] In the debate over the second-century Antonine Plague, for instance, the written sources have meant everything or next to nothing. Cf. Schnurrer, *Chronik*, pp. 90–94; Hecker, *De peste antoniniana commentatio*; Gilliam, 'The Plague Under Marcus Aurelius'; Duncan-Jones, 'The Impact of the Antonine Plague'.

[28] On these claims see below.

[29] Among the alleged evidence for smallpox during the conquests are a few supposedly pox-marked caliphs (Woodville, *History*, p. 25; Moore, *History*, pp. 64–65; Schnurrer, *Chronik*, p. 155) and the south-west Asian epidemic of 638–39, known as the Plague of Amwās. Many now think that that epidemic was an outbreak of *Yersinia pestis*, a Justinianic Plague recurrence, but others have argued it was smallpox. Cf., for instance, Schnurrer, *Chronik*, p. 155 with Dols, 'Plague in Early Islamic History', pp. 376–79, Conrad, 'Tāʿūn and Wabā'', pp. 298–301.

Indeed, better evidence for the disease in late antique Arabia, though itself problematic, is found in the sixth century and as discussed below, may be connected to the evidence from Europe. And while extant texts do give reason to suspect that smallpox or something that looked like it afflicted several European regions decades before the rise of the first caliphate, the current evolutionary biology of VARV is at odds with the idea that smallpox, as we clinically and epidemiologically understand it, was anywhere in the sixth century.

We tackle these issues below. We begin with the written sources, before transitioning first to the origins and trajectory of any interregional smallpox-like outbreak in the sixth century and second to the paleogenomics and evolutionary history of VARV. Throughout, we aim for transparency and to demonstrate the range of ways in which the evidence could be combined and read. In the process, we hope to demonstrate the complications one must overcome when writing histories of late antique disease. One must not only sort through and make sense of the written sources and come to terms with data from paleogenomics and evolutionary biology, but also get a handle on formative work that has influenced the way we think about a disease and outbreaks of it. All considered, it may come as no surprise that the late antique history of smallpox, like the late antique history of other diseases, is in the midst of a period of change. Definitive histories may be out of reach for the time being, but it is for the better, not the worse, that things are unsettled. With closer attention to the range of data at hand and to the foundations of old narratives, new, more reliable disease histories will emerge.

Van Ess proposed this epidemic was smallpox: *Der fehltritt des gelehrten*, pp. 279–85, but see Stathakopoulos, *Famine and Pestilence*, pp. 349–50. Whether smallpox or not, multiple aspects of the Plague of Amwās do not square with the idea that smallpox was native to Arabia or that Arab soldiers introduced it to former Byzantine lands. Notably, Arab forces had been in the region for several years before they took Amwās (Emmaus Nicopolis in Byzantine Palestine), and therefore before the outbreak commenced, and sources claim the epidemic killed at least twenty thousand Arab soldiers: Ibn Wāḍiḥ al-Yaʿqūbī, 'The History (Taʾrīkh)', p. 780. That accounts of this plague say nothing of the disease spreading in the local population also works against the idea that Arab armies then and there introduced it to the region. The claim that Arab armies brought the disease to Egypt, in the conquest of 640 (Freind, *History*, II, p. 189; Woodville, *History*, pp. 5–6, 12, 19; Schnurrer, *Chronik*, p. 155), is likewise problematic. That the 665 epidemic in southern Turkey identified in Arabic sources as 'judari' (Stathakopoulos, *Famine*, p. 352) was in fact smallpox is as well uncertain. The clinical expression of that epidemic is not described and it should not be assumed *judari* was employed systematically in late antiquity to refer to smallpox as we know it (see below). In any case, that outbreak is reported to have occurred within an Arab army.

The Pusulae of Gregory of Tours

Most histories of disease that touch on the sixth century lean weightily on the writings of Gregory of Tours. This history is no different. The well-connected promoter of St Martin is one of the principal witnesses to the epidemics of the 500s and he wrote with some frequency about disease outbreaks. Recent histories of late antique plague, malaria, and bovine epizootics have thoroughly exploited his historical and hagiographic texts.[30] Histories of smallpox once did too.

Jean-Jacques Paulet, physician, *mycologue* and *epizootiques* aficionado, identified smallpox time and again in Gregory's writings in his two-volume history of *petite vérole* of 1768.[31] Paulet was certain the bishop knew smallpox, even though Gregory was not, he stressed, a doctor. In fact, the observations on epidemics that Gregory made in his lengthy ten books of history were enough to permit Paulet to assert firmly that the sixth century was 'le siècle de petite vérole'.[32] This was exceptional, as not only had Paulet failed in 51 previous pages to identify smallpox in any earlier European writing, but previous commentators on smallpox's past had overlooked the Frankish bishop.[33] Creighton claimed a century later that Gregory's contributions to smallpox history were 'much-debated', but he only did so because a few smallpox historians, blinded by their epidemiological orientalism, refused to believe that smallpox could be found in the *Decem libri historarum* (*DLH*) or any European text before the Arab conquests.[34] In any case, over the years, a number of passages from

[30] Biraben and Le Goff, 'La peste'; Horden, 'Disease, Dragons and Saints'; Bachrach, 'Plague, Population, and Economy'; Newfield, 'Early Medieval Epizootics'; Faure and Jacquemard, 'L'émergence du paludisme en Gaule'; Newfield, 'Malaria and Malaria-Like Disease'; McCormick, 'Gregory of Tours on Sixth-Century Plague'.

[31] Paulet, *Histoire*, I, pp. 79–87.

[32] Paulet, *Histoire*, I, pp. 80, 93. Gregory 'ne fût pas médecin'. Early commentators on smallpox's past privileged medical treatises over historical texts. They seem to have thought physicians more likely to recognize smallpox when they saw it or more capable of providing a reliable account of symptoms. Most were physicians themselves.

[33] Paulet, *Histoire*, I, pp. 25–76, sorts through Greco-Roman texts as well as late antique writings predating 500 CE.

[34] Creighton, *History*, p. 440. For a pox-marked *Decem libri historiarum*, see: Smith, 'Preface', pp. xxiv–xxv; Willan, 'Inquiry', pp. 87–93; Moore, *History*, pp. 7–10; Schnurrer, *Chronik*, pp. 138–42; Haeser, *Lehrbuch*, p. 255; Hirsch, 'Blattern', p. 216. Fuelling Creighton's remark, Moore questioned whether Gregory had in fact described smallpox. On this point, Moore was Paulet's Werlhof. His ridding of the bishop's text of variola, however, was greatly influenced

Gregory's principal text have been considered relevant to smallpox's past. They are presented here in full and in sequence.

> **V.34** A most serious epidemic [gravissima lues] followed these prodigies. While the Kings were quarrelling with each other again and once more making preparations for civil war, dysentery [desentericus morbus] spread throughout the whole of Gaul. Those who caught it had a high temperature, with vomiting and severe pain in the small of the back: their heads ached and so did their necks. The matter they vomited up was yellow or even green. Many people maintained that some secret poison must be the cause of this. The country-folk imagined that they had boils [pusulas] inside their bodies and actually this is not as silly as it sounds, for as soon as cupping glasses were applied to their shoulders or legs, great tumors [visicis/vesica] formed, and when these burst and discharged their pus they were cured. Many recovered their health by drinking herbs, which are known to be antidotes to poisons. The epidemic began in the month of August. It attacked young children first of all and to them it was fatal: and so we lost out little ones, who were so dear to us and sweet, whom we had fed and nurtured with such loving care. As I write I wipe away my tears and I repeat once more the words of Job the blessed: 'The Lord gave, and the Lord hath taken away; as it hath pleased the Lord, so is it come to pass. Blessed be the name of the Lord, world without end.' In these days King Chilperic fell ill. When he recovered, his younger son, who had not yet been baptized in the name of the Holy Ghost, was attacked in his turn. They saw that he was dying and so they baptized him. He made a momentary recovery, but then Chlodobert, his older brother, caught the disease. When their mother Fredegund realized that she, too, was at death's door, she repented of her sins, rather late in the day, it is true, and said to the King: 'God in his mercy has endured our evil

by his ideas about the disease's origin and early diffusion. While Moore merely expounded upon the idea popularised by others that smallpox had spread to Europe initially with the Arab conquests, as discussed below, he was among the first to refute the claim that Gregory had documented smallpox in an effort to keep the association between the disease and the spread of Islam intact. Others who proposed Europeans met smallpox and Islam simultaneously either did not know of Gregory's writings or turned a blind eye to them: Freind, *History*, II, p. 190; Woodville, *History*, pp. 18, 22; Monro, *Observations on the Different Kinds of Smallpox*, p. 48. That Gregory knew smallpox has been treated as fact more recently: Zinsser, *Rats*, pp. 124–25 n. 8; Dixon, *Smallpox*, p. 190; De Raymond, *Querelle de l'inoculation*, p. 12; Hopkins, *Greatest Killer*, pp. 24–25; Devroey, *Économie rurale et société*, p. 46; Devroey, 'Catastrophe, crise et changement social', p. 154. 'The earliest undeniable description of smallpox occurring in the West', for Cockburn, 'is that of Gregory of Tours, A.D. 580, who gives a detailed account of cases he has seen': *Evolution*, p. 86. Taking Leven's lead, Stathakopoulos, *Famine*, pp. 97, 313–14, hesitated to see smallpox in Gregory or Marius of Avenches' writings. Leven, 'Zur Kenntnis der Pocken', p. 346; cf. McCormick, 'Gregory', pp. 60, 91. Variola was visible in earlier sources, however: Stathakopoulos, *Famine*, pp. 91–95.

goings-on long enough. Time and time again He has sent us warnings through high fevers and other indispositions, but we never mended our ways, now we are going to lose our children.' [...] Meanwhile their youngest son wasted away before the onslaught of the disease and finally died. With broken hearts they carried him to Paris from their estate at Berny, and buried him in the church of Saint Denis. As for Chlodobert, they placed him on a stretcher and carried him to the church of Saint Medard in Soissons. They set him down before the Saint's tomb and made vows for his recovery. He died in the middle of the night, worn to a shadow and hardly drawing a breath.[35]

V.35 At the same time died Austrechild, King Guntram's Queen, and of the same disease [morbus].[36]

V.36 Nantinus, Count of Angoulême, was another who wasted away with this disease [aegritudo] and then died [...] A few months later he fell ill with dysentery [morbus] [...] his temperature rose higher and higher [...] his bodily strength

[35] Gregory of Tours, *The History of the Franks*, pp. 296–97. 'Sed haec prodigia gravissima lues est subsecuta. Nam et discordantibus reges et iterum bellum civile parantibus, desentericus morbus paene Gallias totas praeoccupavit. Erat enim his qui patiebantur valida cum vomitu febris renumque nimius dolor; caput grave vel cervix. Ea vero quae ex ore proiciebantur colore croceo aut certe viridia erant. A multis autem adserebatur veninum occultum esse, — rusticiores vero coralis hoc pusulas nominabant — quod non est incredibile, quia missae in scapulis sive cruribus ventosae, procedentibus erumpentibusque visicis [visices/vissicis/vesicis/virius], decursa saniae, multi liberabantur. Sed et herbae, quae venenis medentur, potui sumptae, plerisque praesidia contulerunt. Et quidem primum haec infirmetas a mense Augusto initiata, parvulus aduliscentes arripuit lectoque subegit. Per dedemus dulcis et caros nobis infantulos, quos aut gremiis fovimus aut ulnis baiolavimus aut propria manu, ministratis cibis, ipsos studio sagatiore nutrivimus, Sed, abstersis lacrimis, cum beato Iob dicimus: Dominus dedit, Dominus abstulit; quo modo Domino placuit, ita factum est. Sit nomen eius benedictum in saecula. Igitur in his diebus Chilpericus rex graviter egrotavit. Quo convaliscente, filius eius iunior, necdum aqua et Spiritu sancto renatus, aegrotare coepit. Quem in extremis videntis, baptismo abluerunt. Quo parumper melius agente, frater eius senior nomen Chlodoberthus ab hoc morbo correpitur; ipsumque in discrimine mortis Fredegundis mater cernens, sero penetens, ait ad regem: "Diu nos male agentes pietas divina sustentat; nam sepe nos febribus et aliis malis corripuit, et emendatio non successit. Ecce! iam perdimus filios." [...] Post haec infantulus iunior, dum nimio labore tabescit, extinguetur. Quem cum maximo merore deducentes a villa Brinnaco Parisius, ad basilicam sancti Dionisie sepelire mandaverunt. Chlodoberthum vero conponentes in feretro, Sessionas ad basilicam sancti Medardi duxerunt, proicientesque eo ad sanctum sepulchrum, voverunt vota pro eo; sed media nocte anilus iam et tenuis spiritum exalavit'. Gregory of Tours, *Libri historiarum decem*, pp. 238–41.

[36] Gregory of Tours, *History*, p. 298. 'His diebus Austrigildis Guntchramni principis regina ab hoc morbo consumpta est; sed priusquam nequam spiritum exalaret, cernens, quod evadere non posset, alta trahens suspiria'. Gregory of Tours, *Libri historiarum decem*, pp. 241–42.

failed [...] In his dying hours his body became so black that you would have thought it had been placed on glowing coals and roasted.[37]

VI.8 Eparchius, the recluse of Angoulême, was the next to die. He, too, was a person of great sanctity, and God performed many miracles through his agency. He was originally an inhabitant of Périgueux, but, after his conversion and his ordination as a priest, he went to Angoulême and built himself a cell there [...] Eparchius arranged for the freeing of a great number of prisoners by using the alms and oblations of the faithful. By making the sign of the Cross over them he would destroy the poison in malignant boils [pusulae malae], by prayer he would cast out evil spirits from bodies which were possessed [...] When he had been a recluse for forty-four years he fell ill of a fever and died.[38]

VI.14 This year the people suffered from a terrible epidemic [lues]; and great numbers of them were carried off by a whole series of malignant diseases, the main symptoms of which were boils and tumours [pusulis et vissicis]. Quite a few of those who took precautions managed to escape. We learned that a disease of the groin [inguinarium morbum] was very prevalent in Narbonne this same year, and that, once a man was attacked by it, it was all up with him.[39]

[37] Gregory of Tours, *History*, p. 299. 'Hac itaque aegritudine et Nanthinus Equolisinensis comes exinanitus interiit. Sed quae contra sacerdotes vel ecclesias Dei egerit, altius repetenda sunt [...] Post paucos vero menses a supradicto morbo corripitur; qui nimia exustus febre, clamabat, dicens: "Heu, heu! Ab Eraclio antistiti exuror, ab illo crucior, ab illo ad iudicium vocor. Cognosco facinus; reminiscor, me iniuste iniurias intulisse pontifici; mortem deprecor, ne diutius crucier hoc tormento". Haec cum maxima in febre clamaret, deficiente robore corporis, infelicem animam fudit, indubia relinquens vestigia, hoc ei ad ultionem beati antistitis evenisse. Nam exanime corpus ita nigredinem duxit, ut putares eum prunis superpositum fuisse combustum. Ergo omnes haec obstupescant, admirentur et metuant, ne inferant iniurias sacerdotibus! quia ultor est Dominus servorum suorum *sperantibus in se*'. Gregory of Tours, *Libri historiarum decem*, pp. 242–43.

[38] Gregory of Tours, *History*, pp. 338–39. 'Obiit et Eparchius reclausus Ecolesinensis. Vir magnificae sanctitatis, per quem Deus multa miracula ostendit; de quibus, relictis plurimis, pauca perstringam. Petrocoricae urbis incola fuit; sed post conversionem clericus factus, Ecolesinam veniens, cellulam sibi aedificavit [...] Magnam enim catervam populorum de oblationibus devotorum redemit; pusularum malarum venenum crucis signum saepe compressit, daemonas de obsessis corporibus oratione abegit et iudicibus plerumque, ut culpabilibus ignoscerent, dulcedine profusa imperavit potius quam rogavit [...] Sed et alia multa fecit, quae insequi longum putavi. Post XLIII vero annos reclusionis suae parumper febre pulsatus tradidit spiritum', Gregory of Tours, *Libri historiarum decem*, pp. 277–78.

[39] Gregory of Tours, *History*, p. 346. 'Magna tamen eo anno lues in populo fuit; valitudinis variae, milinae cum pusulis et vissicis, quae multum populum adficerunt mortem. Multi tamen, adhibentes studium, evaserunt. Audivimus enim eo anno in Narbonensem urbem inguinar-

VI.15 Felix, Bishop of the city of Nantes, contracted this disease and became gravely ill [graviter aegrotare coepit].[40] Bishop Felix seemed to be recovering somewhat from his illness. His fever abated, but as the result of his low state of health his legs were covered with tumours [pusulas]. He applied a plaster of cantharides, but it was too strong. His legs festered and so he died in the thirty-third year of his episcopate, when he was seventy years old.[41]

VIII.15 [Deacon Vulfolaic] 'When I went home for some food I found my whole body from the top of my head to the soles of my feet covered with malignant sores [pusulis malis], so that it was not possible to find the space of a single finger-tip which was free from them. I went into the church by myself and stripped myself naked by the holy altar. It was there that I kept a flask full of oil which I had brought home with me from Saint Martin's church. With my own hands I anointed my whole body with this oil, and then I went to sleep. It was nearly midnight when I awoke. As I rose to my feet to say the appointed prayers, I found my body completely cured, just as if I had never had any ulcers [ulcus] at all. Then I realized that these pustules [vulnera] had been caused by the hatred which the Devil bore me. He is so full of spite that he does all he can to harm those who seek God.'[42]

VIII.18 Bishop Namatius remained behind in Nantes [...] three malignant boils [pusulae malae] grew on his head. He felt extremely ill as a result and decided to

ium morbum graviter desevire, ita ut nullum esset spatium, cum homo correptus fuisset ab eo', Gregory of Tours, *Libri historiarum decem*, p. 284.

[40] It is initially indeterminate which disease, the *pusulae* or the groin plague, Felix suffered, but later it becomes clear he contracted the former. That Nantes is 650 km north-west of Narbonne, but only 250 km from Angoulême, where we can put *pusulae* in 580, also suggests plague did not catch up with the Ligérien churchman.

[41] Gregory of Tours, *History*, pp. 346–47. 'Felix vero episcopus Namneticae civitatis in hac valitudine corruens, graviter aegrotare coepit. [...] Felix episcopus ab incommodo levius agere videretur. Sed postquam febris discessit, tibiae eius ab humore pusulas emerserunt. Tunc cantaredarum cataplasmam nimium validam ponens, conputrescentibus tibiis, anno episcopatus sui XXXIII, aetate septuagenaria vitam finivit', Gregory of Tours, *Libri historiarum decem*, p. 284.

[42] Gregory of Tours, *History*, p. 447. 'Ipsa quoque hora, cum ad cibum capiendum venissem, ita omne corpus meum a vertice usque ad plantam pusulis malis repletum est, ut locus, quem unus digitus tegerit, vacuus invenire non possit. Ingressusque basilicam solus, denudavi me coram sancto altario. Habebam enim ibi ampullam cum oleo plenam, quam de sancti Martini basilicam detuleram; ex qua propriis manibus omnes artus y perunxi, moxque sopori locatus sum. Expergefactus vero circa medium noctis, cum ad cursum reddendum surgerem, ita corpus totum incolomem repperi, acsi nullum super me ulcus apparuisset. Quae vulnera non aliter nisi per in vidiam inimici emissa cognovi. Et quia semper ipse invidus Deum quaerentibus nocere conatur', Gregory of Tours, *Libri historiarum decem*, p. 382.

return to his own city as quickly as he could. As he was passing through Angers he died.[43]

That is it for the *Decem libri historarum*. Does smallpox lurk in these passages? Paulet was certain it did. Although Gregory's supposedly pox'ed passages have not been interrogated to the degree that the alleged evidence for the disease in the Hippocratic corpus or the writings of Galen have, over the last few centuries relatively few have disagreed with the Ancien Régime physician. There is, nevertheless, room for doubt.

Some 'verbalist ingenuity' is required to secure the diagnosis, as is often the case when looking for a particular disease in prelaboratory writings.[44] Retrospectively diagnosing prelaboratory disease outbreaks with written sources alone is always a precarious venture. Some argue it should not be pursued at all.[45] We proposed it is best not pursued without the support of ancient pathogen DNA or RNA.[46]

That written evidence is generally thin and spotty in late antiquity makes attempts to apply modern medical labels to late antique diseases all the riskier.[47] What can we say without a diagnosis? The 'most grave plague' recounted in V.34, like the prodigies that preceded it, can be dated to 580. It seems to have been characterized not only by pustules and blisters but also by diarrhea, vom-

[43] 'Namatius vero episcopus, dum, receptis villis moraretur, pusulae malae ei tres oriuntur in capite. Ex hoc valde confectus taedio, dum ad civitatem suam reverti cupiret, infra Andegavensis territorii terminum spiritum exalavit'. Gregory of Tours, *Libri historiarum decem*, p. 432.

[44] We borrow this phrasing from Creighton, *History*, p. 440.

[45] Cunningham, 'Identifying Disease in the Past'; Arrizabalaga, 'Problematizing Retrospective Diagnosis'; Stein, '"Getting" the Pox'. Arrizabalaga points out the difficulty entailed in applying a modern medical diagnosis to a historical disease increases the farther back in time the disease occurred, 'Problematizing Retrospective Diagnosis', p. 57. In diagnosing with modern labels and pursuing linear disease histories, we do, as these and other historians stress, lose some of the disease experience of the people with study.

[46] Palaeo- or archaeo-geneticists have understandably observed that only with pathogen aDNA can we obtain confident or confirmed retrospective diagnoses of centuries-old plagues: Andam and others, 'Microbial Genomics'; Spyrou and others, 'Ancient Pathogen Genomics', p. 323. Only aDNA 'unequivocally demonstrated the involvement of *Y. pestis* in the first pandemic': Keller and others, 'Ancient *Yersinia pestis*', p. 12,364. It should be stressed that the detection of aRNA in archaeological remains has been and is still very hard to come by.

[47] Arrizabalaga, 'Retrospective Diagnosis', p. 57, rightly stresses that the difficulty entailed in applying a modern medical diagnosis to a premodern disease increases the farther back in time the disease occurred.

iting and backache too. It began in Gaul in August, at least as far as Gregory knew, but we have no idea as to when it petered out. Gregory has this plague spreading across all of Gaul, undoubtedly an exaggeration, and he twice underscores its toll on children. This is the only plague in his *DLH* that Gregory specifically characterizes as killing many young people.[48] Several prominent Merovingian political figures are identified as having fallen sick, including the Neustrian king Chilperic and his queen consort Fredegund, who Gregory despised. Two of their children, Chlodobert and Dagobert, also fell sick and only they, from that family, died.

Other major figures appear to have suffered the disease too. While Chilperic and Fredegund recovered, Austrechild (V.35), the Queen consort of the Orléanais king Guntram, and Nantinus (V.36), Count of Angoulême, did not. Fortunate as we are to be able to associate an epidemic with particular figures, one might question why Gregory identifies these specific individuals as sickly but not others, assuming there were other pustular victims for Gregory to choose from. That Gregory liked neither Austrechild nor Nantinus might explain why our good bishop drew attention to them and dispel any doubts that the disease did not spread more widely. The consort was a 'wicked woman' who led a 'worthless life' and had an 'unhappy soul'. She had two doctors killed because their remedies had failed her.[49] Nantinus, on the other hand, had done more 'damage' to 'God's bishops and His churches' than Gregory had time or space to relate, or so our bishop reports.[50]

Four consecutive chapters in the *DLH* address the passing of influential people in 580. After Nantinus, Gregory comes to the saintly bishop of Braga, Martin, whose death, Gregory makes sure to emphasize, was very much lamented (V.37).[51] Whether Martin suffered the same disease or any disease, our author does not say. One wonders, though, whether he might have, given the timing of his passing and the placement of the notice of his death

[48] Gregory refers to the odd child or two dying in the context of other outbreaks reported in his magnum opus (for example, Gregory of Tours, *Libri historiarum decem*, pp. 214–15), and kids are occasionally singled out as disease victims (Gregory of Tours, *Libri historiarum decem*, pp. 304–05, 494), but this is the only child plague. It is remiscent of the *mortalitas puerorum* reported in Irish annals a century later: MacArthur, 'The Identification of Some Pestilences', pp. 179–81.

[49] Gregory of Tours, *History*, pp. 298–99; *Libri historiarum decem*, pp. 241–42.

[50] Gregory of Tours, *History*, pp. 299–301; *Libri historiarum decem*, pp. 242–43.

[51] Gregory of Tours, *History*, p. 301. 'Hoc tempore et beatus Martinus Galliciensis episcopus obiit, magnum populo illi faciens planctum', Gregory of Tours, *Libri historiarum decem*, p. 243.

in Gregory's text. It is not beyond Gregory to have principally identified his nemeses, and the ill-reputed, as suffering physical manifestations of disease.[52] Symptoms could be interpreted as the materialization of divine disfavour and an outward expression of sin within. At times, Gregory is painfully clear about this. That the disease left Nantinus blackened was incontrovertible proof, our bishop tells us, that he had been punished for his wrong doings. More than that, the count's charcoaling was a warning to all that 'the Lord will avenge His servants'.

In striking contrast, there is Eparchius, who did more good than Gregory had time or space to convey.[53] Gregory's account of this recluse's miraculous works (VI.8) touches in passing on 'pustules' (*pusulae*). The reader is led to believe the disease affected many near Angoulême, but Gregory does not say whether the miracle worker's own death in 581 followed his own infection with 'pustules'. That the elderly Eparchius died suffering a fever it seems shortly after he had been in contact with 'pustules'-laced miracle seekers makes us wonder whether he too succumbed to the disease.

We must be careful, though, of making too much of the evidence. On a map, one might identify the disease at Berny and Angoulême, and perhaps, if we include Austrechild and Martin among the epidemic's dead, Orléans and Braga. One might therefore argue that the disease was rather contagious, as we are told that it not only spread within a royal family but also that it managed, likely in a single year,[54] to transverse considerable distance (more than 550 km separate Angoulême and Berny, and more than 1500 km lie between Berny and Braga). A few leaps of faith are worked into this conclusion, however.

To begin, one might ask how Gregory established that the Neustrian princes died of the same disease or that their parents suffered it to. That Eparchius was working miracles outside Angoulême offers some backing to Gregory's claim that 'pustules' [pusulae] sickened Nantinus, but by what means did Gregory figure out that the disease also attacked Austrechild and Nantinus? Had Gregory visited each of these individuals on their sickbeds, which he did not, we would still be wise to question the reliability of his diagnoses and the care he took to claim sameness between one illness and another.

[52] Gregory may have not written about Fredegund's symptoms because she recovered.

[53] As Gregory writes, 'Sed et alia multa fecit, quae insequi longum putavi', *Libri historiarum decem*, p. 278.

[54] Gregory does not, of course, explicitly write that the epidemic began and ended in 580.

Relatedly, could Gregory have employed *pusulae* or *vesicae* ('vesicles') systematically? If so, did he intend to? One real risk when hunting for a diagnosis, according to Paulet, is that 'on s'attache à des mots dont on corrompt la signification'.[55] Whether Paulet corrupted their meaning or not, when reading Gregory he certainly attached himself to a couple of nouns: *pusula* and, to a lesser extent, *vesica*.[56] These, for him, were smallpox. The former, 'pustule' or 'blister', and the latter, possibly 'vesicle' or 'blister' from *vesicare*,[57] are both fitting of VARV's maculopapular rash. A number of other observations, of course, shored up the association for Paulet, such as the disease's propensity for killing children and the spread of the 'pustules' or 'vesicles' over the body. Yet, Paulet compiled these details from multiple *pusula* epidemics and 'case reports' collected from Gregory's *History*. His confidence in the *pusula*-petite vérole linkage, in other words, hinged both on one disease underlying Gregory's pustular eruptions and also on Gregory's ability to apply (and intention of applying) *pusula* discriminately.[58] This is tricky. Some have questioned rightly whether any particular disease identifier was used in a strict sense to refer to a particular disease in late antiquity.[59]

Lastly, if Gregory sought to use *pusula* discriminately, and if we can depend on his diagnoses, how certain can we be that he did not conflate concurrent outbreaks or that one and the same epidemic sickened those in Angoulême,

[55] Paulet, *Histoire*, I, p. 57.

[56] Paulet, *Histoire*, I, pp. 79–89. Cf. Haeser, *Lehrbuch*, p. 255.

[57] Certainly not 'great tumours' as in Gregory of Tours, *History*, pp. 296–97.

[58] Notably, not all of Gregory's *pusulae* seem very smallpox-like. See discussion of Namatius below, n. 63.

[59] Even if the intention was there, how successful could our authors have been? For instance, Willan, 'Inquiry', pp. 86–87, proposed *lues inguinaria* referred originally to 'the actual pestilence', but that others 'adopted this distinctive title' for 'specific notice of other forms of contagious disease'. Schnurrer emphasizes Gregory's unsystematic terminology when discussing what could have been smallpox: *Chronik*, pp. 138–39, 142. Also, Newfield, 'Human-Bovine Plagues', p. 22. Of course, historians of Latin late antiquity have long thought *lues inguinaria / inguinalis plaga / inguinarium morbum* were used exclusively for plague (*Y. pestis*) across several late antique centuries. Biraben and Le Goff identified this as a problem but popularised the approach nonetheless: 'La Peste', pp. 1491–92. Cf. Conrad, 'Tāʿūn and Wabā'', pp. 271–73, 305, 306. While not all plague outbreaks may have been described in these terms, all 'groin plagues' were *Y. pestis*, or so many assume, regardless of whether or not a late antique author supplemented an account of *inguinaria* with telling symptomatologic descriptions. Significantly, several late antique authors, including Gregory of Tours, often do not provide us with any symptomatology, but leave us with bald accounts of groin plagues.

Berny and Orléans? That certainty on this point eludes us should prevent us from making too much of the distribution of 'pustules' in 580 or, by extension, attempting to infer the vastness of any one pustular epidemic or the speed with which the disease was capable of diffusing.[60]

Yet, *pusula* could clearly take on epidemic proportions. Crucially, the epidemic of 580 was not an isolated event. Gregory records a second plague characterized by *pusulae* and *vesicae* in 582 in VI.14. This epidemic is not geographically fixed, but we might presume, at a minimum, central or north-western Gaul were hit, respectively the region with which Gregory was principally concerned and the region where Gregory identifies a *pusula*-sufferer. Whether this plague was a continuation of the epidemic of 580 is uncertain, but that is perhaps likely.

The epidemic of 582, which claimed the lives of many but was not inescapable, coincided with an outbreak of 'groin disease' (*inguinarium morbum*), traditionally diagnosed as plague (*Y. pestis*) in Narbonne. Crucially, Gregory is careful to differentiate this groin disease from the 'pustules' by means of their whereabouts, their manifestation, and their virulence. The 'groin disease', Gregory implies, killed a greater proportion of its victims than did the 'pustules'. Among the victims of the latter, Gregory draws attention to Felix, the bishop of Nantes (VI.15), who languished with *pusulae* on his legs.

Whether the deacon Vulfolaic, who Gregory paraphrases in VIII.15, fell ill in the outbreak of 580 or 582 is unclear and clearly unimportant to our bishop. Gregory's insouciance on this point could be taken to mean that the two epidemics were, in fact, one or that *pusulae* were circulating in Gaul throughout the early 580s. Gregory visited Vulfolaic, a fellow promoter of Saint Martin, in 585 outside the town of Carignan in the Ardennes.[61] We learn that Vulfolaic was struck with the disease after he tore down an old pagan statue next to which he had erected his monastery in or soon after 580. The proximity of Carignan to Berny, where 'pustules' overwhelmed the Neustrian princes in 580, suggests the deacon may have come down with the disease the same year.

Again, we might ask, why Felix and Vulfolaic? That Gregory had reason to single out these churchmen as disease victims supports the notion that he used illness deliberately and shores up Gregory's own assertion that the pustular dis-

[60] As historians of early medieval disease have occasionally done, Maddicott, 'Plague in Seventh-Century England', pp. 27–28.

[61] The monastery was built outside of Carignan, but which in the text is described as being situated in the territorium Trevericae, 'the territory of Trier'.

ease could spread widely. In other words, 'pustules' were more noteworthy in some circumstances than others. They were especially noteworthy when they manifested on his nemeses.

To say Gregory was not a fan of Felix is putting it lightly. 'A man of unlimited greed and boastfulness', Felix wrote Gregory a letter 'full of insults' and backed a 'vile and wicked' attempt to remove Gregory from his office.[62] Although Vulfolaic was, like Gregory himself, a champion of Saint Martin, it has been argued that our bishop cared little for the deacon. Gregory was careful to note that Vulfolaic was a mere Lombard by birth and only a recent convert to Catholicism. Vulfolaic was also, it seems, unaware of Gregory's efforts to promote their patron saint. That Gregory had to promise Vulfolaic that he would not tell his story in order to get him to share it and then paraphrased the deacon at length, is itself revealing. Were this not enough to coat Vulfolaic in *pusulae*, Saint Martin had proved vital to the Lombard's convalescence, something Gregory was sure not to omit.[63]

Whether or not the 580 and 582 epidemics were one and the same, or the outbreaks in north-eastern and western Gaul in 580 connected, it is clear, if we trust Gregory's diagnostic talents, that *pusulae* were somewhat widely diffused across a large area of Gaul in the early 580s. It likewise seems as though they afflicted people indiscriminately of age, sex or wealth, and that many of the sick appear to have recuperated perhaps with the notable exception of the young and elderly. Gregory stresses the toll 'pustules' took on children and he names two children who succumbed to the disease. Of course, if the plague of the early 580s was smallpox, that it killed many children might suggest that the disease had spread recently in the regions of Gregory's concern, as smallpox survivors are usually conferred life-long immunity to reinfection.[64] In this detail one might see support for identifying Marius' *variola*, discussed below, as smallpox or something like it.[65] Equally notable, many of those Gregory has the disease killing were relatively long in the tooth. Austrechild, Eparchius, Felix, Martin and Nantinus were all advanced in age. Martin had serviced his flock for 30 years, Felix was 70, and Eparchius was ordained in 542. Surely had the

[62] Gregory of Tours, *Libri historiarum decem*, pp. 200, 262, 284; *History*, pp. 346–47.

[63] Gregory of Tours, *History*, p. 447; *Libri historiarum decem*, p. 382. Winstead, 'Vulfolaic the Stylite', p. 68.

[64] Fenner and others, *Smallpox*, pp. 51–52, 167–68.

[65] But see below.

disease been more acute than smallpox, Eparchius would have had more difficulty neutralizing it.[66]

We can see why Paulet thought there were grounds here for a smallpox diagnosis. Both the presentation of the disease and the epidemiology of the outbreak, as far as we can discern them, fit smallpox. Gregory stresses *pusulae*, 'pustules', and *vesicae*, 'vesicles', characterized the disease. With a couple of exceptions,[67] these were neither few nor restricted to a particular body part. Rather, they manifested, like smallpox, on the head, feet, legs and shoulders. That Gregory does not once draw attention to a pustular trunk or abdomen, also hints of smallpox and its classic centrifugal rash.[68] That the *pusulae* and *vesicae* were fluid-filled is likewise very reminiscent of smallpox.

What of Gregory's dysenteric illness that spread through Gaul at the start of his account? Some orientalists, convinced Arab armies introduced smallpox to the late antique Europe, clung to this diarrhea.[69] While most may not think of loose bowels when they think of smallpox, upwards of 10 per cent of the victims of the disease may present with it.[70] The vomiting is a less rare complication of VARV than some might suppose, manifesting as much as 50 per cent of the time.[71] Both loose bowels and emesis typically set in before the onset of smallpox's defining pustular rash, which might possibly account for Gregory initially describing the epidemic of 580 as a 'dysenteric disease' (*desentericus morbus*).[72] Severe backache, which our bishop also observes, may have occurred

[66] Elsewhere it has suggested that diseases of miracles were best distinct (or memorable) and widespread (or well-known), but also benign enough that they could appear to have been cured: Newfield, 'Malaria', pp. 274–75.

[67] The lethal *pusulae* Namatius endured (VIII.18) in 587 do not read like smallpox. They may have been located on his head, where a smallpox rash typically starts (Fenner and others, *Smallpox*, pp. 4, 9, 11, 19, 21, 56), but Gregory specifies there were only three of them.

[68] Fenner and others, *Smallpox*, pp. 8–21, 56, 130, 167. Of course, it could be said that most had their abdomens covered most of the time. An article on Gregory's use of *pusula* throughout his many works is in progress.

[69] Moore, *History*, pp. 8–10, 12. Paulet, *Histoire*, I, p. 80, thought smallpox and dysentery irrupted simultaneously: 'ce n'étoit autre chose que la petite vérole jointe à la dysenterie'.

[70] In line with the foregoing discussion, one might as well wonder how Gregory established that the *desentericus morbus* and *pusulae* were related. Could he not have conflated the symptoms of distinct diseases?

[71] Fenner and others, *Smallpox*, pp. 5–6. Schnurrer was comfortable with Gregory's mention of vomiting in the context a putative smallpox epidemic: *Chronik*, p. 141.

[72] Fenner and others, *Smallpox*, pp. 5–6, 48–49. That said, Gregory wrote about this epidemic after the fact.

in upwards of 90 per cent of smallpox cases, making it a canonical symptom of the disease.[73] Even the overcooked Nantinus could be construed as a victim of haemorrhagic smallpox or as coated with smallpox scabs.[74] At the same time, Gregory may have conflated multiple co-occurring diseases.

Meanwhile the inkling of a U-shaped mortality pattern fits epidemic smallpox too. Indeed, smallpox was not plague: even in 'virgin' lands, it was far more likely to kill the young and the old than those in the prime of their life.[75] That *pusulae* killed many but were hardly always fatal as well lines up with well-documented pre-variolation/vaccination smallpox epidemics, for which case fatality rates rarely exceeded 25 per cent.[76] More generally, the epidemics Gregory describes were widespread and not limited to a particular region or, as far as can be told, environment or season, which suggests person-to-person transmission, which also fits smallpox as we know it. Missing, however, is mention of classic smallpox sequelae.

None of this confirms the diagnosis, of course. Considering what little faith could be put in clinical diagnoses of smallpox leading up to its extinction or of bubonic plague now,[77] we might question whether *pusula* or *inguinaria* were dependably put to use in late antiquity to characterize disease outbreaks. Some linguistic acrobatics are unavoidable on our part when we try retrospectively to diagnose epidemics 15 centuries after the fact, but our authors had more than lexical hurdles to overcome. Still, it is hardly outside the realm of the possible that, on the basis of Gregory's writings, a disease related to smallpox spread through parts of Gaul in the early 580s.

[73] Fenner and others, *Smallpox*, pp. 5–6, 37, 39, 200.

[74] Fenner and others, *Smallpox*, pp. 17, 35.

[75] This naturally works against the proposal advanced above that Gregory's notice of many children dying suggests smallpox had recently spread in the same regions. The evidence is such that either argument can be made.

[76] Fenner and others, *Smallpox*, pp. 164, 227, 244, 246, 271, 790–91.

[77] Fenner and others, *Smallpox*, pp. 55–56; Stenseth and others, 'Plague', p. 3; Baril and others, 'Can We Make Human Plague History?', p. e00,1984; Rajerison and others, 'Performance of Plague Rapid Diagnostic Test', p. 90.

The First Variola

There is more decent evidence for smallpox-like disease in mid-sixth-century continental Europe, or at least Paulet thought so.[78] In the so-called *Chronica* of Marius of Avenches, Gregory's contemporary and another bishop, there are two brief passages, which have since Paulet's time been occasionally deployed to support the claim Europeans suffered smallpox in the 500s:[79]

> In this year [570], virulent disease greatly afflicted Italy and France with a flow of the bowels and *variola*, and beef animals died especially through the aforementioned places.[80]
>
> In this year [571], an unspeakable illness of the *glandula*, named the *pustula*, devastated innumerable people in the aforementioned regions.[81]

These passages present a number of difficulties despite their brevity. To start, the dates may be off. Marius' chronology — or at least some of his entries — are a year late from 566.[82] As such, the *variola*, the first extant usage of the term, might better be assigned to 569 and the *pustula* to 570. But there are a number of more pressing issues with each passage. About the first: what was meant by *variola*, what are we to do with the cows and whose bowels were loose?

While many have assumed the Avenchene bishop understood *variola* as affecting humans alone and that his *variola* was smallpox, some have expressed doubt. Early smallpox historians convinced that smallpox fanned outward from Arabia with Arab armies in the seventh century have speculated that *variola* was not coined by the 'old chronologist' Marius, but a later copyist. To shore up this idea, Moore wagered that there 'could be no term for [smallpox] in the Greek or Latin languages', as it was not a Greco-Roman disease.[83]

[78] Paulet, *Histoire*, I, p. 87. Orientalists who argued that the disease only reached Europe with the Arab conquests discounted both the passages mentioned here and above by Gregory and Marius.

[79] Paulet, *Histoire*, I, pp. 78–79, 85–86.

[80] 'Hoc anno morbus validus cum profluvio ventris et variola Italiam Galliamque valde afflixit et animalia bubula per loca supra scripta maxime interierunt', Marius of Avenches, *Chronica*, p. 238.

[81] 'Hoc anno infanda infirmitas, quae glandula, cuius nomen est pustula, in supra scriptis regionibus innumerabilem populum devastavit', Marius of Avenches, *Chronica*, p. 238.

[82] So, Justinian here dies in 566: Marius of Avenches, *Chronica*, p. 238.

[83] Moore, *History*, p. 88.

Unfortunately for Moore, this is as baseless as his assertion that the term could only have been coined in the tenth century.[84] The physician Constantinus Africanus, who wrote a hundred years later, has been said for centuries to have been the first author to use *variola* in specific reference to smallpox, but as we make clear smallpox historians have not shied away from seeing smallpox in Marius' *variola* for centuries.[85]

Whether or not Marius coined the term, *variola* is thought to derive either from the adjective *varius*, meaning 'changing', 'varying', or 'different', but often translated in histories of smallpox as 'blotchy' or 'spotted', or from the noun *varus*, 'spot' or 'pimple', which has, in smallpox histories, evolved into 'pustule'.[86] That the disease was cutaneous, visibly altering the appearance of its victims, and spread widely, infecting many, is hardly a sure bet that smallpox lurks in this *variola*, but smallpox remains a good guess. As it stands, though, a number of diseases could have been to blame, including measles and typhus which manifested cutaneously.[87]

Then there are the cows. Paulet deliberately ignored them. In his history of petite vérole, we encounter *variola*, but no bovines. He saved the cattle for his *Recherches historiques et physiques sur les maladies épizootiques* published in 1775.[88] As no disease known to modern science spreads widely in cattle and

[84] Moore, *History*, pp. 88–90, 98. This was picked up by later writers, for instance: Schnurrer, *Chronik*, p. 138. Although it is not impossible that Marius' text was admended here, there is no evidence for it. There is also no reason to believe Latin terms could not be coined in the sixth century. Marius' chronicle survives in one manuscript dated to the mid- or late ninth century (British Library Add MS 16974, *variola*: 112^{v}). There it is found alongside one of three extant texts of the Chronicle of 452. The compiler of the manuscript appears to have 'transcribed [...] faithfully' the Chronicle of 452 and copied Marius' text 'without making any obvious admendments': Wood, 'Chains of Chronicles', pp. 67, 69.

[85] Woodville thought the claim that Constantinus was the first to refer to smallpox with *variola* 'mistaken': *History*, p. 31. That Constantinus refers to *variola* as an 'ordinary disease' seems to have shored up Moore's belief that *variola* was coined earlier. In any case, Constantinus Africanus' description of *variola* will fail to convince everyone: Constantine the African, *Constantini Africani post Hippocratem et Galenum*, pp. 152–53.

[86] Edwardes, *A Concise History of Small-Pox*, p. 3; Dixon, *Smallpox*, p. 187; Biraben and LeGoff, 'La peste', pp. 1493–94; Tucker, *Scourge*, p. 2; Hopkins, *The Greatest Killer*, p. 25; Devroey, *Economie rurale*, p. 46; Carrell, *The Speckled Monster*, p. xv; Kotar and Gessler, *Smallpox*, pp. 3–4. Dixon, *Smallpox*, p. 187, was careful to point out that without a description of the disease, we cannot be certain the first *variola* was smallpox.

[87] Newfield, 'Human-Bovine', pp, 8–9, 27–29, 36.

[88] Paulet, *Recherches historiques*, p. 73.

people, causing severe sickness and death, it has been posited that premodern accounts like Marius' conflated concurrent disease outbreaks, one in people with another in livestock.[89] A real possibility this may be, but a few have nevertheless contended that a single disease was to blame.

Both a smallpox-cowpox ancestor and a measles-rinderpest ancestor have been put to work in Marius' time. The former is by some measure the more problematic of two. Cowpox may be zoonotic, and its genetic affinities with smallpox were essential for Jennerian vaccination, but molecular clock and phylogenetic analyses, discussed below, seriously problematize the suggestion.[90] That cowpox is neither a rapidly spreading nor virulent disease of cattle itself complicates the association. The closest relative of measles, on the other hand, is both of those things. Rinderpest is a highly contagious and acute bovine disease. Furthermore, although, as noted, a percentage of smallpox victims present with diarrhea, debilitating loose bowels rank among rinderpest's primary symptoms and diarrhea is a common complication of measles.[91]

Unfortunately, the bishop was not exactly clear about whether the loose bowels were human, bovine or both. That the only other account of a cow die-off about 570 says nothing of a flux, could suggest Marius' bovines were not diarrheal. That said, a minuscule percentage of post-Roman authors took pains to document the symptoms of non-human diseases.[92] The passage in question — the 'in the fifth year of Emperor Justin II [570] there was a pestilence of oxen and a destruction everywhere' (*pestilentia bovum et interitus ubique fuit*) encountered in the *Liber pontificalis ecclesiae ravennatis*, composed about 840 by Agnellus of Ravenna from earlier sources — might only confirm that the cattle mortality was not a local phenomenon.[93]

[89] To quote one veterinary historian, 'there was a tendency', in the distant past, 'to observe a "simultaneity" of epidemics in man and animals [...] and to attribute the same cause to them [...] now we know that such epidemics [...] had different causes [...]': Mantovani, 'Notes on the Development of the Concept of Zoonoses', p. 41.

[90] As discussed below, smallpox's closest known relatives are camel- and tatera-poxes. That an ancestral VARV was zoonotic is not, however, out of the question. Indeed, extinct VARV strains may have had less host specificity than the VARV known to modern science. See below.

[91] Wohlsein and Saliki, 'Rinderpest and Peste des Petits Ruminants', pp. 68, 70–72, 79; Perry, Orenstein, and Halsey, 'The Clinical Significance of Measles', pp. S6–S7.

[92] Newfield, 'Human-Bovine', p. 8.

[93] Agnellus' known sources do not refer to this epizootic, and although it is indeterminate whether Agnellus consulted Marius' *Chroncia*, his writing shows no verbal parallels with it. Agnellus may have drawn on two non-extant texts he cites: Maximian of Ravenna's *Chronicon*,

Potentially crucial here, the most recent attempt to date the emergence of the measles virus has pushed back earlier such attempts by more than 1000 years.[94] While independent studies in 2010 and 2011 estimated the date of the measles-rinderpest divergence to about 800–1100 CE, the new earliest possible point of divergence is closer to around 600 BCE.[95] In other words, it seems less likely now than it did just a few years ago that an ancestral morbillivirus, a zoonotic pathogen capable of causing disease in both people and their cows, could have spread through Marius' Italy and Gaul.[96] Paulet may have been right to treat the sick people and cows separately. If so, there is more reason to think smallpox or something related to it lurks in this *variola*.

Marius' second passage is equally problematic. It is sometimes interpreted as evidence for an outbreak of the Justinianic Plague.[97] The text itself, though, causes us to doubt that. Neither 'glands' (*glandula*) nor 'pustule' (*pustula*) are sure indicators of *Y. pestis*.[98] They might just as well point to smallpox, though that too is uncertain. That Gregory of Tours writes nothing of *pusulae* or *vesicae* in 570, but does refer to a 'great' plague-like 'destruction' of people in central and western Gaul,[99] in an epidemic commonly, but hardly definitively, dated to 570 or 571, supports the *Y. pestis* identification of Marius' second epidemic.

which possibly concluded in the 570s, and an unknown 'annalistic source' which seems to have wrapped up in 573. Agnellus of Ravenna, 'Liber Pontificalis Ecclesiae', p. 337; *The Book of Pontiffs*, pp. 49, 205.

94 Düx and others, 'Measles Virus and Rinderpest Virus', pp. 1367–70. Crucially, this new molecular clock analysis integrates the first measles genome reconstructed from human remains (a century-old lung). The two decade-old molecular clock studies, in contrast, relied exclusively on younger measles genomes.

95 Furuse, Suzuki, and Oshitani, 'Origin of Measles Virus', p. 52, proposed an eleventh- to twelfth-century divergence. Shortly thereafter, Wertheim and Kosakovsky Pond, 'Purifying Selection', pp. 3355–65, proposed a split around the year 900. Düx and others, however, date the most recent common ancestor for measles and rinderpest to the sixth century BCE. The pre-peer-review version of the article dates the divergence two centuries later: Düx and others, 'The History of Measles'.

96 As proposed in Newfield, 'Human-Bovine Plagues', pp. 8–9, 26–29. At the same time, that the two could have divereged as early as 600 BCE does not mean they did. A later divergence remains on the table.

97 Biraben and Le Goff, 'La peste', pp. 1493, 1495, 1499.

98 Notably, Biraben and Le Goff, 'La peste', p. 1491, thought the *pustula* 'laisse subsister un doute'. Later (p. 1495), however, they conjectured that Marius referred to plague in one passage and smallpox in the other.

99 Gregory of Tours, *History*, pp. 225–27; *Libri historiarum decem*, pp. 164–66.

Water-tight, however, this is not. Not only is the date questionable, contingent as it is on Marius' chronology, but how Gregory's plague passage might help us explain Marius' second epidemic is unclear. That Gregory seems to have his plague disseminating not far from Marius' familiar territory (Chalon-sur-Saône lies less than 150 km from Avenches) and around the same time could mean that Marius's second epidemic, which does date to 570 or 571, was one and the same. That yet more support for a plague outbreak about then and around there is found in Paul the Deacon's late-eighth-century *Historia Langobardorum* would seem to confirm this suspicion. Paul's 'greatest plague' (*maxima pestilentia*), sometimes dated to 570, spread in Liguria, on the opposite side of the Alps from Marius. It has long been identified as *Y. pestis*, as it caused *glandulae* to swell, Paul is careful to point out, in the groin and other 'delicate places'.[100] Yet, the date of this outbreak, like Gregory's, is not certain. Both Paul and Gregory's plagues occurred sometime in the 560s or early 570s, and both have sometimes been dated to 570/71 on the basis of Marius' passage. That these three authors refer to a single interregional plague outbreak is a real possibility, but it is not the only possibility. Two or all three of the authors may be referring to different outbreaks.

That Marius' epidemics were geographically sandwiched between two at least roughly concurrent plague outbreaks more-or-less might assure us that one of the Avenchene outbreaks was plague and that in one year plague spread widely in these areas of Europe alongside another disease, which Marius stresses also diffused broadly.[101] But if the *glandula* is plague must the *pustula* also be? Marius is clear that the first of the epidemics he records was not the same as the second. The outbreaks are reported in consecutive entries, not one, and unique terms are employed for each. Yet, might the bishop refer to two epidemics in his second passage? Could the *pustula* have been the *variola*? If so, why would Marius have not simply said so? At the risk of attaching ourselves to particular terms, we propose that Marius records one year of plague, which Gregory and Paul also observe, though possibly in a different year, and two years of another epidemic, quite possibly variolous, following the origins likely of *variola* and the mention of 'pustule'. As in the 580s under Gregory's watch, so in the 570s under Marius', plague and a smallpox-like disease may have co-occurred.

[100] Paul the Deacon, 'Historia Langobardorum', p. 74.

[101] Or possibly his first, considering the dating uncertainties of the key passages.

A Death-Dealing Cloud and the Bolggach

Unknown to Paulet, there are yet more indications for something similar to smallpox in mid-sixth-century Europe. Some of it made an appearance in post-Paulet inquiries into smallpox's origins,[102] but none of it has received proper consideration in the context of the continental material.

Sometime around 690,[103] the abbot Adomnán composed a hagiography of his cousin, the sixth-century Irish saint Columba, on the Hebridean island of Iona. Among a series of miracles and prophecies in his *Vita sancti Columbae*, one comes across a disease outbreak among people and cattle in the area of modern-day Dublin.[104] Adomnán reports that Columba, already an accomplished healer, set sail from Iona for his native Ireland shortly after the disease struck, to attend to the 'very many' it afflicted. Allegedly transmitted through 'morbiferous rain' (*pluviam … morbiferam*) from a 'morbiferous cloud' (*nubis … morbifera*), the disease caused great sickness and death, and manifested as 'grave and purulent ulcers [or sores]' (*gravia et purulenta … ulcera*) that formed 'on human bodies and cow udders' (*humanis in corporibus et in pecorum uberibus*).[105] Although the disease-carrying rain 'devastated' people, as Columba purportedly forecasted it would, the miracle worker did not disappoint. Adomnán tells us that the saint's efforts proved remarkably efficacious: people and cows convalesced almost instantaneously once sprinkled with holy water.

With some verbalist ingenuity this outbreak could tentatively be identified as smallpox. The fluid-filled pustules and fairly wide spread fit the disease. So does the hint that this plague did not kill most of its victims. That Columba, like Eparchius against the *pusulae*, managed to work cures implies that the disease was not always fatal. The cows, though, do not suggest smallpox. Unsurprisingly, an ancestral smallpox-cowpox orthopoxvirus has been put forward for this plague too. The ulcered udders certainly bring cowpox to mind.[106] Considering how unlikely it is that such an orthopoxvirus once existed, might

[102] For instance, Willan, 'Inquiry', pp. 114–15.

[103] Brüning, 'Adamnan's Vita Columbae, pp. 227–29; Adomnán, *Life of Columba*, pp. 5, 94, 96.

[104] The affected region is identified as lying 'ab illo rivulo qui dicitur Ailbine usque ad Vadum Clied'.

[105] Adomnán, *Vita S. Columbae*, pp. 73–75.

[106] See Adomnán, *Vita S. Columbae*, p. 74 n. 4; Adomnán, *The Life of Saint Columba*, n. 217; Shrewsbury, 'The Yellow Plague', p. 39. Cowpox'ed udders, of course, were central to Jennerian vaccination: Jenner, *Inquiry*, pp. 3–4, 6, 74.

Adomnán, like Marius, have conflated two plagues? Or might an extinct variety of VARV have been zoonotic and caused disease in people and their cattle?

When the death-dealing cloud formed is indeterminate. Columba left for Iona in 563 and died there in 597. The plague must have occurred between those dates. The Irish Annals, notably, report only one epidemic during Columba's Hebridean residence. While the chronology of epidemics these annals give is undoubtedly fragmentary, it is feasible that the epidemic reported in the mid-570s is Adomnán's *pluviam morbiferam*.

Different sets of Irish annals give different years for the epidemic in question, but it has been argued that the proper date is 574.[107] Each of the extant passages clearly stems from a single entry in a non-surviving text very likely written, like the saint's vita, at Iona.[108] The *Annals of Tigernach* reports in 574 'a glimmer of *lepra* and an unheard-of abundance of nuts' (*scintilla leprae et abundantia nucum inaudita*). The *Annals of Ulster* sets the same passage at 576, when the *Annals of Inisfallen* observes 'a plentiful crop of nuts' (*cnómes imda*). A year later, in the latter text, we find 'people afflicted with bolggach' (*bolggach for doenib*).[109]

Whether the original entry was made roughly when people were dying is debated. Some hold that the lost text was kept current at the monastery of Iona from just about the time Columba founded it in the 560s;[110] others argue it provides a contemporary witness from Adomnán's time.[111] Either way, the surviving annals may provide a contemporary witness to the pestilential rain. If they do, they notably omit cattle.

[107] Mc Carthy, 'Chronological Synchronisation'; Ludlow, 'The Utility of the Irish Annals', II, p. 23.

[108] It is thought the non-extant Iona Annals, written at the monastery of Iona, was a significant source before 740 for the so-called Chronicle of Ireland, another non-extant text, which informs pre-912 entries in the surviving Irish annals. Not everything included in the Iona Annals, though, was written in Iona. It is possible the 574 mortality was an item included from another Irish text. Bannerman, *Studies in the History of Dalriada*, pp. 9–26; Evans, *The Present and the Past*, pp. 2–3, 12–14; Evans, 'The Irish Chronicles', p. 23.

[109] These translations are from *CELT: The Corpus of Electronic Texts*. The dates assigned to these passages vary across editions and translations of these texts. For example, in the 'The Annals of Tigernach: third fragment, AD 489–766', p. 151, the passage is dated to 575, as is the passage in *Annala Uladh*, pp. 64–67. These passages are near identical. A similar passage does not appear in the *Chronicon Scotorum*.

[110] Smyth, 'The Earliest Irish Annals', pp. 9–12; Charles-Edwards, *The Chronicle of Ireland*, I, pp. 8–9.

[111] Hughes, *Early Christian Ireland*, p. 142.

Unsurprisingly, there is no consensus on a diagnosis. Editors and translators have proposed several options, some more plausible than others. Certainly, *lepra* is unlikely to have been slowly progressing and faintly contagious leprosy/Hansen's Disease, as some have it, and whether *bolggach* is modern and medieval Irish for smallpox, as other scholars have proposed, is difficult to say.[112] This is the earliest dated usage of *bolggach* and problematically the annalist seems to have made no attempt to supplement it with a description of the disease's symptoms.[113] The parallels with Marius' *variola* are striking.

Although *bolggach* appears later in Ireland's annals, those later usages are no more informative. We can only say that at least some medieval annalists thought *lepra* and *bolggach* equivalent.[114] Like the former, the latter implies the disease manifested cutaneously, possibly causing blisters, a rash or lesions. As such, *bolggach* could have been smallpox or a number of other infections, including again measles and typhus. Alternatively, the disease might have been

[112] MacArthur memorably considered the identification of this *lepra* and other epidemics mentioned in the Irish annals as leprosy 'absurd'. Instead, he argued that *bolggach* was smallpox: MacArthur, 'Identification', pp. 183, 184. Others agreed: Shrewsbury, 'Yellow Plague', pp. 25, 38–39; Bonser, *The Medical Background of Anglo-Saxon England*, pp. 60, 63, 66–70. MacArthur rightly questions whether a leprosy-like disease and smallpox could have been confused on account of the 'extensive scabbing which accompanies the drying of [smallpox] pustules', 'Identification', p. 184. *eDIL* defines *bolggach* as a 'name of disease(s) characterized by eruptive spots or pustules on the skin, smallpox'. As helpful as this may seem, it is seriously problematic, as the definition hinges in part on modern translations of Irish annals.

[113] The earliest extant copy of any set of Irish annals, the *Annals of Inisfallen*, dates to the late eleventh century. It is near certain that the compiler translated the Latin encountered in the text which the Annals of Ulster and the Annals of Tigernach too depended on, into Irish, and that he equated *lepra* with *bolggach*. The *bolggach* here, in other words, is likely a later translation of the *lepra* still encountered in other annals. Nevertheless, it is uncertain whether *lepra* or *bolggach* were used systematically or for a number of different diseases. Unaware of Adomnán's text and the possible link between the outbreak Adomnán recorded and the *bolggach* encountered in other early medieval Irish sources, Dixon lamented that 'bolgach' was never defined clinically. He thought it smallpox nonetheless: *Smallpox*, p. 190.

[114] At 680, in the 'Annals of Ulster', 'Annals of Tigernach' and the 'Chronicon Scotorum', we find: *lepra grauisima in Hibernia que vocatur bolgacach* ('a most severe *lepra* in Ireland which is called bolgacach', Ulster); *lepra grauissima in Hiberniam quae uocatur Bolgach* (Tigernach); *lepra grauissima quae uocatur bolgach* (Chronicon). In the *Annals of Ulster*, at 743 [...] & in *bolgach* [...] ('the *bolgach* was rampant') and at 779 [...] *in bolggach for Erinn huile* [...] ('[...] the *bolggach* throughout Ireland [...]'), and, in the *Annals of Tigernach* at 1061 [...] *teidm mor a Laignib .i. in bolgach & treghaid, cor'ladh ár daíne sechnón Laighen* [...] ('great pestilence in Leinster, to wit, the *bolgach* and colic, so that there was a great destruction of people throughout Leinster').

gastrointestinal, as *bolg* can mean blister or boil.[115] This said, the association with *lepra* does suggest *bolggach* often or always affected the skin.[116] Adomnán writes of *ulcera* (ulcers or sores), painful, purulent and possibly raised, and *lepra* undoubtedly refers to a disease that affected the look and feel of one's skin, perhaps making it 'scabby, scaly or crusted'.[117]

There might be something in the nuts too. It has been proposed that these *nuces* were not nuts at all, but a corruption of the initial, perhaps contemporary, description of the disease's symptoms: nut-sized buboes, carbuncles, pustules or ulcers. If so, this epidemic was unknown to the Irish, or at least the generation affected in 574, as in the extant annals the nuts are described as 'unheard of' (*inaudita*). Arguing that this plague was yersinial and Justinianic, Woods proposed *habundantia nucum inaudita* is a misreading of *magna pestis glandularia*. 'Etymologically', he stressed, 'the noun *glandula*', often translated as 'glands', 'does mean "little nuts"'.[118] In other words, Woods proposed a copyist mistook a description of the disease to mean there was a good nut crop. If so, need those 'glands' have been the size of plague buboes? Might the nuts have been smaller?[119]

In any case, there is reason to think the plague Columba extinguished was the *bolggach*. While there are no cows in the annals,[120] both epidemics were just that, disease outbreaks, and both occurred in near spatiotemporal proximity. Furthermore, the *morbifera nubes* and *bolggach* manifested cutaneously. There is also a chance that both were recorded at Iona. On that note, were these plagues the same it is unlikely that the community at Iona would have escaped unscathed. According to Adomnán, Columba saw the deathly cloud pass qui-

[115] In 'Annales Inisfalenses', p. 8: *bolggach for doenib* is translated *fluxus ventris in populo*.

[116] That we find 'diseases of the leaprosie did abound and knobbes this year', in *Mageoghagan's Book*, an early-seventeenth-century English translation of an earlier Clonmacnoise-based Irish text, provides more support that *bolggach* affected the skin: *Annals of Clonmacnoise*, p. 89. This passage may not be precisely dated, but it is found between entries affixed dates of 569 and 579.

[117] Shrewsbury, 'Yellow Plague', p. 25.

[118] Woods, 'Acorns, the Plague, and the "Iona Chronicle"', pp. 498–99.

[119] The annals do not specify the species of nut, but oak mast (acorns), prized for pigs, was common in Ireland and features often in later Irish annals. Where the nut-sized lesions were located is not known; nor is it said how lesions one typically had. We might also add that most acorns (if the *nuces* were acorns) are small for a plague bubo.

[120] It may be noted that the Irish annals do not observe a cattle mortality for more than 100 years after 574. While later sections of the annals report epizootics, this section does not.

etly over the monastery he had founded, and where Adomnán was abbot when the vita was composed, before it rained death in eastern Ireland. Yet, the lost set of Ionan annals, kept as it was within the territory of Dál Riata, possessed an outlook that was Irish, western Scottish and Hebridean. That text may reference epidemics from multiple areas of the insular north-west. At the same time, those annals were connected with events in the vicinity of Iona. It is reasonable to suggest, therefore, that the annals and vita refer to an epidemic in 574 that sickened and killed in parts of Ireland, Dál Riatan territory in western Scotland and the Hebrides.

Virgin-Soil Smallpox in the Sixth Century?

The practice of retrospective diagnosing centuries-old plagues is fraught with peril. Neither Gregory's *pusulae* nor Marius' *variola* were necessarily smallpox. The same holds true of Adomnán's *morbifera nubes*, the *lepra* and *bolggach* of the Irish annals or the Avenchene *pustulae*. At the same time, we cannot claim that smallpox or something related to it does not lurk in these texts. That might not seem like much, but it is something. Consider that there is little chance, perhaps with the exception of the nut-riddled *bolggach*, that these epidemics were plague (*Y. pestis*).[121] Most other demographically significant diseases known to modern medical science would likewise seem to fit poorly with the evidence at hand. Measles and typhus should be considered for a differential diagnosis, but perhaps not, for instance, cholera, influenza or malaria.[122] Smallpox may just be the best match. Of the passages considered, Gregory's seems the most variolous.

[121] Hopkin's aforementioned proposal that Marius' variola could have been plague is doubtful: *Greatest Killer*, p. 25. Of note, this paper ignores the widely doubted speculation of Shrewsbury, 'The Yellow Plague', pp. 23, 38, that the Irish *blefed* of the 540s was also smallpox. His treatment of *sámthrosc* a decade later is another issue, but also remains unambiguous. Few question the plague diagnosis of that epidemic, although it is based primarily on its chronological fit with the Justinianic plague and now the isolation of *Y. pestis* in sixth-century Edix Hill: for example, Stathakopoulos, *Famine*, p. 166; Horden, 'Mediterranean Plague', pp. 138, 139; Keller and others, 'Ancient Yersinia pestis', p. 12,366. Earlier: MacArthur, comments on Shrewsbury's 'The Yellow Plague', pp. 214–15.

[122] That the lead author previously proposed that *variola*, with Marius' dead bovines and loose bowels, was an ancestral morbillivirus related to measles and rinderpest, indicates just how thin our evidence is, how easily our readings of the sources can be influenced by state-of-the-art, yet sometimes fleeting, results from evolutionary biology and palaeogenomics, and the uncertainty of the smallpox-*variola* association prescribed here.

That the outbreaks considered here were connected is likewise a possibility. While we face particular, insurmountable problems in attempting to tie the *bolggach* to the *variola*, or one outbreak of *pusulae* to another, that we can establish a few striking similarities between them raises the possibility that we are dealing with a series of interrelated outbreaks. It is notable that each epidemic visibly altered the appearance of their victims; that each may have sickened many more than it killed; that each occurred in relatively close proximity (within Western Europe); that each occurred within a roughly twelve-year period, and that each was, in fact, an epidemic, an outbreak of a disease distinct from the endemic disease baseline.

There are also indications the disease(s) underlying these plagues were novel or, at the very least, our authors seem to have been unfamiliar with them. Gregory might not claim the *pusulae* were 'unheard-of', but the Ionan annalist may say as much about the 574 epidemic. Notably, *pusulae* appear for the first time in Gregory's ten books of history, which span centuries, in 580. We are also dealing with the first appearance of *variola* and of *bolggach*.[123] Marius and Adomnán's bovine die-offs would as well appear to link the continental and insular epidemics. Although that would contradict a smallpox diagnosis, there is a chance, following Paulet's lead, that cattle epizootics were erroneously conflated with human epidemics.

With some verbalist ingenuity, there are grounds to propose that smallpox or a disease similar to it afflicted parts of Europe in the mid-sixth century. Those grounds are not as firm as we would like, but late antique epidemics are usually built on less. To be sure, it is less certain that the Antonine Plague (156/65–182/90) and most outbreaks of the first plague pandemic (541–750) were widespread than it is that the evidence consulted here represents a series of smallpox-like outbreaks across Western Europe.[124] We very much hesitate,

[123] Or if *variola* and *bolggach* were inserted later (nn. 83 and 84 above), the first recorded instance of a disease familiar to or known in the time of the complier or copyist. *Lepra* makes an earlier appearance in the Annals of Ulster, in 553 or 554 (cf. *Annala Uladh*, p. 55 with the translation on *CELT*), but there it refers to something different, as far as the annalist was concerned, the disease called *sámthrosc*, which some think is smallpox.

[124] From the Antonine Plague to the recurrences of the Justinianic Plague, the written evidence for epidemics in late antiquity is more often scarce than not. Consult, on the second-century 'pandemic': Gilliam, 'The Plague'; Duncan-Jones, 'The Impact of the Antonine Plague'; Greenberg, 'Plagued by Doubt'; Bruun, 'La mancanza di prove'. On the construction of first-pandemic plagues after 545, see Biraben and Le Goff, 'La Peste', pp. 1494–97; Stathakopoulos, *Famine*, pp. 113–24.

though, to argue that the sources surveyed here evidence a pandemic, virgin-soil or not.

It could be argued that a disease related to smallpox affected many parts of Western Europe between 570 and 582, including regions of modern-day Belgium, France, Ireland, Italy, Scotland, and Switzerland. Whether the disease spread earlier or later in these areas, or elsewhere, we cannot say. No late antique author provides us with a complete account of the epidemics that occurred within their purview and for many regions in late antiquity we have no authors at all.[125] That the *bolggach*, *pusulae*, and *variola* took on epidemic proportions does imply that the disease or diseases underlying them were rather communicable and unlikely to have been restricted to the years and regions when and where epidemics are reported, unless they were the result of one or more local spillovers from a nonhuman host. That Gregory stresses that children died primarily in the plague of 580 also suggests that if VARV were to blame, it was not spreading for the first time. That the old suffered as well indicates that when it did previously spread, its spread was not universal. For these smallpox-like epidemics, commencing before 569/70, persisting after 582 and travelling farther afield are all possibilities. But that they were discrete outbreaks, circumscribed and locally emergent is possible too. The evidence frustrates.

Home Grown or Imported? Debunking Early Smallpox Orientalism

Working with the evidence at hand, and assuming that each of these epidemics were variolous, one might argue that smallpox spread between Italy and the Hebrides between 569/70 and 574, possibly, as we could infer from Marius and Adomnán, by way of France and Ireland. It then reappeared (or caught the attention of our extant authors again) in France in the early 580s. At the same time, we should refrain from making too much of any epidemic geography the

[125] The story of smallpox's patron saint, St Nicasius, deserves mention. Vandals or Huns (it depends on who you read) beheaded the Rémoise bishop in the early or mid-fifth century. Latin supplications, likely ninth-century or later, against a disease identified as *variola* and interpreted as smallpox — but not described clinically — invoke Nicaise's name and describe him as both suffering smallpox and also looking out for victims of the disease. No earlier references to the bishop, however, mention smallpox (or any disease). Moore thought this claim just another 'one of those pious frauds' and he may have been right: Moore, *History*, pp. 95–98. Others have expressed less doubt (like Willan, 'Inquiry', pp. 96–97, Hopkins, *Greatest Killer*, p. 23), but the evidence for Nicasius having smallpox is very weak. As Moore stressed (*History*, p. 97), 'not a syllable is mentioned' of smallpox in Flodoard's tenth-century life of Nicasius: 'Historia Remensis ecclesiae', pp. 417–20.

sources permit, considering how thin the evidence is and that we do not know, for example, whether the *variola* or *bolggach* was in Germany in 565 or Spain or Scandinavia a decade earlier than that. For all we know, the disease followed the path of the *morbifera nubes* and reached Ireland from Scotland.

Where these epidemics originated is another problem. There is no indication, internal to the evidence, to suggest where they emerged. The long-standing approach would be to assume that they did not start where they are reported. Greco-Roman authors popularized the narrative that epidemics begin somewhere else, and the epidemiological othering of early smallpox historians ensured the smallpox-like outbreaks addressed in this paper were often early on fixed extra-European origins. Twentieth-century theories about VARV's emergence that will be addressed below did not change that. But that thinking is no longer robust.

It could be argued that had smallpox already emerged somewhere by the mid-sixth century, the epidemics observed in our sources would have originated in regions where the disease was endemic. However, to accept that argument and presume that the *bolggach*, *pusulae* and *variola* did not originate in the regions in which they are known to have spread is to conjecture the recorded epidemics were not the product of one or more local spillovers. Naturally, a local spillover of a pathogen hosted in wild animals, and possibly amplified in domesticated animals, which the human population, because of its density and distribution, could not sustain, is an option we cannot completely discount. It has been estimated for the better part of a century that to maintain smallpox a well-connected population must exceed 200,000.[126] Urban centres of that size, or networks of closely connected smaller cities, did not exist in sixth-century Europe, outside of central Italy or western Turkey. If smallpox was introduced to European populations or locally born, therefore, we must assume it eventually died out nearly everywhere.

But let's momentarily forget these possibilities and return to the eighteenth century. Early commentators on smallpox's past assumed that the disease reached Europe from somewhere, and many soon enough turned their attention to Arabia and East Africa. They did so in part because of what the physician-philologist Johann Reiske reported in 1746. Reiske had stumbled upon a text in a medieval Arabic codex in Leiden's public library that was, he supposed, of little value. Its only redeeming use was that it put smallpox, or

[126] Fenner and others, *Smallpox*, pp. 117–18, 1325.

so he surmised, at Mecca in the same year as Mohammad's birth.[127] The text referred to the arrival of two new diseases in Arabia, *al-nawāṣel*, which Reiske understood to be smallpox, and translated into Latin as *variola*, and *al-hasbah*, which Reiske read as measles and translated as *morbilli*.

Reiske tied the arrival of these new diseases in Arabia to the Axumite empire, named after its capital Axum in modern-day northern Ethiopia, and its attempt then to 'subjugate' Arabia north of Himyar, 'invade' Mecca, and 'destroy the great temple of Caaba'.[128] That an army from Ethiopia could have introduced smallpox to Arabia then required no stretch of the imagination: it was already suspected that an 'eastern or other remote people' initially introduced smallpox to the world.[129] The text itself put the blame on invading *Isräelitas*, undoubtedly Christian Ethiopians (*Aethiopes*) according to Reiske, and, as Reiske pointed out, Ethiopia was a known source of infectious disease, having introduced 'plague' (*pestis*) to Europe 'in the time of Hippocrates'.[130] Reiske did not doubt any of the text's claims. Not only were *al-nawāṣel* and *al-hasbah* at Mecca in the year of Mohammad's birth, but this was also, as the text asserted, their first appearance in Arabia. Further, the *Isräelites* were already familiar with the disease.[131] Yet, Reiske could not assign the text, as one nineteenth-century historian complained, 'any collateral authority',[132] for he knew nothing of the text's author or date.

Four decades passed before the 'adventurous traveler' James Bruce published an account of his encounter with a set of 'Abyssinian Annals' in the early 1770s

[127] Reiske, *Miscellaneas*, pp. 9–10. This was perhaps the most stunning of Reiske's observations on medieval Arabic medical texts.

[128] Reiske, *Miscellaneas*, p. 9.

[129] As, Reiske noted, pioneering medical historian Freind had speculated twenty years earlier: *Miscellaneas*, p. 9.

[130] In pinning this sixth-century smallpox epidemic on Ethiopia, Reiske was also following the lead of Thucydides, who had identified 'Ethiopia' as a disease hotbed in the mid-fifth century BCE. Several Greco-Roman and late antique authors as well pointed to Ethiopia as a source of epidemics. Reiske seems to have been aware of this Thucydidean thinking, as the Ethiopian-delivered plague in Hippocrates' age he refers to was quite possibly the Athenian plague which Thucydides thought born in 'Ethiopia': *Miscellaneas*, p. 10.

[131] Hirsch noted how unlikely it is that the first account of smallpox in south-west Asia would actually mark the first occurrence of the disease in the region: 'Blattern', p. 216.

[132] Willan appears to be one of very few to question Reiske's contribution: 'Inquiry', p. 1 (footnote). He had to, of course, if he was going to push anew Hahn's ridiculed theory that ancient Mediterraneans knew smallpox.

in the Red Sea town of Massuah. The annals seemed to corroborate Reiske's report.[133] The text, attributable to an El Hameesy, tells of seabirds with 'faces like lions' liberating the besieged Meccans and protecting the Ka'bah. This flock of birds, literally *abābīl*, let rain pea-sized stones, clutched in their claws, that 'destroyed' the Axumite army. But these were no stones, Bruce's author was certain — they were the marks of smallpox and measles.[134]

Accounts of this event — stone-slinging birds and all — we now know survive in diverse late antique and medieval sources. They are principal elements of the Year of the Elephant miracle, so-called for not only did an *abābīl* save Mecca, but also the lead (or only) Axumite war elephant (named Mahmūd in some accounts) miraculously abandoned the cause. The provenance of the passages Reiske and Bruce came across remains to be sorted, but it seems very unlikely that they provide independent witnesses, as was long claimed. They share notice not only of the initial arrival of smallpox in Arabia, but also of the first appearance of measles.

The Abbasid historian, al-Ṭabarī, captures several elements that surface in the texts Reiske and Bruce consulted. His voluminous early-tenth-century history observes two interrelated versions. First, a God-sent 'flock of birds, like swallows' flew in from the sea, 'each bird bearing three stones, like chickpeas and lentils, one stone in its beak and two in its claws'.[135] When the birds dropped the stones, 'everyone whom the birds hit perished, although all of them were not in fact hit'. Later we encounter the 'Abyssinian troops [...] continually falling by the wayside and perishing at every halting place' and their leader Abraha 'smitten in his body' and 'with his fingers dropping off one by one'. 'As each finger dropped off', he continues, 'there followed a purulent sore in its place'.[136]

In another take, we read that 'the birds [...] hurled down the stones upon the Abyssinian [Abraha's] troops and everyone who was hit suffered either a wound or else a spot that erupted into blisters and pustules'. Al-Ṭabarī then continues that 'the stones snuffed them [Abraha's army] out completely', before remarking that the 'survivors', Abraha among them, 'took to flight'. Not long afterwards, however, Abraha's 'limbs began to drop off one by one'.[137]

[133] Bruce, *Travels*, I, pp. 513–14. His adventurousness: Woodville, *History*, pp. 13–14.

[134] The text specifies smallpox and measles, though few might consider pea-sized stones emblematic of measles.

[135] Al-Ṭabarī, *History*, pp. 942, 229.

[136] Al-Ṭabarī, *History*, pp. 942, 231.

[137] Al-Ṭabarī, *History*, pp. 945, 235.

Could this be smallpox? Abraha's loss of appendages does not, of course, sound like smallpox, or any other highly infectious disease for that matter, but the blisters and pustules do and it has been proposed that the loss of fingers or arms was poetic for malignant smallpox skin shedding.[138] The mention (poetic?) in some of the sources (addressed below) of survivors having lost their sight also fits with smallpox.[139] That there were survivors also fits well with what we know of the disease, though most epidemics have survivors and some versions of these events purport everyone died.[140] That al-Ṭabarī underscores that this marked the first time 'smallpox' had appeared in Arabia adds some support to the diagnosis.[141] But it was also supposedly the first showing of 'measles' (not to mention of a particular sort of foreign shrubbery, another detail Reiske caught). Later in his text, however, the Abbasid-era polymath refers to an earlier outbreak of 'pustules', among the Khuzā'ah who had earlier ruled Mecca. The epidemic 'was likely to wipe them out' and consequently caused them to abandon Mecca, allowing the Qusayy tribe, into which Mohammad would be born, to assume control of it.[142] The definitive smallpox diagnosis is out of the question, but this at least sounds more smallpox-like than plague-like.[143]

What of other sources? The oral traditions collected by Ibn Isḥāq in the mid-eighth century, which survive in a number of recensions, relate much the same, in part because al-Ṭabarī drew on them. Details, though, vary between editions. After Mahmūd refused to do as he was told, we read that 'God sent upon them [Abraha's army] birds from the sea like swallows and starlings; each bird carried three stones, like peas and lentils, one in its beak and two between its claws. Everyone who was hit died, but not all were hit'.[144] And again we later learn

[138] That said, Dixon wagered accounts of Abraha losing appendages (arms or less) could be read as skin shedding following a case of malignant smallpox: *Smallpox*, p. 189.

[139] Fenner and others, *Smallpox*, pp. 49–50.

[140] One text also claims a one hundred per cent case mortality rate, which is hardly fitting of any epidemic diseases known to modern science, with the possible exception of pneumonic plague. See below.

[141] This 'smallpox', like Reiske's 'smallpox' and Bruce's 'smallpox', very much requires more attention, both the translation and the history of the translation. It has been noted that smallpox has explained Abraha's failed campaign since the work of 'Ikrimah, who died in 723/24 CE: Al-Ṭabarī, *History*, p. 229, n. 563.

[142] Al-Ṭabarī, *History*, p. 21.

[143] Following Dols, Selassie argues this epidemic was one of *Y. pestis*: 'Plague as a Possible Factor', pp. 50–51.

[144] Ibn Isḥāq, *The Life of Muhammad*, XXXVI, p. 25.

that as Abraha's force retreated they were 'continually falling by the wayside, dying miserably by every waterhole' and that Abraha himself 'was smitten in his body, and as they took him away his fingers fell off one by one' and that 'where his fingers had been, there arose an evil sore exuding pus and blood [...]'.[145] And later yet, we find 'Ya'qūb ibn'Utba told me that he was informed that that year [the year of Abraha's campaign] was the first time that measles and smallpox had been seen in Arabia'.[146] Additional details include 'the conductor of the elephant and its groom' subsequently 'walking about Mecca blind and crippled begging for food' and that 'sixty thousand men [from Abraha's army] returned not home, nor did the sick recover after their return'.[147] And yet the relevant sūrat in the Qur'ān (al-Fīl, 105.1–5) says nothing of the epidemic:

> Have you not considered how your Lord dealt with the possessors of the elephant?
> Did He not cause their war to end in confusion,
> And send down (to prey) upon them birds in flocks,
> Casting against them stones of baked clay,
> So He rendered them like straw eaten up?

Not hung up on the absence of an explicit reference to a smallpox-like disease in the Qur'ān, early smallpox historians were convinced that the flock of birds was 'symbolical of a pestiferous contagion' and the stones they threw 'emblematic of variolous pustules'.[148] One of the more flagrant orientalists, who preferred smallpox to have spread out of Arabia with Arab armies in the seventh century than without them in the sixth, went so far as to suggest that the birds are 'an easy explanation' for the epidemic and the whole Axumite army an 'eastern allegory of the origin of smallpox' or, more likely, 'a parable invented by Mahomet to excite veneration for the city in which he was born'.[149]

This supposed smallpox outbreak at or near Mecca became a central feature in early commentaries on the disease's past. How the sūrat, which underpinned the narrative, came to evidence an epidemic went unassessed, and for generations this element of Islamic exegetical tradition was not only accepted

[145] Ibn Isḥāq, *The Life of Muhammad*, XXXVI, p. 26.

[146] Ibn Isḥāq, *The Life of Muhammad*, XXXVI, p. 27.

[147] Ibn Isḥāq, *The Life of Muhammad*, XXXVI, p. 28.

[148] Woodville, *History*, p. 15. Schnurrer was one exception. He thought the stones possibly a rare Arabian hailstorm: Schnurrer, *Chronik*, p. 144. Powerful hailstorms are hardly unknown in the region, however. More recently, Tucker, *Scourge*, p. 6, remarked smallpox was 'described metaphorically in the Koran'.

[149] Moore, *History*, pp. 53–55.

matter-of-factly but turned to Islamophobic ends: to other smallpox and shore up the idea that Arabia, Arabs and Islam played a vital role in the early history of the disease. As indicated, however, commentators disagreed on whether smallpox spread beyond Arabia in the sixth century or the seventh. By Freind's time, it was commonplace that 'Arabians brought it from their own country' to Europe, as he claimed in his influential and oft-translated medical history of 1727,[150] and most, like Freind, supposed, despite an utter lack of evidence, that smallpox was the secret weapon of the Arab conquests.[151] Epidemiological orientalism was pervasive and influenced greatly the way narratives of smallpox were written.

> In less than thirty years [the Arab people] did propagate its religion and empire, so did it no less this modern evil [smallpox], not only through Aegypt, but Syria, Palestine, and Persia, and a little while after along the Asiatick coast through Lycia, and Cilicia: and in the very beginning of the next century farther into the Maritime parts of Africk, and cross the Mediterranean even into Spain itself.[152]

According to another early authority, 'from India to the Atlantic Ocean' Arab armies 'did not fail to widely disseminate the distemper'. Both soldiers and their families 'contaminated with disease' the 'surrounding nations',[153] adding 'to the devastation of the sword that of the fatal pestilence'.[154]

[150] Freind, *History*, II, pp. 188–90; *Histoire*, II, pp. 99–100; *Histoire*, pp. 202–04; *Historia*, pp. 137–38. That Paulet equated the 'irruption des Sarrazins' with the irruption of smallpox did not raise an eyebrow in the eighteenth century. Paulet, *Histoire*, I, p. 96. As discussed, Paulet thought smallpox spread widely in the 570s, from Arabia to continental Europe, but he also ventured that the disease diffused anew with Arab armies from the seventh century, albeit after they contracted the disease in Egypt.

[151] In this regard, Hirsch noticeably differed. He relied not on Arab conquerors for the spread of smallpox into or through Europe. Drawing on the writings of Marius and Gregory, he thought the disease had already spread in southern Europe before it arrived in Mecca: 'Blattern', p. 216. Like Hahn and Willan before him, Hirsch thought it possible the disease circulated in the ancient Mediterranean.

[152] Freind, *History*, II, pp. 188–90; *Histoire*, II, pp. 99–100; *Histoire*, pp. 202–04; *Historia*, pp. 137–38.

[153] Moore, *History*, pp. 44–45, 57, 63.

[154] Woodville, *History*, p. 22. While he admitted that there was 'no notice' of any of this in historical sources, he was certain that the effects were devastating: the 'ravages' of the disease 'must have unquestionably accompanied the general spread of the small-pox': Woodville, *History*, p. 25.

The dearth of evidence for any of this was typically overcome with an abundance of confidence.[155] It was asserted that 'greater multitudes' than those reported to have fallen to the armies of the caliphates had 'silently perished by disease'.[156] Although 'no direct proof' was had, 'circumstantial evidence' proved 'sufficiently conclusive'.[157] Time and again, the following imaginary geography of smallpox in Europe was sketched: with Arab armies the disease reached Spain, Portugal and France in the early eighth century after crossing North Africa in the seventh; Arabs introduced it as well to Sicily and Italy; and eventually, via these many points of entry, smallpox made its way to Germany and England.[158]

All of this thinking, questionable as it is and was, hinged on the epidemic at Mecca. Without it, smallpox could have fanned out of Arabia neither in the sixth century nor in the seventh. Not only was the Meccan epidemic the best-known evidence for smallpox in Arabia before Rhazes, but there seemed to be no other reports of disease outbreaks in or around the Year of the Elephant, in or anywhere near the region, that could have been smallpox. Early small-

[155] In this regard, Hirsch also stood out. He considered post-sixth-century medieval European medical and historical texts too vague about epidemics to identify smallpox in them: 'Blattern', p. 216. Cf. Schnurrer, *Chronik*, p. 159. As already stressed, smallpox had been repeatedly tied to Islam before 1800, though early commentators, like de Fonseca, Porchon and Voltaire, did not always explicitly write that the Arab conquests spread the disease: de Fonseca, *Consultationes*, p. 170; Porchon, *Nouveau*, pp. 8–9; Voltaire, *Histoire*, p. 74.

[156] Moore, *History*, p. 58.

[157] Moore, *History*, pp. 77–78. For some of this circumstantial evidence, see n. 29 above.

[158] Monro, *Observations*, p. 48; Moore, *History*, pp. 74–77. Additional to the works already cited, see Schnurrer, who has Arabs introducing rice, sugar and smallpox to Spain and Sicily, *Chronik*, pp. 144, 162: 'War aber nun die Krankheit den Arabern mitgetheilt, so musste sie durch diese im Verlauf der Zeit auch ins südliche Europa gelangen, da sie bald darauf als weit sich verbreitende Eroberer auftraten und disseits des mittellandischen Meers in Sicilien und Spanien eindrangen' and 'Aber für die Geschichte der Krankheiten ist sie sehr wichtig, weil ausser dem Bau des Reis und des Zukerrohrs durch die Araber von Spanien, vielleicht aber auch von Sicilien und Unter — Italien aus viel früher als durch die Kreuzzüge die Poken und, wie es scheint, auch der Aussaz ins westliche Europa gebracht wurden'. More recently, Dixon followed the centuries-old narrative of smallpox reaching Europe with Arab conquerors with their eighth-century campaigns in Spain and France. For Dixon, though, this was not the first time the disease had reached Europe. In Gregory's writing he also saw smallpox: Dixon, *Smallpox*, pp. 189, 190. Zinsser too: *Rats*, pp. 124–25 n. 8. Tucker likewise associated the diffusion of smallpox across North Africa and Iberia with the Arab conquests: *Scourge*, p. 6. For Hopkins, *Greatest Killer*, p. 26: 'Islamic armies spreading their new faith carried smallpox across north Africa and into Spain, Portugal and France before they were halted in 732'.

pox commentators appear to have missed Alexandrian John of Nikiû's account of an epidemic around the year 570, but it would have made little difference. Like some of the European evidence surveyed here, John's record is painfully brief and of limited use: 'a plague in all places and a great famine'. Without any mention of symptoms, John's plague eludes diagnosis. Further, the date of this outbreak is uncertain, though there is a possibility it occurred shortly before Byzantine Emperor Justin II abdicated in December 574.[159]

It was not lost on Paulet, however, that the Meccan epidemic roughly correlated with the Avenchene *variola* and it seemed perfectly reasonable to Paulet, even before Bruce's discovery, that a smallpox outbreak in Europe could have come from Arabia, as the extra-European origins fit neatly with prevailing theories about the disease's origins. While there was no doubting what direction the smallpox travelled, how precisely the disease made its way to Italy from Arabia was never clearly sorted, though Byzantines were sometimes implicated.[160]

The earliest of the potential European pox epidemics surveyed here, the *variola*, can be assigned to 569 (or 570), and placed, unlike the later *pusulae* and *bolggach*, in Italy. At no other point, can we identify another one of our smallpox-like epidemics in southern Europe, excusing Marius' *pustulae* of 570 (or 571). That the sixth-century outbreaks are reported in southern Europe before northern Europe could suggest the disease arrived in Italy after crossing the Mediterranean or spreading quickly overland through the Byzantine empire. Unless John of Nikiû was a few years off, the Meccan epidemic would seem to be the only recorded plague to have possibly preceded the *variola* by a year or so, and if John were a few years off it could still be said that the epidemic which caught Marius' attention arrived from the south-eastern Mediterranean region. Yet, to argue that smallpox docked in Italy after emerging in Arabia, we would have to both acknowledge the centuries-old orientalism that underlies the argument, overlook large uncertainties in our sources, and assume that smallpox, as we know it, then existed.

Let us return to the Axumite campaign. Its traditional date did not, though it should have done, raise issues for the smallpox histories Paulet and others sketched. The Year of the Elephant was identified as 568, 569, 570, and 572 at

[159] John of Nikiu, *Chronicle*, p. 150, XCIV.18. John appears to address the plague and famine after the Samaritan revolt of 572–73 but before the retirement of Justin II.

[160] Although there is no Byzantine connection in the sources themselves, it was proposed that Byzantine troops introduced *variola* to Italy after arriving there to repell the Lombards and that Byzantine troops sent to fight alongside Abraha brought the disease back with them to Byzantium, for instance: Paulet, *Histoire*, p. 94; Hopkins, *Greatest Killer*, p. 25.

one time or another in early accounts of the disease.[161] Paulet himself dated the *variola* to 570 and the Year of the Elephant to 570 or 572, following Reiske.[162] Were we to accept these dates, the possibility would exist, of course, that the epidemic Marius documented preceded the epidemic that routed the Axumite army outside Mecca and that the Meccan plague in fact arrived from Europe. Perceived truths about where smallpox could have emerged and initially spread explain why this possibility was overlooked repeatedly. Things have changed since the eighteenth century, however: Abraha's campaign has new dates.

Few specialists now think the Year of the Elephant corresponds to 570 or thereabouts. Epigraphs discovered in 1951 and 2009, and an offhand remark in the writing of Procopius, have led several historians and archaeologists of late antique south-west Asia to redate the Year of the Elephant or, more specifically, Abraha's failed assault on Mecca, to the mid-550s, sometime likely between 552 and 558, though a date as late as 565 has been proposed also.[163] These new dates would change the traditional narrative, but leave the orientalist thinking intact. The new date, in fact, better fits with the narrative Paulet wanted to tell. There is no doubting now that the campaign preceded the smallpox-like epidemics in Europe. Although a 15-year interval between the Meccan epidemic and the *variola* is hard to account for, historians of late antiquity have sketched histories of allegedly great plagues with less and there is a chance, if the outlier date of 565 is correct, that the interval was considerably shorter, which fits well with the old narrative.[164] Importantly, neither the inscriptions nor Procopius

[161] Gibbon went with 569 in his influential history of Rome's fall, which worked well with the narrative that the European epidemics originated in Arabia: Woodville, *History*, pp. 13, 16.

[162] Paulet, *Histoire*, I, pp. 77, 78.

[163] On the inscriptions, redating and the life Abraha, see: Kister, 'The Campaign of Hulubān', pp. 425–28; Conrad, 'Abraha and Muhammad' (p. 227 for Procopius' particularly notable comment); Robin, 'The Peoples Beyond the Arabian Frontier'; Robin, 'Himyar', pp. 151–53; Fisher, *Rome, Persia and Arabia*, pp. 140–42. Hirsch, 'Blattern', p. 215, uniquely, among early commentators on smallpox's past, suggested Abraha's campaign may have dated to 558. Zinsser, *Rats*, pp. 124–25 n. 8, oddly assigned it to the fourth century. Note that the historicity of Abraha's 'semi-mythical' campaign has been called into question recently, but that most specialists seem to think it plausible, though its traditional date and association to the year of Mohammad's birth are unlikely: Robin, 'Himyar', p. 152; Fisher, *Rome*, pp. 141–42.

[164] The Antonine plague has long been said to have comprised outbreaks documented years apart, in 165 and 189 for instance: Hecker, *De Peste Antoniniana Commentario*, pp. 1–24; Littman and Littman, 'Galen and the Antonine Plague', p. 243. Unlike the epidemics discussed here, outbreaks commonly thought to comprise the Antonine plague cannot on be linked, even tentatively, on symptomologic grounds: Gilliam, 'Plague Under Marcus Aurelius', p. 227.

mention the Meccan epidemic, but if we are to redate Abraha's campaign we should redate the pox-slinging flock too.

Had the *variola* Marius describes originated in Arabia, we should also ask where the Meccan plague originated. Here, too, there are problems with the traditional view. Early smallpox historians, following Reiske, did not argue for a local emergence near Mecca but instead wagered that the disease arrived from the Kingdom of Axum, which then spanned modern-day Eritrea, Ethiopia and south-western Arabia, though Reiske focused specifically on *Aethiopes*.[165] That Abraha's army itself fell victim to the disease outside Mecca and that the disease followed Abraha's force home, however, suggests otherwise.[166] Whether the population of the capital of Axum was then sufficiently large or connected to other sizeable cities to sustain smallpox is unclear.[167] Importantly, however, we now know that the army did not return to Axum; it retreated to Himyarite Arabia — most likely it returned to Ṣanʿāʾ.[168] Abraha had been an Axumite general and the Axumite viceroy of southern Arabia, but the campaign in question was more Himyarite than Axumite, as by the 550s Abraha ruled independently from Axum in Himyar, in south-western Arabia.[169]

That Abraha's force was southern Arabian and did not cross the Red Sea is an important detail. It derails some of the popular thinking about Ethiopia's place in smallpox's past. This, though, is perhaps besides the point, as it is unlikely that those on campaign with Abraha would have suffered smallpox, assuming it was indeed the disease in question, were they from a region where the disease

That the Meccan and European plagues can be considered smallpox-like strengthens the assertion the two were connected.

[165] Schnurrer, *Chronik*, p. 4, thought 'kamen die verheerendsten Krankheiten über Aethiopien und Aegypten'.

[166] Bruce concludes the Aksumite army was smallpox's first victim: *Travels*, p. 514.

[167] Hirsch stressed it was yet to be proven that smallpox's cradle was African. He favoured India or China: 'Blattern', pp. 215–16. Axum's population is said generally to have numbered in the tens, not hundreds, of thousands.

[168] It does not seem as though Axumite soldiers travelled to join Abraha's campaign, but it has been argued that Mahmūd and any other elephants Abraha may have had could have been African bush elephants, a species native to East Africa. Charles, 'The Elephants of Aksum', especially p. 173. How long Mahmūd had been in southern Arabia is unclear. Curiously, Charles suggests the elephant(s) also died in the epidemic, 'The Elephants of Aksum', p. 171.

[169] Abraha sought to destroy the Kaʿbah in order to redirect the pilgrim traffic to his newly built great church at Ṣanʿāʾ.

had already focalized. Where endemic, smallpox is primarily a disease of the young and those who recover are usually immune to reinfection.[170]

What of the Meccans? Accounts of the plague at Mecca suggest that the local population was entirely or largely unscathed. This could mean they simply were not exposed to the disease — it is not claimed, after all, that the disease spread within Mecca — or that Meccans were familiar with the disease. Al-Ṭabarī's remark about the pustular, pre-Mohammad Khuzā'ah reinforces that thinking. So does the detail that Abraha and his army suffered the disease on their way home and not when en route. At the same time, Meccans suffering alongside Abraha and his *Isräelites* would have undermined the miracle. That smallpox could have persisted and focalized in Himyarite Arabia and not travelled beyond Himyar along lines of trade and conflict, reaching not only Axum but also Mecca, before the mid-sixth century seems doubtful. Perhaps the disease had just arrived in Himyar. But if smallpox had just touched down, how could Abraha have thought to undertake a campaign and why would the disease have only irrupted in the army after it set out for Mecca?

That several urban centres populous enough to sustain smallpox dotted the shores of the eastern Mediterranean in the mid-sixth century is well known. Even had the region's largest cities — Constantinople, Alexandria and Antioch — fallen below the 200,000 mark by the 570s, which seems very doubtful, commerce and communication between these metropolises and many smaller centres with populations in the tens of thousands would have been sufficiently kinetic to sustain the disease. Were this the case, the Meccan epidemic could have been imported through Red Sea ports, by way of Trajan's Canal or carried overland. For one lone smallpox historian two centuries ago, this was 'an easy and natural explanation' for an event 'hitherto deemed so extraordinary'.[171]

The simplest interpretation this may be, but it is not problem-free. As emphasized, the European disease outbreaks surveyed above were of epidemic proportion and quite plausibly caused by a disease unfamiliar to the populations they afflicted or, at the very least, the authors who documented them. As large urban centres lined the shores of the eastern Mediterranean and trade and communication between them remained relatively frequent into the mid-sixth century, we must ask how a disease hosted in Mediterranean cities could remain unknown to populations in Western Europe. One might wager that the disease,

[170] Fenner and others, *Smallpox*, pp. 51–52, 167–68.

[171] Smith, 'Preface', p. xvi.

endemic in the populous eastern Mediterranean, broke out and burnt out on occasion in the areas Gregory, Marius and Adomnán were concerned with, but before their time. Yet, how could earlier authors in Western Europe have not documented a disease that manifested like smallpox? To blame the European epidemics on Arabia or Ethiopia is shortsighted, but to claim the disease was maintained in eastern Mediterranean cities is to overlook these questions and, more importantly, the dearth of evidence for endemic smallpox around the late antique Mediterranean; it is also to turn a blind eye to scattered accounts of smallpox-like epidemics in the eastern Mediterranean region in late antiquity that pre-date the *variola* and *abābīl*.

There is no reason to return to the Greco-Roman texts Hahn combed. More pertinent for our purposes is the account of a smallpox-like epidemic in the Syriac chronicle of Pseudo-Joshua Stylites. The future Vatican librarian Giuseppe Assemani made the disease outbreaks reported in this text available to European scholars in 1719, but his version of the text was heavily truncated, and the symptoms of epidemics left out. These epidemics apparently remained unknown to early smallpox historians well into the nineteenth century.[172]

Two epidemics, seemingly interconnected, in the city of Edessa (now Şanlıurfa in south-eastern Turkey) and the wider region of Byzantine Mesopotamia are observed and dated to the mid-490s. The recorded symptomatology encourages a tentative smallpox diagnosis.[173]

> **XXVI** 494/95 Our whole country [...] as encompassed with health at this time, but the diseases and sicknesses of our souls were numerous. Since God wills that sinners should repent of their sins and be saved, he made our body like a mirror and completely filled it with sores, so that by our outside we might see what our inside was like, and by the marks on our bodies we might learn how foul were the marks on our soul. As all the people had sinned, they all fell victim to this disease. Swellings and tumours appeared on all our citizens, and the faces of many became puffed up and filled with pus, making a fearful sight. Some had sores or pustules over their whole body, even to the palms of their hands and the soles of their feet, while others had great fissures on every single limb. But by the grace of God which protected them, the disease did not last long with anyone, and no damage or injury was done to the body. Although the marks of the afflictions were (still) evident after they were restored to health, the necessary bodily functioning of the limbs was preserved.[174]

[172] Assemani, *Bibliotheca Orientalis*, II, p. 267.

[173] Stathakopoulos, *Famine*, pp. 248–49.

[174] Psuedo-Joshua the Stylite, *Chronicle*, XXVI, pp. 23–24.

> **XXVIII** 496/97 The disease of tumours took a firmer hold on the population, and many in the city itself and in the (outlying) villages lost their eyesight.[175]

These passages would have thrilled Paulet but confounded his claim that European smallpox was imported from Arabia. Of course, for the orientalists who had smallpox spreading initially through south-west Asia with seventh-century Arab armies, this would have been devastating. Pseudo-Joshua, with symptoms and sequelae, might even offer better evidence for smallpox than does Gregory, Marius or Adomnán. The pustular rash, its distribution and scarring, the disease's relatively short course and implied quick onset, not to mention, as observed already, the effect of the disease on the sight of its victims, all suggest smallpox.[176] The vague focus on sinners implies the disease did not primarily attack children, as did Gregory's *pusulae* of 580, which in turn implies the exposed population was naive to the disease. That Pseudo-Joshua's epidemics were epidemics, however, indicates, again, that the disease was introduced from elsewhere, that it was not endemic in the eastern Mediterranean, or that it had only recently spilled over within the area.

Earlier yet there is Eusebius' account in his ecclesiastical history, published initially about 313, of a smallpox-like epidemic that, in concert with a food shortage, reportedly killed widely and fast in 312 in parts of the eastern Empire:

> another severe disease consisting of an ulcer (or sore), which on account of its fiery appearance was appropriately called a 'carbuncle'. This, spreading over the whole body, greatly endangered the lives of those who suffered from it; but as it chiefly attacked the eyes, it deprived multitudes of men, women and children of their sight.[177]

While less convincing of a smallpox diagnosis, Eusebius' passage, like Pseudo-Joshua's, forces us to treat gingerly claims that mid-sixth-century smallpox-like epidemics in Europe arrived from Arabia.

These passages also cast doubt on the idea that smallpox or something similar to it was endemic in the late antique eastern Mediterranean. Neither passage refers explicitly to an outbreak in a large city (that which Pseudo-Joshua refers to was Edessa in south-western Turkey), which leaves room to suggest

[175] Pseudo-Joshua the Stylite, *Chronicle*, XXVIII, p. 26.

[176] Fenner and others, *Smallpox*, pp. 219, 225–26, 231.

[177] Eusebius, *Church History*, p. 368, IX.8.1–12. That Eusebius here documented a smallpox epidemic has been suggested often: Baron, *Life*, p. 188; Baas, *Grundriss*, p. 148; Dixon, *Smallpox*, p. 189; Hopkins, *Greatest Killer*, p. 23.

that what is reported here are large outbreaks in smaller centres, where the disease could not have focalized, that were seeded by larger centres, where the disease was endemic. We might argue instead that the disease was reintroduced to the eastern Mediterranean in Pseudo-Joshua's time, though why it would have required a reintroduction considering how populous the region was, and why it would have taken more than seven decades to then spread widely in Western Europe, is unclear.[178]

If we adopt the old argument that smallpox is discernible in accounts of the Justinianic Plague in Constantinople and Antioch,[179] we might propose that the disease began to spread in other parts of the Mediterranean, after the epidemic Pseudo-Joshua records and before the 550s, and possibly that it had become endemic in large cities, assuming the smallpox-like disease discernible in accounts of the Justinianic Plague was endemic and not epidemic smallpox.[180] Yet, then we would have to explain why the initial dissemination of such a demographically significant and clinically severe disease did not attract more attention before the early 540s and, again, why it took so long to appear further west. In the 540s, but not before, plague could have overshadowed a smallpox-like disease, as it does now in the eyes of modern scholars. In sum, the European and Meccan epidemics may have been connected, so too the plague at Alexandria, but where they started, like where and when VARV initially spilled over, is anyone's guess.

The Rug Pulled Out? Insights from Evolutionary Biology

Paulet was not exceptional for thinking that *petite vérole* could have caused epidemics in the sixth century. That smallpox existed then somewhere, as stressed, was not in doubt in his lifetime. Little changed in the nineteenth century or

[178] As Hopkins underscored, smallpox may have been introduced anew multiple times to Europe in antiquity and the Middle Ages before there were cities there large enough to sustain it: Hopkins, *Greatest Killer*, pp. 20, 22, 26. We cannot say the same about the eastern Mediterranean region.

[179] Hahn, *Variolarum antiquitates*, pp. 74, 119; Smith, 'Preface', p. xvi; Willan, 'Inquiry', pp. 16, 86–87; Schnurrer, *Chronik*, p. 126; Baas, *Grundriss*, p. 193; Zinsser, *Rats*, p. 147; Hopkins, *Greatest Killer*, p. 23.

[180] Procopius and Evagrius mention smallpox-like symptoms but do not give their readers an indication of whether those symptoms were previously unknown or if any population cohort, such as children, suffered them most: Procopius, *History of Wars I*, II.XXII.30, pp. 462–63; Evagrius, *Ecclesiastical History*, IV.XXIX.178, p. 231.

the twentieth. More than three centuries worth of epidemic catalogues and smallpox studies agree: the disease has been around for a long time.[181]

This did not change when smallpox entered the laboratory, became pathogenic and was reframed as a 'souvenir' pathogen of the agricultural revolution.[182] Sequencing and molecular clock studies over the last decade and change, however, have reframed VARV again, considerably cutting VARV's age and undermining the idea that it diverged directly from a disease of livestock (cowpox).[183] In lieu of ancient or medieval smallpox strains reconstructed from bone, dental pulp or soft tissue, most attempts to pin down the emergence and untangle the evolution of the smallpox virus until recently have relied exclusively on the application of molecular clocks, which measure genetic distance between modern sequences, of the virus and closely related pathogens, and transform that genetic distance into time via a calibration rate (an estimate of the quantity of changes considered predictable over a unit of time).[184] Strains of smallpox assessed alone, without including other othropoxviruses, can speak to the origins of known VARV diversity. Studies that include other othropoxviruses can tell us something, some evolutionary biologists argue, about when VARV emerged. Importantly, a pathogen and its near-relatives must have a 'steady' temporal structure or molecular clock for its age to be reliably estimated. Smallpox possesses such a structure.[185]

[181] As discussed above.

[182] For example, Downie and Dumbell, 'Pox Viruses', p. 247; Cockburn, *Evolution and Eradication*, pp. 84–87, 197–98; Cockburn, 'Infectious Diseases in Ancient Populations', pp. 52–53; Fiennes, *Zoonoses*, p. 28; McNeill, *Plagues and Peoples*, p. 69; McKeown, *The Origins of Human Disease*, p. 38; Fenner and others, *Smallpox*, pp. vii, 216; Dobson and Carper, 'Infectious Diseases and Human Population History', p. 118. Earlier Fenner had doubts, see 'The Impact of Civilisation on the Biology of Man', p. 58. Long before the mid-twentieth century, however, some saw cowpox in Marius' variola: Maunoir, 'Lettre aux Rédacteurs', pp. 102–04; Schnurrer, *Chronik*, p. 138.

[183] Some recent studies, such as Babkin and Babkina 'Origin of the Variola Virus', pp. 1100, 1101, 1106, and Zehender and others, 'Bayesian Reconstruction', pp. 1134, 1135, have assumed that the cowpox virus is the ancestor of the smallpox, taterapox and camelpox viruses, but this is conjectural. This thinking is based solely on the fact that cowpox has 'the broadest [known] geographical range of susceptible hosts' and the lengthiest genome of known othropoxviruses.

[184] On molecular clocks: Rodriguez-Trelles, Tarrío, and Ayala, 'Molecular Clocks: Whence and Whither?'; Lemey and Posada, 'Molecular Clock Analysis'; Furuse, Suzuki, and Oshitani, 'Origin of Measles', p. 3.

[185] Duggan and others, '17th century Variola Virus', pp. 3407–12.

Smallpox's rate of evolution is fast, but not unlike other DNA viruses.[186] Just how fast it is, though, is debated. A 2016 molecular clock analysis, the first to integrate a pre-twentieth-century VARV genome, one reconstructed from a mid-seventeenth-century smallpox victim in Lithuania, estimated that the diversity of known VARV had a much more recent origin than previously thought.[187] Using a relaxed clock, which allows for different rates of evolution along different phylogenetic branches, that study concluded that all sequenced VARV strains, including the reconstructed seventeenth-century genome, share a common ancestor dating to between 1530 and 1654. The study, which has withstood criticism and the integration of additional premodern VARV sequences into studies of VARV's evolutionary history,[188] consequently proposed a faster rate of evolution for smallpox than earlier analyses.[189]

Known genetic diversity of VARV, following these studies, might originate after 1600, but when smallpox emerged is a different matter. Estimates of emergence require the assessment of other othropoxes. Yet, twentieth-century strains of smallpox and related othropoxviruses, suffered and sequenced, can only tell us so much about smallpox's deeper past. Earlier sequence data, reconstructed genomes, are needed to get a better picture of when smallpox initially diverged, and they remain very, very few. Like measles and tuberculosis, among other pathogenic diseases, estimates of smallpox's emergence have varied widely. For example, in 2007 VARV was reported to have diverged from

[186] Peck and Lauring, 'Complexities of Viral Mutation Rates', pp. e01021–17.

[187] Duggan and others, '17th Century Variola Virus'.

[188] Pajer and others, 'Characterization', introduced two additional 'ancient' VARV genomes and argued that known VARV strains shared an ancestor roughly two centuries earlier (about 1350) than Duggan and others, '17th Century Variola Virus', had proposed. The age of two genomes, however, had been incorrectly calibrated — both were more recent than alleged, as demonstrated in Porter and others, 'Comment', p. 276. Notably, Smithson, Imbery, and Upton, 'Re-Assembly and Analysis', p. 253, used Pajer's erroneously dated genomes to erroneously push back Duggan and others' estimated divergence by about 100 years; Porter and others, 'Comment', p. 276. More recently, Ferrari and others, 'Variola Virus Genome Sequenced' reported a late-eighteenth-century VARV sequence. Integrated with the correctly dated samples from the Czech Republic and Lithuania, it dated known VARV diversity to 1639–1662.

[189] Notably those proposed in Y. Li and others, 'On the Origin of Smallpox', and Babkin and Babkina, 'Origin', pp. 1100–12. The latter, used in Harper, *Fate*, pp. 91–92, 107, 327 n. 52, 329 n. 75, were repeated from Babkin and Babkina, 'A Retrospective Study', and Babkin and Schelkunov, 'Molecular Evolution of Poxviruses'.

a rodent-hosted taterapox between 16,000 and 68,000 years ago.[190] A year before it had been estimated to have spilled over six thousand years past,[191] and a year later, those ten plus thousand years of 2007 were shortened to three or four thousand years, an estimate re-published in 2012 and 2015.[192] Calculated rates of evolutionary change, sample size, and the data used to calibrate a clock, account in large part for this diversity of dates. Problematically, supposedly incontrovertible archaeological and historical evidence for smallpox has sometimes been used to constrain molecular clocks or confirm clock outputs, and thereby to greatly influence emergence estimates. The texts used here have yet to be employed to this end, but the alleged smallpox epidemic recorded by Ge Hong/Ko Hung in fourth-century China is partially responsible for the 2007 pre-Holocene estimate of smallpox's spillover.[193]

In an effort to establish an emergence date for VARV, the aforementioned papers, like more recent studies, tend to apply a single evolutionary rate to smallpox, camelpox and taterapox viruses, a perilous approach, as recently argued. Indeed, it has been proposed that molecular clocks cannot be used to date the divergence of VARV from its ancestors or, in other words, to demonstrate the emergence of smallpox.[194] Pinpointing a specific time when smallpox diverged from its closest known relatives, taterapox and camelpox, is infeasible as VARV's rate of evolution is unlike that of non-human orthopoxviruses. In other words, the different rates of evolutionary change known for the smallpox, camelpox and taterapox viruses muddles molecular-clock dating. Further, spillovers across species may result in a change in evolutionary rates that could 'confound molecular-clock dating'.[195]

In current reconstructions, a long phylogenetic branch connects VARV to camelpox and taterapox, one much, much longer than those currently associ-

[190] Li and others, 'On the Origin of Smallpox'. Contrast this with a more recent assessment of smallpox evolutionary history which argued the entire genus of othropoxviruses are a common ancestor about 10,000 years ago: Zehender and others, 'Bayesian Reconstruction', pp, 1134–41. In 2008, meanwhile, Babin and Babkina, 'Molecular Evolution', pp. 895, 903, 907, estimated the ancestor of the orthpoxviruses dated to 131,000±45,000.

[191] Babkin and Shchelkunov, 'Time Scale of Poxvirus Evolution', p. 18.

[192] Babkin and Babkina, 'Origin', pp. 1100–12; Babkin and Babkina, 'Retrospective', pp. 1601–02, and Babkin and Schelkunov, 'Molecular Evolution', pp. 895, 903.

[193] Dixon, *Smallpox*, p. 189; Fenner and others, *Smallpox*, p. 210; Babkin and Babkina, 'Origin', p. 1102; Babkin and Babkina, 'Retrospective', p. 1603.

[194] Porter and others, 'Comment', p. 3.

[195] Duggan and others, '17th Century Variola Virus', p. 5.

ated with sequenced smallpox strains.[196] The spillover to humans could have taken place at any point along that branch. It is also unclear whether this lengthy limb represents evolution in humans or other mammals. Not knowing that problematizes attempts to assign a divergence date or identify a specific point along that branch when smallpox leapt to humans or became smallpox as we understand it. Notably, the studies that have posited that VARV emerged thousands of years ago tend to work non-human orthropoxvirus genomes into their estimates. A recent report that twentieth-century VARV strains share a common ancestor in the fourteenth century does the same and is consequently problematic.[197] Confounding things further, the species we now know to harbour the orthropoxviruses most closely related to smallpox may not have served as reservoirs for those viruses in the past.[198] This warns against assigning the emergence of VARV to a specific region based on the current host geography of its closest relatives.[199]

That these key parts of the story — where and when smallpox emerged and the reservoir of the ancestral pox — remain unknown, seriously complicates efforts to place smallpox in Europe or Arabia in the sixth century. At present, the best evidence available indicates that smallpox as we understand it may be considerably younger than previously thought and, quite possibly, that VARV capable of causing disease we would associate with smallpox was not anywhere in the sixth century. This conclusion still stands despite the recent report of 11 VARV sequences and the reconstruction of four near-complete VARV genomes from early medieval northern Europeans.[200] The earliest of the

[196] Duggan and others, '17th Century Variola Virus', Figure 3; Porter and others, 'Comment', Figure 1.

[197] Zehender and others, 'Bayesian Reconstruction', pp. 1134–41. Here a mean substitution rate is worked out for a number of othropoxviruses and used to date important moments in smallpox's past.

[198] Duggan and others, '17th Century Variola Virus', p. 5. Rodents were central to the emergence of smallpox, but precisely what species is unclear. Cf. Babkin and Babkina, 'Retrospective', p, 1602.

[199] That VARV emerged 3000 to 4000 years ago in east Africa, as Babkin and Babkina proposed, and Harper adopted in his account of the Antonine plague, is, thus, problematic: Babkin and Babkina, 'Origin', pp. 1106–08; Harper, *Fate*, pp. 91–92, 107, 327 n. 52, 329 n. 75; also Green, 'Climate and Disease' . A more recent attempt to assign smallpox a cradle, using other means, has focused on South Asia (refered to as 'Southern Asia', but see Figure 4), albeit with a limited dataset potentially unrepresentative of past transmission: Adam, Scotch, and MacIntyre, 'Bayesian Phylogeography'.

[200] Mühlemann and others, 'Diverse Variola Virus'. This paper appeared after this essay

four genomes dates to the first half of the seventh century, the latest to the last half of the tenth. While upwards of a millennium older than the aforementioned Lithuania genome, the gene content of these VARV sequences contrasts greatly both with already reconstructed prelaboratory genomes and also with the VARV eradicated in the twentieth century, which in turn do not contrast nearly as greatly between themselves. Known VARV diversity has a common origin now very tentatively dated,[201] with the new early medieval VARV genomes, as early as the third or fourth century CE, but those new genomes, as noted, are separated from the early modern and modern VARV genomes by a long evolutionary branch, represent a previously unknown VARV sister clade, and, on account of the remarkable contrast in their gene content, very possibly did not cause a disease we would associate with or recognize as smallpox. The VARV responsible for the disease clearly discernable, on symptomologic and epidemiological grounds, as smallpox may have evolved long after these early medieval strains diverged from the common ancestor of later known VARV.[202] In short, there is no genetic evidence yet for smallpox, as we know it, before the early modern period.

The contrast with centuries of thinking about the early history of smallpox is striking. Yet, historian Ann Carmichael observed more than 30 years ago that prior to the sixteenth century there is little evidence, in Europe at least, for epi-

was submitted for publication, but an effort has been made to carefully integrate its findings throughout where appropriate. Some have misinterpreted these findings in papers also published after this paper was submitted. For example, McCormick, 'Gregory', p. 60, writes that smallpox's 'early history has been controverted until July 2020' when Mühlemann and others identified 'aDNA of the virus' in early medieval remains, and that VARV phylogeny now suggests VARV 'originated by *c.* AD 300'. As stressed here, and by Mühlemann and others, there are multiple VARV viruses and the VARV genomes published in 2020 are of a VARV that is very distinct genetically from the VARV known to cause smallpox. The current phylogeny also suggests the VARV we know to cause smallpox and the VARV identified in early medieval remains had a common ancestor around 300 CE, but a great deal remains unknown and a smallpox-causing VARV could have emerged, on the basis of current data, at any point after 300 CE, before early modernity. Cf. Newfield, Duggan, and Poinar, eLetter response to Mühlemann and others, 'Diverse Variola Virus'. Importantly, as emphasized in McCormick, 'Gregory', p. 60 n. 114, the early medieval remains which turned up the VARV sister clade may need to be redated (closer to the present) to take into account marine reservoir effects (Mühlemann and others, 'Diverse Variola Virus', SI, 5, 8).

201 As stressed in Mühlemann and others, 'Diverse Variola Virus'.

202 Newfield, Duggan, and Poinar, eLetter response to Mühlemann and others, 'Diverse Variola Virus'.

demics of virulent smallpox-like disease.[203] If smallpox was known in the Middle Ages and earlier, it could be that it was caused by lineages of the pathogen not known to modern science, like the aforementioned early medieval VARV, or no longer circulating at the point of its eradication. It is certainly plausible that many lineages of VARV appeared and died off. Some strains may have had increased host-specificity, perhaps accounting for the deaths of animals other than people, like the cattle Marius reports, while others may have been more virulent. At this point, we cannot be certain. Also not entirely certain is the significance of the pattern of gene inactivation found in the reconstructed early medieval and early modern VARV genomes. Importantly, gene degredation shared between early modern and modern VARV, and therefore predating 1650, is not present in other othropoxviruses, like vaccinia virus.[204] That degradation seems to have occurred, therefore, before 1650. It may have enhanced virus transmission, creating a 'natural host species barrier', limiting the virus to humans, or augmented pathogenesis. Or it may not have. Unfortunately, with so few data points much of smallpox's evolutionary history remains murky.

Implications and Ways Forward

Where does this leave us? Not much written in histories of late antique disease is known for sure. There is nothing new about that. In the preceding pages, we have surveyed the evidential basis for smallpox-like epidemics in the mid-sixth century. The caveats are many, as stressed, but it is not completely inconceivable that our sixth-century authors refer, in their own terms, to interconnected smallpox outbreaks between roughly 570 and 582. Whether a one-off pandemic lasting just over a decade, or a collection of distinct epidemics in different regions spread out over years, the texts, Gregory's in particular, do give reason to suspect that smallpox, or something like it, was present, at least for a time, in late antique Europe. The epidemics may have begun there or reached Europe from the eastern Mediterranean, Arabia or elsewhere, we cannot tell. On the basis of the evidence in hand, no particular region can be identified as the origin of smallpox, in the sixth century or altogether. Further analysis of the extant textual tradition is required to determine whether a smallpox-like disease was present anywhere in Arabia circa the Year of the Elephant and propogated by Abraha and his army, whether around 555 or 570. That a VARV

[203] Carmichael and Silverstein, 'Smallpox in Europe'.

[204] Duggan and others, '17th Century Variola Virus', pp. 1, 2.

capable of causing smallpox as we know it appeared for the first time in Europe or anywhere in the sixth century is yet less certain. None of the evidence is altogether indisputable.

Current evolutionary biology and paleogenomics of smallpox does not upend the theory that smallpox was in Europe in the sixth century, but it does not exactly support it either. Known VARV diversity has a common origin that has now been dated tentatively to as early as late antiquity. That will change as more prelaboratory VARV genomes are reconstructed, but how much it will change remains to be seen. Crucially, the VARV now known to have existed in Europe in the early Middle Ages is genetically very distinct from the VARV that caused smallpox as it is understood, epidemiologically and symptologically, by modern medical science. At present, that the science of smallpox's past will ever come to support age-old thinking about the timing and emergence of the disease, as we recognize it, appears unlikely. Rather, it seems probable that late antique strains of VARV, like the abovementioned early medieval VARV, were not the VARV we associate with smallpox; all varieties of VARV currently known that caused disease clinically and epidemiologically recognisable as smallpox are much younger than previously thought. While this could explain the apparent conflict between smallpox's molecular timeline and recorded sixth-century smallpox-like symptomatology, the discrepancy between smallpox, as experienced in early modern and modern centuries, and the sixth-century epidemics assessed here (from an ability to cause disease in cattle to a capacity to persist in thinly populated regions), may, in fact, lie in the gene content contrast identified in known early medieval VARV. That the smallpox-like epidemics observed by Gregory, Marius and others were caused by the VARV recently isolated from early medieval northern Europeans, however, is hardly certain.[205]

Nevertheless, the textual evidence is enough to unravel the reckless claim that smallpox spread, for the first time, around the Mediterranean and into Europe with the seventh- and eighth-century Arab conquests. Smallpox-like epidemics were recorded in Europe and the Mediterranean region long before

[205] The association was advanced in Mühlemann and others, 'Diverse Variola Virus', pp. 5–6, but VARV would need to be uncovered in populations known to have been afflicted by the epidemics discussed here before the two could be connected even tentively. See, Newfield, Duggan, and Poinar, eLetter response to Mühlemann and others, 'Diverse Variola Virus'. Of course, one might rightly regard identifying an account of an epidemic, like those reported by Gregory or Marius, as being caused by pathogen the clinical expression and epidemiology of which are very little understood as careless.

the armies of the first caliphates won Byzantine lands in the Mediterranean region or reached Visigothic Iberia or Merovingian France. Paulet knew this, but other early historians of the disease, misguided by their epidemiological othering, saw smallpox appearing first in Arabia and spreading outward, not via commerce and communication in the mid-sixth century, but via conflict after 630.[206] But that smallpox did in the Mediterranean region in late antiquity what it is claimed to have done in the Americas in early modernity, or that smallpox, as has been proposed of first-pandemic plague, weakened Greek or Persian empires ahead of, or amidst, the Arab conquests,[207] is entirely unsubstantiated. Not only is evidence for the smallpox-like disease in those conquests wanting, but the written evidence for smallpox, or something like it, in Europe and Arabia in the sixth century, decades before the Arab conquests, is far more robust than it is in the seventh or eighth centuries anywhere. At the same time, we must stress that while smallpox may not have cleared the way for the first caliphates, in most regions of Europe the disease could have been introduced or spilled over as a virgin soil epidemic in the sixth century, as well as in the seventh, eighth, ninth and tenth, as the population there was generally not populous enough to maintain it. Indisputable evidence for smallpox in Europe in the immediate centuries after the sixth is lacking, but the disease, at least as we know it, would have had to be re-introduced were it present.[208]

Insofar as certainty is within our reach, it appears absolutely certain that sixth-century populations suffered epidemics caused by a number of infectious diseases. If disease helped transform the ancient world, *Y. pestis*, as the Justinianic Plague, did not do it alone. That a smallpox-like disease and plague appear to have cocirculated in multiple instances is notable. Not only may plague and something like smallpox have co-occurred in the 570s and 580s, but physicians and historians alike have suggested since at least the eighteenth century that smallpox overlapped with plague during the initial outbreak of the first plague pandemic, the Justinianic Plague, in the 540s.[209] Whether or not

[206] The earliest of the northern European VARV sequences would now seem to indicate this as well, Mühlemann and others, 'Diverse Variola Virus', p. 5.

[207] First-pandemic plague as helping to pave the way for the Arab conquests: Russell, 'That Earlier Plague', pp. 174, 182–83; Biraben and Le Goff, 'La peste', p. 1508; Kennedy, *The Armies of the Caliphs*, p. 2; Sarris, 'The Justinianic Plague', p. 173.

[208] Again, this recalls that the VARV recently identified in abovementioned early medieval northern Europeans seems to differ considerably from the VARV that causes the disease we recognise as smallpox.

[209] On smallpox-plague co-occurrence, see above. Naturally, the confirmation of plague

VARV was new in late antiquity, or a smallpox we would recognize existed in the sixth century, there is still the potential for a smallpox-like disease to have left a demographic mark. How widespread and deleterious any late antique VARV may have been, whether its presence at all shaped economic or political outcomes, requires more study. As recently argued for plague, a smallpox-like disease may have taken a greater toll in some regions than others and may not have contributed much to the reconfiguration of the ancient world.[210] It is too soon to say.

As in Paulet's time, there are still more questions than answers. Further, many of the questions raised and debated in the eighteenth and nineteenth centuries remain unsolved. Although recent advances and multidisciplinarity have pulled the rug out from beneath several perceived truths in smallpox's early history, raising yet more unknowns, going forward, collaborative, multi-disciplinary efforts will be essential. No one discipline can uncover smallpox's cradle, date its emergence, chart its evolution, identify where it first spread, or map the course of its globalization. That much is certain.

in late antique human remains neither invalidates the hypothesis that multiple diseases cooccurred nor proves that plague took more lives in late antiquity than other infectious diseases.

[210] Mordechai and others, 'Justinianic Plague'.

'AWAY WAS SONS OF ALLE AND BREDE': THE DECLINE OF THE MEDIEVAL CLIMATE ANOMALY AND THE MYTH OF THE ALEXANDRIAN GOLDEN AGE IN SCOTLAND

Richard Oram*
History, Heritage and Politics, University of Stirling

Quhen Alexander our kynge was dede,	When Alexander our king was dead,
That Scotland lede in lauche and le,	Who led Scotland in law and security,
Away was sons of alle and brede,	gone was abundance of ale and bread,
Off wyne and wax, of gamyn and gle.[1]	Of wine and wax, playfulness and joy.

Introduction

Amongst the deepest ingrained popular perceptions of Scotland's medieval past, neatly encapsulated in the stanza from Andrew of Wyntoun's early fifteenth-century *Orygynale Cronykil of Scotland* above, is that of what has come to be called the 'Golden Age' of the second half of the thirteenth century. What Wyntoun and subsequent chroniclers presented was a tale of how the kingdom had flourished through an era of unprecedented political stability and economic prosperity under the benign rule of King Alexander III (1249–1286).[2] Down to the 1990s, that narrative was accepted almost uncritically in a Scottish historiographical tradition which viewed the thirteenth century as marking the developmental climax of the medieval kingdom.[3]

* This paper was written to complement the research developed by Alasdair Ross on the Scottish economy and taxation in the later thirteenth and fourteenth centuries, presented in Chapter 5 of this collection.

[1] Androw of Wyntoun, *Orygynale Cronykil of Scotland*, II, p. 266.

[2] For the most recent analysis of the reign and its significance, see Reid, *Alexander III*.

[3] That trend culminated in a volume of essays by leading Scottish historians of the 1980s,

'With Our Backs to the Ocean': Land, Lordship, Climate Change, and Environment in the North-West European Past. Essays in Memory of Alasdair Ross, ed. by Richard Oram (Turnhout: Brepols, 2023)
EHNAW 5, pp. 83–107 BREPOLS PUBLISHERS 10.1484/M.EHNAW-EB.5.131271

Underpinning that climax, it was argued, was the final episode in Scotland's experience of a pan-European era of prolonged economic growth which, starting in the eleventh century, had first triggered and was then sustained by a cycle of colonization, land-clearance and agricultural expansion that spanned the following century and a half. From the 1980s, an increasing body of published scientific data revealed significant long-term trends in historic climate change and in particular of a sustained period of what was believed to be more benign conditions in the northern hemisphere throughout the twelfth and thirteenth centuries. This episode, first labelled the Medieval Warm Period and, as understanding of these climate processes has become more refined and nuanced, more recently re-styled the Medieval Climate Anomaly (MCA), was seen as providing ideal environmental conditions to support the kind of sustained growth posited in traditional historical narratives. In Scotland, that traditional historiography saw the end of this 'golden' era at the very end of the thirteenth century as the result of human factors; dynastic failure and subsequent descent into a war for independence with England, and a civil war between rival contenders for the crown of Scotland. From the 1980s onwards again, climate change was added as a further agent in ending that period of growth, with the fourteenth century witnessing a transition from the benign conditions of the MCA into the colder and wetter episode that was initially styled the Little Ice Age. The newly-available scientific data, however, was rarely taken up by Scottish historians or archaeologists; but where it was utilized it was deployed uncritically and largely deterministically as corroboration of the traditional documentary record; the 'Medieval Warm Period' almost by definition meant benign conditions and, therefore, growth; the Alexandrian Golden Age was validated.

The resilience of the traditional narrative was grounded in largely untested assumptions and what can be seen as *a priori* determinations; the weight of authority given to the suite of historical sources upon which it was built; the apical place given amongst generations of historians to the idea of the Golden Age; and its passing into the meta-narrative of deeply-ingrained Scottish cultural identity. It must be said that it was long recognized amongst academics that both constructs were founded on the version of history presented by a narrow group of surviving later fourteenth- and fifteenth-century Scottish

arising from events to commemorate the 700th anniversary of Alexander's death, which presented a re-evaluation and affirmation of the reign's significance as a 'high point' in the kingdom's medieval political, cultural, and economic life: Reid, ed., *Scotland and the Reign of Alexander III*.

chronicles and praise poems,[4] but the absence of other forms of historical record evidence provided few alternatives against which to test their authority. From the 1990s, however, revisionist historiography began a critical re-evaluation of these sources and of the narrative built upon them, exposing through textual criticism, access to new and more accurate source editions, and techniques such as discourse analysis, their propagandist nature; all were written from the perspective of the victorious post-Wars of Independence Bruce and Stewart political elite.[5] The juxtaposition of an Alexandrian 'Golden Age' with the subsequent decades of crisis, followed in turn by the restoration of the security and prosperity of the Alexandrian age under the victor of the civil war has become transparent as a contrivance to legitimize that man, King Robert I's rule.

Although the motivation underlying the chronicle discourse has been exposed and questioned for nearly a quarter of a century, the perception of an Alexandrian 'Golden Age' which had formed as a concomitant dimension of the propaganda construction has been more resilient. To Nicholas Mayhew, in his contribution to the ground-breaking volume of essays celebrating Alexander III's reign published in 1990, it was:

> a period of prolonged relative peace which fostered a growing population who brought more land than ever into cultivation. It was a period of great trading prosperity, founded chiefly on wool, which stimulated and sustained the development of town life, drew Scotland into the mainstream of European trade and won for Scotland a bigger share of growing European silver supplies.[6]

This summary characterization of the reign of King Alexander III and the state of Scotland's economy in the late thirteenth century was founded on three pillars: Mayhew's close examination of the fragmentary financial records of the reign; analysis of the archaeological evidence for the material wealth and social state of the kingdom (especially the evidence for the volume of coin in circulation); and regression modelling from the position of the better-documented late fourteenth century. Reflecting the limited impact that environmental thinking had then yet had on mainstream social and economic history, it made no reference to the early climate models for the period. He

[4] These comprise the chronicles of John of Fordun (*Chronica Gentis Scotorum*), Wyntoun's *Orygynale Cronykil*, Walter Bower's *Scotichronicon*, and Barbour's *The Bruce*.

[5] See especially Boardman, 'Chronicle Propaganda'.

[6] Mayhew, 'Alexander III — A Silver Age?', p. 66.

thus qualified it as a 'Silver Age', in reference to the abundance of bullion as coin in circulation, but essentially confirmed the Alexandrian 'Golden Age' construction first set out by Patrick Fraser Tytler in the early nineteenth century.[7] Reinforced by Mayhew's analysis of the economic state of Alexander's kingdom, subsequent discussions of later thirteenth-century Scotland have continued to present a broadly positive view, although some of the rosier tints of earlier interpretations have become more subdued as awareness of the wider climate change debate has penetrated historians' consciousness. The available scientific data, however, has often been incorporated into the old narrative in ways that are largely teleological and still very deterministic.[8] This present essay argues that the traditional narrative requires to be revised more radically and offers the alternative hypothesis that far from being a 'Golden Age' the second half of the thirteenth century in Scotland was a time of progressive decline and incipient crisis in which systemic stresses were exposed in regional economies and societies. Recent revisions of the traditional narrative of Scotland's later thirteenth-century socio-economic and more general environmental experience have been dependent largely on the use of climate proxy data which permit the proposition of generalized models for impacts on a regional scale.[9] For the medieval period and with applicability across the entire country, few reliable long-sequence climate proxies that are of direct Scottish provenance are as yet currently available.[10] We are, thus, reliant on sample replication and all the question marks presented by that, on geographically distant proxies or on short-run or more regionally specific sequences.

Reconstructions of weather- and wider climate-related impacts on Scotland have been dependent hitherto on the synthesis of data from mul-

[7] Tytler, *The History of Scotland*, i.

[8] See, for example, Brown, *The Wars of Scotland 1214–1371*, pp. 57–60. For a rather sombre overview of the socio-economic position of Scotland in the Middle Ages, see Ditchburn and Macdonald, 'Medieval Scotland, 1100–1560'.

[9] Oram, 'From "Golden Age" to Depression'; Oram, 'Estuarine Environments and Resource Exploitation'; Hindmarch and Oram, 'Eldbotle'; Oram and Adderley, 'Lordship, Land and Environmental Change'; Oram and Adderley, 'Re Innse Gall'; Oram and Adderley, 'Lordship and Environmental Change'.

[10] See, for example, Charman and others, 'Paleohydrological Records'; Proctor and others, 'A Thousand Year Speleothem Proxy Record'; Mills, 'Historic Pine and Dendrochronology'; Wilson and others, 'Reconstructing Holocene Climate'; Mills and Crone, 'Dendrochronological Evidence'; Rydval and others, 'Reconstructing 800 Years of Summer Temperatures'.

tiple climate proxies but principally from Greenland ice-core, oceanic sediment, and Siberian tree-ring sequences. The problems of using such geographically remote data to explore the Scottish context have been discussed elsewhere by the present author and their value will be reviewed below. There has, however, been no significant parallel discussion and reappraisal of the medieval written record for this period, with the narrative founded on late fourteenth- and fifteenth-century chronicles remaining largely unchanged from that advanced in the 1800s. In large part, this situation reflects the relative paucity of illustrative texts of medieval Scottish provenance or concerned with Scottish domestic conditions. Conventional historical records such as have been used to excellent effect in England, particularly the long sequences of manorial accounts and court rolls, is almost entirely lacking and there is a very limited range of contemporary chronicles, the majority derived from a single set of now lost earlier medieval narratives. Only two examples of rental material survive and are available in published form and, while it is known that other now lost late thirteenth-century chronicle material is embedded in the *Gesta Annalia* sections of the later fourteenth-century chronicle attributed to John of Fordun and in the fifteenth-century works by Wyntoun and Bower, only two full chronicle texts of thirteenth-century compilation have come down to us.[11] Alongside these sources, however, we can set the extensive corpus of Irish annal evidence, whose synchronous accuracy and relevance as a reflection of wider climate and weather trends down the whole Atlantic seaboard of the British Isles can be tested against British — primarily English — sources and climate proxy data. The first part of the following discussion is an environmental narrative for the half-century from 1250 synthesized from that annal and chronicle material, followed by an examination of the climate proxy evidence and an interpretation of what it and palaeoenvironmental data-sets suggest about conditions through Alexander III's reign and down to the end of the century.

[11] The rental material relates to the monastic estates of Coldingham Priory and Kelso Abbey: see *The Correspondence, Inventories, Account Rolls and Law Proceedings of the Priory of Coldingham* and *Liber S. Marie de Calchou*, II, pp. 455–73, 'Rotulus Reditum'. The chronicles are *The Chronicle of Melrose* and the so-called *Chronicon de Lanercost*, much of which was composed in a southern Scottish context.

The Chronicle-based Narrative[12]

Scottish chronicle sources for the early 1250s are preoccupied with the struggle for political power in Scotland which followed the death of King Alexander II (reigned 1214–1249) and the accession of his seven-year-old son, Alexander III, in July 1249. Due to that preoccupation, we are reliant largely for record evidence for the climatic conditions and weather events of this period upon a closely-related group of Irish monastic annals. The weather observations made in these sources are still generalized and of widely varying levels of detail. Let us take one example. The north-eastern source, the Annals of Ulster [AU], has a simple entry for 1252 which records laconically that the summer was hot but the Roscommon-based Annals of Loch Cé [ALC] for that same year provide a more detailed account, reporting:

> Great heat and drought in the summer of this year, so that people used to cross the Sinuinn without wetting their feet; and the wheat was reaped twenty nights before Lammas [1 August], and all the corn was reaped at that time; and the trees were burning from the sun.[13]

Qualitative commentaries of this kind are a poor substitute for quantitative data that would permit analysis of seasonal temperature variations, rainfall levels or crop yields amongst other measures, but they do offer insight onto the perceptions of general human experience through the eyes of a privileged elite which, though insulated from the worst impacts, witnessed the consequences of climate variation.

From such records we can begin to build a narrative of observed trends. At the start of our period, accounts for 1249 and 1250 are dominated by political events; this suggests that the unexceptional weather which had been noted through the 1240s after a period of violent extremes in the 1230s continued into the new decade. 1251, however, was the first of a series of years in which comment was made on the heat of the summer, which was peppered by thunderstorms and torrential downpours, as recorded in ALC and the Annals of Connacht [AC].[14]

[12] For a detailed discussion of the thirteenth- to fifteenth-century position as reflected in the contemporary sources, see Oram, '"The Worst Disaster Suffered by the People of Scotland in Recorded History"'.

[13] 'Annala Uladh', *s.a.* 1252.1 [hereafter AU]; 'Annals of Loch Cé', *s.a.* 1252.9 [hereafter ALC].

[14] ALC, *s.a.* 1251.9, 1251.10, 1251.12; 'Annals of Connacht', 1251.10, 1251.11, 1251.13 [hereafter AC].

According to the Kerry-based Annals of Innisfallen [AI], 1253 saw a similarly hot summer to that experienced in 1252, with both AU and ALC reporting fruitfulness of trees. The latter source again provides the more expansive account of what the annalist saw as 'the best year that had ever come for nuts, and the produce of the earth, and of cattle, and of trees and herbs'.[15] It is ALC again that provide the most information for the productivity of 1254, reporting 'an excellent year' in which a bumper crop of acorns provided first-class pannage opportunities, and there were excellent milk yields from herds, and an abundance 'of all other good things besides'.[16] The run of good years, however, ended with 1254 and the absence of comment on weather or on the harvest and yield levels for 1255 suggests at best unexceptional conditions throughout the Atlantic-facing zone of the British Isles in that year.

Irish sources fall silent in respect of environmental matters for five years after 1255 but the narrative is here filled by Scottish and wider British accounts. The most relevant for Scottish conditions is the so-called Lanercost chronicle, which in this period was probably still being composed chiefly in a northern English Franciscan context and had access to south-eastern Scottish information from the friary at Haddington in East Lothian.[17] For 1256, the chronicler reported that:

> In this year there was so great corruption of the air, and inundation of rain, throughout the whole of England and Scotland, that both crops and hay were nearly all lost. And some men's corn rotted in the fields from the day of harvest; some men's corn, shaken out by the wind, grew again under the straw; some men's harvest was so late that they did not reap it until about the festival of St Martin [11 Nov] or later [...][18]

15 'Annals of Innisfallen', *s.a.* 1253.3 [hereafter AI]. See also AU, ALC, *s.a.* 1253.

16 ALC, *s.a.* 1254.13.

17 Little, 'The Authorship of the Lanercost Chronicle'.

18 *Chronicon de Lanercost*, *s.a.* 1256. Independent corroboration of Lanercost's claims of widespread devastation is provided by Matthew Paris, who records extreme wind and rain storms beginning on 19 June and running on through the summer, noting especially catastrophic flooding of the Ouse around Bedford and extensive structural damage affecting mills, bridges and houses. In his summary for the year, Paris claimed that it rained incessantly from the Feast of the Assumption of the BVM (15 August) to the Feast of the Purification (2 February), rendering roads impassable and fields sterile, and at the end of autumn grain rotted in the ears: Paris, *Chronica Majora*, v, pp. 561, 600.

With a certain air of inevitability, it reported for the following year that:

> There was a great dearth of grain throughout this whole island, for there was scarcely enough flour to shape into the meanest bread, nor malt to make liquor suitable to drink. [Followed by description of peasants fighting over horse carcases, etc.][19]

Sadly, the non-survival of Scottish royal exchequer, manorial accounts or teind[20] records from this period renders it impossible to verify the scale and extent of either the weather-related harvest failure of 1256 or the consequent dearth of 1257. A similar lack of long sequences of such data from England from before 1275 leaves us to conjecture the degree of human misery caused by this episode.

The 1260s was a decade of wide year-to-year variations in general weather and saw a succession of poor harvests, food shortages and episodes of epidemic and epizootic disease.[21] In Scotland, 1260 brought dearth followed by another poor harvest before a descent into a stormy winter, with a famine recorded in 1261–1262 due to harvest failures.[22] AU and ALC record that Ireland experienced a summer of extreme heat in 1263, but Scottish sources for that year are preoccupied with the events of the Scoto-Norwegian war; the contemporary Melrose chronicle, however, did attribute the Norwegian defeat on 2 October to divinely-sent pestilence and storms.[23] A sharp deterioration from an already poor position began mid-decade beginning with AU's recording of plague and famine in 1265.[24] In Britain, 1267 began benignly but sources record first infestations of worms and caterpillars which destroyed the vegetable crops. Autumn, however, saw extreme weather events commencing in late October, with a

[19] *Chronicon de Lanercost*, *s.a.* 1257. Paris records poor weather in south-eastern England from February to May 1257, with rain, strong winds and coastal flooding, and that as a consequence many farmers were obliged to re-sow their fields: Paris, *Chronica Majora*, v, p. 630. Heavy rains continued to affect southern England into late summer, prompting the monks of St Albans to stage a religious procession to secure saintly intercession: Paris, *Chronica Majora*, v, pp. 644–45.

[20] Teind is the Scottish equivalent of ecclesiastical tithe in England.

[21] Matthew Paris did record prevailing conditions in England in 1258 in equally bleak terms to what had preceded: *Chronica Majora*, v, p. 728.

[22] Bower, *Scotichronicon*, v, pp, 325, 335.

[23] AU, *s.a.* 1263.7. AU's 1263.11 account of mortality through famine and pestilence belong to 1265 rather than 1263. ALC, *s.a.* 1263.15; *Chron Melrose*, *s.a.* 1263; Bower, *Scotichronicon*, v, p. 341.

[24] AU, *s.a.* 1263.11 (recte 1265, the dates of the annal entries down to the early fourteenth century are two years out of sequence).

major sea-flood caused by northerly winds, which created storm surges affecting the east coast of Scotland from the Tweed to the Tay, with the following winter described as 'very harsh with continual whirlwinds of hail, snows and rain'.[25] It may be coincidental that Lothian in 1268 was struck by a possibly pulmonary cattle epizootic, the so-called 'Lungessouth', but the spring of that year was recorded as particularly wet and windy, with the summer remaining wet, cold and windy.[26] The thirteenth-century source used by the fifteenth-century chronicler Walter Bower records high livestock mortality that year, especially of sheep, but with wild animals also badly affected. Ireland, in the meanwhile, suffered a severe famine.[27] The late autumn and winter of 1268–1269 brought no relief. Scottish sources record extreme conditions from late November to early February that prevented ploughing and preparation for the new season's crop, English accounts speak of a spoiled harvest followed by a wet, tempestuous and pestilential winter, whilst Irish sources speak of severe storms in January 1269.[28] The decade evidently ended as it had begun, with a catastrophically bad harvest in 1269, for Scottish and Irish sources identify 1270 as a year of 'dearth and much rain' and 'hunger and great destitution'.[29] This protracted period of unfavourable weather conditions in the north-east Atlantic region continued into the early 1270s, with winter 1270–1271 recorded as severe and cold for Scotland and in Ireland poor weather accompanied a further episode of famine which promised a repeat of that experience the following year.[30] A crescendo was reached in 1272, for which Bower's thirteenth-century source recorded:

> In the year 1272 there was a great lack of productivity on the land and unfruitfulness at sea, as well as turbulence of the air, as a result of which many people fell ill and many animals died ... a great famine hit France, England, Scotland and many areas, for the cattle mostly died, the crops failed, and the poor died of their poverty.[31]

[25] Bower, *Scotichronicon*, v, pp. 359, 369.

[26] *Chronicon de Lanercost*, *s.a.* 1268; Bower, *Scotichronicon*, v, p. 369.

[27] AU, *s.a.* 1268.3.

[28] Bower, *Scotichronicon*, v, p. 373; William of Newburgh, 'Continuation of the Chronicle', p. 555; AI, *s.a.* 1269.

[29] Bower, *Scotichronicon*, v, p. 381; AI, *s.a.* 1270; ALC, *s.a.* 1270.6.

[30] *Chron Melrose*, *s.a.* 1271; AI, *s.a.* 1272.2.

[31] Bower, *Scotichronicon*, v, p. 385.

Bearing in mind that most of the Scottish material was composed in monastic centres in east central and south-eastern Scotland, districts buffered from the most extreme Atlantic conditions by the mountainous regions to the north and west, we can assume that the Atlantic seaboard experienced severe weather events across this period.

After the nadir of 1272, there is a decade-long lull from reference to extreme weather events in the Irish and Scottish records until the beginning of the 1280s. Irish sources indicate that winter 1281/2 was extremely poor, AI labelling the weather that year as generally bad, beginning with heavy snow in February and early March which caused widespread cattle deaths and high mortality and general hardship for the human population.[32] AU corroborates this record but starts the episode of extreme weather earlier, from around Christmas 1281.[33] Winter 1282/83 was similarly poor, as reported in AC, AI and ALC, the former in particular noting extreme weather through December and January and observing that continuing poor conditions thereafter prevented 'any useful work being done'.[34] The middle part of 1283 appears to have been characterized by stronger winds than usual, which resulted in widespread damage to essential hay-ricks as well as to built infrastructure. Early 1284 again saw heavy snowfall with its associated consequences for animal and human populations, and the early part of 1285 was noted for extreme storminess.[35]

Winter 1285/86 passed without noteworthy weather events beyond the seasonal gales recorded in contemporary accounts of Alexander III's death on 19 March 1286.[36] Irish sources, however, recorded the first of a new series of blows to the economic regime upon which both Gaelic Irish and upland Scottish society was founded; the spring months of 1286 were described in AC and ALC as 'the spring of the cattle-plague' or 'cow-mortality'.[37] Over the next decade epizootic disease recurred regularly, affecting cattle and, in lowland Scotland, sheep also. From the late 1270s into 1330s, disease outbreaks seem to have occurred regularly in ovine populations in north-western Europe.[38] Cattle were the mainstay of the complex exchange mechanisms throughout the Scottish and

[32] AI, *s.a.* 1281.2, 1281.8.

[33] AU, *s.a.* 1279.7 [there is a three-year error in the dating of the annals at this point].

[34] AC, *s.a.* 1282.9; AI, *s.a.* 1282.2; ALC, *s.a.* 1282.8.

[35] AI, *s.a.* 1284.2, 1285.2.

[36] *Chronicon de Lanercost*, *s.a.* 1286.

[37] AC, *s.a.* 1286.3; ALC, *s.a.* 1286.2.

[38] Slavin, 'Mites and Merchants'.

Irish Gaedhealtachd upon which lordship structures were founded and, while the impact of these murrains is quantified in no surviving source, the fact that the outbreaks were so widely and prominently reported indicates a likely level of associated socio-economic stress. In Scotland, the early 1290s saw episodes of what was labelled colloquially as 'the pluk' and 'pilsoucht', which appear to have been some form of scab, amongst sheep flocks in the eastern Southern Uplands and North East. The severity of these outbreaks and recognition of the potential economic consequence of inaction is reflected in 1292/93 in the stringent measures taken by the administration of the new Scottish king, John Balliol. The king's clerks sent out formulaic precepts to royal officials around the kingdom ordering the culling of all infected flocks, a formula repeated under King Robert I in subsequent outbreaks in the 1320s.[39] Bower's late thirteenth-century source, perhaps reflecting a central eastern lowland provenance, makes no reference to these events, focusing instead on the poor arable yield in 1293 arising from a prolonged episode of snow and winds that affected mainland Britain.[40] This event is corroborated by Lanercost, which recorded that the weather brought 'such a lack of produce throughout the whole of England and Scotland that in some parts a quarter of grain sold for thirty shillings'.[41] Lanercost goes on to record late and heavy snowfall in East Lothian into mid-April 1294 and violent rainfall across Southern Upland Scotland around the start of harvest time in early August. The latter episode most directly affected the hilly interior of the region but brought a rapid and violent spate of the River Teviot which swept away mills and flooded ripening crops in the arable lowland east of Teviotdale, and caused the lower reaches of the River Tweed to flood between Roxburgh and Berwick, where the bridge was destroyed.[42] In Ireland, after an apparent lull in 1294 and 1295, the poor weather resumed in 1296 with AI's report of 'very stormy weather this year, with wind, snow, and lightning, and a great murrain of cattle and loss of life also'.[43] Similar poor conditions afflicted Scotland through 1296 and 1297, exacerbated by the outbreak of war with England; in autumn 1297 Bower's thirteenth-century source

[39] *The Records of the Parliaments of Scotland to 1707*, *s.a.* 1293/8/1. Duncan, ed., *Formulary E: Scottish Letters and Brieves, 1286–1424*, no. 16.

[40] Bower, *Scotichronicon*, *s.a.* 1293.

[41] *Chronicon de Lanercost*, *s.a.* 1293.

[42] *Chronicon de Lanercost*, *s.a.* 1294 (pp. 155, 157).

[43] AI, *s.a.* 1296.2, 1296.7.

recorded a threat of major dearth 'since there was shortage of grain resulting from inclement weather'.[44]

In summary, the surviving annal and chronicle records for the northern and western parts of the British Isles for the half-century after 1250 present a picture of an intermittent but generally downward trajectory in weather conditions. There were two sustained periods within which conditions appear to have been at least passably acceptable with no comment on either weather events or crop yields; 1273–1280 and summer 1286 to the winter of 1290/91. In total, twenty years across the half-century can be so categorized. In only two years, 1253 and 1254, however, is there consensus that harvests were particularly good and the weather especially favourable. There is, on the other hand, also consensus that there was a general harvest failure in only two years, 1256 and 1260, but with famine resulting in the years following. In ten out of the fifty years, famine or dearth was recorded, with seven of those years falling in the period 1261–1272. In eighteen out of the fifty years, extreme weather events are recorded, chiefly periods of severe cold, snowfall or rain, with ten of the worst years falling in the period 1260–1272 but with four out of the five years down to Alexander's death in 1286 witnessing a return to the negative pattern of the 1260s. On the basis of the written record alone, therefore, there seems hardly to be a compelling case for the second half of the thirteenth century and Alexander III's reign especially to be characterized as the era of 'wyne and wax, of gamyn and gle' of the Wyntoun tradition.

Climate Proxy Data

As has been discussed extensively elsewhere, proxy environmental measures are widely recognized as offering a long-term perspective on the changes in climate that runs in parallel to the weather-event data contained in the historical accounts discussed above. It is generally accepted that such measures can give context for these reported events. In recent decades, it has been research into current global climate change that has driven forward most discussion of the historic trends reflected in palaeoclimate records. Various models of increasing refinement have been advanced as a greater body of data has been gathered and climatologists and environmental geographers have developed their understanding of how the trends evident in their data sources would have been physically manifest and experienced as weather phenomena. A variety

[44] Bower, *Scotichronicon*, *s.a.* 1297.

of different proxies are employed, but for the North Atlantic these are principally ice-core, ocean sediment and tree-ring data.[45] It is accepted that these data are region-specific but it is recognized that synthesis of different sources permits modelling of geographically wider and temporally longer episodes of climate change and the human experience of those changes. Bearing always in mind that the picture produced from the data is at best a generalized, headline view that cannot be used to reconstruct fine-grained local experiences, for the thirteenth century there are climatic proxies which can deliver data with annual or seasonal resolution; in effect, they can permit us to see the episodes recorded historically as periods of intense cold or heat, above average precipitation or drought. Ural and Siberian tree-ring data[46] and Greenland ice core data[47] are the chief proxies employed as summer and winter temperature indicators. None of these proxies lies in geographical proximity to Scotland and therefore none can be used as absolute indicators of climatic variability affecting the country.[48] For the British Isles, therefore, the most common method is to contrast Northern Hemisphere/North Atlantic summer temperatures from dendrochronological analyses with stable isotope records that provide indices of relative winter 'severity' from ice core data, which together can be used to produce an annualized long-multi-proxy mean.[49] The method and the interpretation of its results have been much-discussed and that will not be rehearsed here. A generalized summary of the climate and weather trends recoverable from such annualized multi-proxy data supports the view of the MCA as a long period of higher annual mean temperatures that commenced before 1000 and continued into the later thirteenth century. In agreement with the historical

[45] For example, Baillie and Brown, 'Dendrochronology and the Reconstruction of Fine Resolution Environmental Change'; Briffa, 'Annual Climate Variability'; Briffa and others, 'Influences of Volcanic Eruptions'; Briffa and others, 'European Tree Rings'; Dawson and others, 'Ice Core and Historical Indicators of Complex North Atlantic Climate Changes'; Oppenheimer, 'Ice Core and Palaeoclimatic Evidence'; Strothers, 'Climatic and Demographic Consequences'.

[46] Briffa and others, 'Low-Frequency Temperature Variations'.

[47] For example, Adderley and Simpson, 'Soils and Palaeo-Climate Based Evidence'; Dawson and others, 'Ice Core and Historical Indicators'; Vinther and others, 'NAO Signal', p. 1387.

[48] Only recently has a speleothem-based record for precipitation appeared for northwestern Scotland: Baker and others, 'A Composite Annual-Resolution Stalagmite Record', p. 10,307.

[49] Crowley and Lowery, 'How Warm Was the Medieval Warm Period?'. For its application in a specific case-study: Oram and Adderley, 'Lordship and Environmental Change in Central Highland Scotland'.

accounts, however, they also indicate that this period was not an unbroken time of stable and, from a human perspective, benign conditions; instead it was punctuated regularly by short, cold episodes and also by extreme seasonal weather events.

For the purposes of this present study, synthesis of the data for the thirteenth century indicates that generally milder conditions experienced across the eleventh and twelfth centuries continued down to the early 1250s before yielding to several decades of much greater variability. Across the thirteenth century, the trends in the proxy data can be characterized as follows. First, although generally more stable, the period down to *c.* 1250 was not without significant temperature dips and was also punctuated with episodes of extreme weather that diverged substantially from the mean across this era. Indeed, two of the coldest decades of the thirteenth century, spanning the 1220s and 1230s, seem to have fallen into this period.[50] It is in this context that we can place saga accounts of extensive ice floes in the Denmark Straight between Iceland and Greenland in the second quarter of the thirteenth century, which in turn contextualize episodes of low winter temperatures and increased storminess across the North Atlantic region that affected the British Isles in the 1230s and 1240s. Then the 1250s brought recurring drought conditions alternating with saturation for much of the British Isles. A trend towards more unstable conditions was evidently already established but profound and lasting change occurred in the mid-1250s. High concentration of volcanic sulphates found in ice cores from Greenland and Antarctica indicate that a major volcanic eruption, now identified as a 1257 eruption of the Samalas vent of the Gunung Rinjani volcanic complex on the island of Lombok in Indonesia, expelled vast amounts of material into the atmosphere.[51] This event was at first seized upon by archaeologists and environmental historians as being sufficient to trigger the marked episode of climatic cooling which extended through the late 1250s and 1260s,[52] but the Scottish and Irish chronicle accounts, corroborated by southern British and mainland European texts, indicate that the downturn had already commenced before 1257; the transitional year appears to have been 1255 with 1256 seeing exceptionally poor weather and a consequent harvest failure. Rinjani, rather

[50] Rydval and others, 'Reconstructing 800 Years of Summer Temperatures', p. 2960.

[51] Lavigne, 'Source of the Great A.D. 1257 Mystery Eruption Unveiled'.

[52] Briffa and others, 'Influences of Volcanic Eruptions'; Dawson and others, 'Ice Core and Historical Indicators'; Oppenheimer, 'Ice Core and Palaeoclimate Evidence'; Strothers, 'Climatic and Demographic Consequences'.

than triggering the shift, may have served more to accelerate and prolong an episode of cooling that was already underway. With the cooling came a rapid southward extension of sea ice off Greenland, replicating the impacts of the 1230s to again introduce poor summers and severe winters to the British Isles in the 1260s, 1270s and early 1280s.[53] The spike of poor conditions in the early 1290s apart, the last decade-and-a-half of the century appeared to see a return to the more stable conditions that had prevailed before 1250 but the proxy data indicate that there was a continuing deterioration which was manifested violently and disastrously in the well-known events of 1315–1322, the Great European Famine.[54]

To understand the nature of the direct and indirect impacts of these changes in the climate on the human population of the northern and western British Isles, and the potential for mortality crises as a consequence, some awareness of the dynamics within the climate processes and their physical manifestation is important. A combination of atmospheric and oceanic circulation systems drive climatic shifts generally and a major systemic transition produced the conditions experienced in Scotland after *c.* 1250.[55] It is a well-known fact that changes in ocean surface temperatures affect the pattern of east-flowing Atlantic weather systems reaching the British Isles. The main driver in this process is the interplay of what is known as the Atlantic Meridional Overturning Circulation [AMOC] of oceanic water, which transports warmer water north and east across the Atlantic, with the North Atlantic Oscillation [NAO] in the atmospheric pressure systems (between the permanent Iceland low pressure and Azores high pressure cells). The interplay of AMOC and NAO was a key factor in the establishment of the MCA and the LIA that followed, through their influence on westerly wind conditions and eastwards movement of atmospheric moisture. A 'positive' NAO through the tenth to mid-thirteenth centuries brought the generally more benign weather regime to the northern and western British Isles that is reflected in both the climate proxy data and the documentary record.[56] From the 1250s, however, climatological modelling sug-

[53] Rydval and others, 'Reconstructing 800 Years of Summer Temperatures', pp. 2959–60, where it is noted that 1282, 1284 and 1285 may have seen the hottest summers of the century. Hot, of course, does not necessarily mean productive.

[54] Kershaw, 'The Great Famine'; Jordan, *The Great Famine*; Campbell, 'Nature as Historical Protagonist'; Slavin, *Experiencing Famine*.

[55] See, e.g., Mayewski and others, 'Changes in Atmospheric Circulation'; Dawson and others, 'Icecore and Historical Indicators'.

[56] The recent NAO-sensitive speleothem record from northwestern Scotland reveals the

gests that the NAO weakened with a consequent shift to greater variation, but principally to cooler and wetter summers and intensely cold and stormy winters. This is the switch that is reflected in Irish annal and northern British chronicle accounts, which record a fairly abrupt transition in the 1250s from a period of stability into one of wide-ranging and primarily negative variation. The effects of wider northern hemisphere climatic cooling mid-century were reflected in the experience of the northern and western British Isles, which saw pulses of extreme weather in the 1260s/70s and again in the 1290s. These pulses were the most highly visible manifestations of that long-term trend towards progressively cooler and wetter conditions which extended into the mid-fourteenth century, possibly associated with the Wolf solar minimum, referred to above.

Ocean-surface warming increased atmospheric moisture and triggered increases in storminess with some extreme summer wind and rain events, and periods of extreme heat or coldness often seen in records of hot, wet summers and prolonged snow and ice in late winter and early spring.[57] Additionally, what happened in Greenland affected Scotland indirectly, and a descent into a period of intense winter coldness in the north-west Atlantic islands brought an extension of the southward range of sea ice which delivered a consequential increase in the incidence of winter storms as the cold surface temperatures led to the high surface air pressures associated with anticyclonic circulation. Such an interaction sees cold air drawn further south to collide with warmer moist air moving east across the Atlantic. Depending on the track of the high pressure cells, which is affected by the AMOC/NAO interaction, the result can be stormy winters and cold, late springs, adverse summer growing conditions, and wet and delayed harvests. That trend was already evident across Alexander III's so-called 'Golden Age', especially during the 1260s and 1270s, but from around 1300 the shift towards colder annual temperatures in the northern hemisphere became entrenched, signifying the final 'end' of the MCA.

Expressed in terms of generalized climate processes, the trends reflected in the climate proxy data and climate models are largely detached from human experience. The dehumanized scientific discourse can be related only partly to the human experience of the events which formed the physical manifestation of the climate processes at work through the surviving historical record.

decline of a positive MCA-NAO beginning in the fourth quarter of the thirteenth century. A more dramatic decline occurred in the early 1300s and the shift to a negative NAO took place a century later: Baker and others, 'Composite Annual-Resolution Stalagmite Record', Figure 4.

[57] Dawson and others, 'Icecore and Historical Indicators', p. 431.

Those records, however, provide data that is itself extremely generalized, impersonal and rarely providing any level of detail for how the population in general were affected by or responded to events other than as passive recipients of the hand being dealt to them by Nature. Reference to 'famine', 'dearth' and 'mortality' in historical accounts from across this period do not provide evidence for the scale, extent, duration or indeed nature of such phenomena and, with the almost non-existence of record evidence to provide quantitative data, we again have to turn to proxies. A total of eleven wheat, four barley, one oats (grain) and nine oatmeal prices survive from Scotland from the whole of the thirteenth century;[58] this is hardly a firm statistical basis upon which to model trends in arable yields or the impact of weather events on the late thirteenth-century agrarian regime of Lowland Scotland. For upland or largely pastoral zones we are equally poorly served, with one mart (slaughtered beef carcase) price, one ox price, eleven cow prices and seven sheep prices surviving across the same period, and for the products of a livestock-based rural regime two hide prices, no wool prices, three cheese prices and no butter prices. Other resources that could be used effectively as proxy measures of climate impacts on the human population, especially prices for migratory fish species, are similarly served; there are no surviving salmon prices and only three herring prices for the whole century. Finally, two commodities whose production in Scotland was highly weather-dependent, salt (produced by boiling brine over mainly peat-fuelled fires at this period)[59] and beeswax, have only four and two prices respectively surviving.

Palaeoenvironmental Data: Towards a Synthesis

This lack of quantitative record evidence can be compensated for in part by a growing body of palaeoenvironmental evidence. Within that generic label a number of diverse primary indicators are covered, from long-established subjects of specialist study like pollen to water-deposited sediments and the environmentally-sensitive biota which they contain, to secondary, more qualitative and less absolute measures such as changing volumes of animal, bird or fish bone by age, species, sex and size recovered from archaeological contexts, the interpretation of which cannot but be subjectively coloured by the circumstances of

[58] See Gemmill and Mayhew, *Changing Values in Medieval Scotland*.

[59] For salt manufacture in medieval Scotland, see Oram, 'The Salt Industry in Medieval Scotland'.

their recovery, on-site sampling and post-excavation processing. Interpretation of this data has been further influenced by the growing awareness of what the potential impacts of the weather trends indicated by the climate proxy data might have been on the species and materials that are the focus of palaeoenvironmental analysis. Human agency, principally social and economic dislocation arising from the disturbance of established political and commercial structures, has also been advanced as a causal factor in the visible changes in the palaeoenvironmental record. For example, excavations in the Western Isles have produced evidence for the apparently commercialized exploitation of marine fish resources from the later tenth to thirteenth centuries, then a collapse in that activity at the end of the thirteenth century, seen in the dramatic decline in the volume of bone recovered at excavated sites.[60] The collapse, if it is reflective of a profound change in resource availability that in turn led to a cultural shift in food consumption patterns rather than being an artificial construct of poorer survival and recovery of bones from later medieval excavation contexts, has been linked in past discussion to political upheaval in the region in the third quarter of the thirteenth century after nearly three centuries of relative stability under Norse rule. It is now argued that such a change was more likely consequent on shifts in ocean temperatures and changes in bottom-water temperatures in inshore waters affecting the migratory behaviour of pelagic species such as herring and breeding and feeding behaviour of both inshore species and the organisms on which they were themselves sustained.[61] Resource availability was also potentially affected by an increase in storminess which heightened the risks attached to deep-water fishing for gadid species and, hence, driving changed approaches to the fishing techniques and wider cultural practices of the human population.

Individual excavation reports produce evidence of highly localized trends in resource availability and exploitation, and human social and cultural responses to them, but analysis of the data from multiple sites across a widely dispersed geographical area is permitting more general models for such transitions to be advanced. Published results from the Outer Hebrides, e.g. the sites at Dun Vulan, Cille Pheadair and Bornais in South Uist, and the Udal in North Uist,[62] all located on the storm-exposed Atlantic west coast of those islands

[60] Serjeantson, *Farming and Fishing*, pp. 73–74, 77–79; Sharples and others, *A Norse Farmstead*, pp. 169, 192–94, 195–96.

[61] Cage and Austin, 'Marine Climate Variability'.

[62] Parker Pearson and others, *Between Land and Sea*; Parker Pearson and others, eds, *Cille*

and heavily dependent on the ocean for both subsistence and commercialized fish exploitation, can be set alongside data from more sheltered Inner Hebrides sites, e.g. Iona and Finlaggan in Islay,[63] and the more intensely-studied late Norse cod fishery in the Northern Isles and east-coast Caithness.[64] All show the same broad trends of development and intensification through the eleventh and twelfth centuries then decline across all exploited species from the mid-thirteenth century onwards. What that decline meant in human terms is unknown. We can suggest greater losses of gear, boats and crews in fishing communities as sea-conditions deteriorated after *c.* 1250, coupled with loss of livelihoods as the diminishing returns on the labour and resource investment drove shifts into alternative activities, but with alternative exploitation niches available it is unlikely that the decline in fish-catches produced anything more than a short-lived economic dislocation; subsistence crises and attendant mortality crises required similar shocks across a broader basket of resources.

Increased storminess affected terrestrial as well as oceanic conditions. Modelling of storminess in the North Atlantic region across the last 12,000 years has suggested that the steady decline in storm frequency that occurred through the first millennium CE began to be reversed after *c.* 1250 but reached unprecedented levels of intensity only in the early fifteenth century.[65] A shift towards a higher frequency of storm episodes was not a smoothly rising curve but one disturbed by phases of greater or lesser intensity and, when related to the evidence from the historical accounts and modern observations, can be seen to have had significantly different impacts depending on the season of their occurrence; Bower's record of gales in October 1267 mentions a greater impact on woodland, which would still have been in leaf, than his January 1272 windstorms, which he records as chiefly damaging to built structures. Such headline-event wind-impacts, however, were less damaging in the long term than

Pheadair; Parker Pearson and others, 'Cille Pheadair'; Sharples and others, *A Norse Farmstead*; Serjeantson, *Farming and Fishing*. For a detailed overview, see Cerron-Carrasco, *'Of Fish and Men'*.

63 Coy and Hamilton-Dyer, 'The Bird and Fish Bone'. For Finlaggan, D. Caldwell (pers. comm.).

64 For example, Morris and others, *Freswick Links*; Barrett, 'Fish Trade'; Barrett, Nicholson, and Cerron-Carrasco, 'Archaeo-Ichthyological Evidence'. The Scottish east coast fisheries are poorly understood in the pre-modern period and the only large-scale analysis of fish-bone evidence is from the urban context of sites like Perth: Jones, 'The Fish Bone'.

65 Mayewski and others, 'Changes in Atmospheric Circulation'; Mayewski and others, 'Holocene Climate Variability'.

localized impacts which affected agricultural regimes and especially where the interplay of storm events and particular agricultural practices resulted in dramatic losses of crops and, more drastically, of cultivated ground. In the Uists, arable cultivation of the calcareous shell-sands of the so-called *machair* that runs down the Atlantic sea-board of the islands had intensified through the MCA, moving off the stabler 'blackland' areas where most cultivation had been located in earlier centuries. This shift can be seen as largely consequent on the Scandinavian settlement of the region from the later ninth century and the culturally influenced high demand for barley and other crops which they brought with them, which the blackland zone could not meet. The intensive cropping of the machair and the resultant concentration of settlement down that highly fertile strip declined dramatically in the later medieval period almost certainly as a consequence of their exposure to wind-erosion in winter and early spring before the crops began to grow and also to coastal storm-surge.[66] Increasing storminess with attendant episodes of dune deflation and sand-blow rendered that regime progressively more unsustainable through the fourteenth and fifteenth centuries but OSL dating at Baleshare in North Uist suggests that major incidents of this kind had been occurring with rising frequency from the mid-1200s. At the Udal in North Uist a major episode of sand-blow buried this high status site in the early fifteenth century and a further episode in the seventeenth century led to its final abandonment, but excavation results indicate that it was already enduring a period of protracted decline marked by episodes of sand drift from the late 1200s.[67]

Dune deflation and sand-drift overwhelming agricultural land was not restricted to the Outer Hebrides and, while again the main episodes occurred from the fifteenth century onwards, events are recorded from the later thirteenth century at sites from Orkney and down the eastern sea-board of mainland Scotland.[68] At Sands of Forvie on Aberdeenshire's Buchan coast and at Eldbotle on the East Lothian coast, in two regions where a cereal-based rural economy was long-established, two important and ancient settlements were abandoned progressively from the later fourteenth century as sand encroached on and then overwhelmed both agricultural land into which centuries of soil-

[66] Dawson, Dawson, and Jordan, 'North Atlantic Climate Change'; Serjeantson, *Farming and Fishing*, p. 31; Sharples and others, *A Norse Farmstead*, pp. 195–96; Gilbertson and others, 'Sand-Drift and Soil Formation'.

[67] Crawford, 'Structural Discontinuity', pp. 23–24.

[68] Oram, 'From "Golden Age" to Depression'.

enrichment had been invested and the settlements themselves.[69] Evidence from the 1230s for the exploitation of the vegetation cover that stabilized the coastal dune systems at Eldbotle highlights how peasant land-management practices perhaps triggered and then accelerated erosion in a disastrous interplay of anthropogenic and climate-driven factors.[70] These sites serve as possible exemplars for the experience from the later thirteenth century onwards of agricultural settlements on coastal links down the east coast from Caithness, through Moray and Aberdeenshire, Angus and north-eastern Fife, to East Lothian, the regions of most intensive arable exploitation in pre-Wars of Independence Scotland.

As Bower's record of inundation in 1267 indicates, storm events brought other threats to low-lying coastal districts. His particular emphasis on the impact on the district between the Tweed and Tay estuaries suggests that easterly or north-easterly storms were at work, causing a storm surge up the narrowing funnels of the main east-coast river estuaries which overwhelmed agricultural land that had been reclaimed from estuarine carseland in the twelfth and earlier thirteenth centuries. In parts of the Carse of Stirling, especially around the in-flow of the River Carron into the Firth of Forth, this episode appears to have marked a limit in medieval reclamation efforts and the beginning of a gradual abandonment of the hard-won cultivated land along the south side of the firth.[71] On the opposite side of the country in the Solway Firth, south-westerlies delivered storm surges up that great marine inlet which overwhelmed the similarly reclaimed districts on the Scottish and English sides of the firth.[72] One such event is illustrated by the Lanercost chronicler's description from 1292 of how

> such a furious wind arose as destroyed all vegetation, and either overthrew travellers afoot or on horseback or drove them easily out of their right course. There was also such a tremendous inroad of an unusually high tide as to overflow the ancient landmarks of the country in a degree beyond all memory of old people, overwhelming beasts pasturing along its shores and destroying sown crops.[73]

[69] Kirk, 'Prehistoric Sites'; Hindmarch and Oram, 'Eldbotle'.

[70] Hindmarch and Oram, 'Eldbotle', pp. 278–80, 281–82.

[71] Oram, 'Estuarine Environments and Resource Exploitation', pp. 371–72.

[72] Tipping and Adams, 'Structure, Composition and Significance'.

[73] *Chronicon de Lanercost*, *s.a.* 1292.

Single episodes, especially ones such as this which were unprecedented within living memory, may not have triggered precipitate action on the part of the communities affected; it would have taken a switch to regular similar events (such as happened in the later fourteenth and fifteenth centuries), or a gradually awakened awareness of change having occurred almost imperceptibly over an extended period, to force radical rethinking of their exploitation strategies. Under the repeated impact of such episodes, peasant cultivators eventually withdrew from the exposed coastal margins, only to return when modern agricultural technologies coupled with robust coastal engineering permitted after *c.* 1800.

Another dimension of the increasing severity of the weather in the second half of the thirteenth century was its impact on livestock. Such impact needs to be thought of in indirect as well as direct terms; actual mortalities caused by extreme cold, suffocation in deep snow or drowning, while immediately visible, were only the proverbial tip of the iceberg. As damaging to the pastoral regimes that prevailed through most of upland Scotland and in the Hebrides as to the predominantly arable regimes of more lowland districts were extended periods of low day-time temperatures in the late autumn or spring. Grass growth requires average day-time temperatures of 6°C and above. Low-lying and especially coastal grazing was probably little-affected even during the extremely cold periods of the 1260s and 1270s as sea-temperatures kept the day-time temperatures in adjacent maritime zones above that critical threshold, but in parts of the country where transhumance was practiced across a wide altitudinal range (especially in the east, central and north Highlands and the Southern Uplands), upland pastures were incapable of sustaining the same numbers of livestock across the same seasonal range as had previously been the case.[74] In the later thirteenth century the recorded episodes of prolonged and unseasonably cold weather, as with the storm impacts on the coastal districts, probably did not stimulate any immediate or systemic change in animal husbandry practices as it was only from the mid-fourteenth century onwards that the downward trend in annual temperatures appears to have had a long-term suppressant effect on pastures. Nevertheless, the year-on-year consequences during the worst period of the late thirteenth century for a peasant society where dairying was the basis upon which the entire rural regime — and the subsistence of the human population within it — was sustained were potentially catastrophic.

[74] Serjeantson, *Farming and Fishing*, p. 30; Oram, '"The Worst Disaster Suffered by the People of Scotland in Recorded History"'.

There has been extensive discussion of the dairying economy of the Western Isles and parts of Highland Scotland in the medieval period and later, and the role that both beef-production and the circulation of dairy products had in the maintenance of lordship structures throughout Scotland's Gaedhealtachd from the twelfth to seventeenth centuries is well known.[75] Livestock, too, played a significant part in the rural economy of most of sub-Highland Scotland north of the Forth and throughout the Southern Uplands, where cattle and sheep for meat, hides and wool and woolfells were as important in the more complex market economies of these regions as they were for milk and milk-products elsewhere.[76] The cold and wet of the later thirteenth century affected both sectors, for not only was the cold a potential inhibitor of grass-growth but cold and wet conditions reduced cattle and sheep's weight-gain or milk-production (as calories were expended on simply maintaining body-heat) and increased the chances of infertility or foetus miscarriage. Once again, the threat of these outcomes in the stuttering conclusion to the MCA should not be overstated but the extended episodes of extreme weather, especially towards the end of the third quarter of the century, probably provided a sharp corrective to assumptions about the carrying capacity of pastures and the stocking-levels or *soums* associated with them, and about flock and herd replenishment through breeding calculations based on the more benign pre-1250 conditions.

When the unquantified mortalities from the cattle and sheep epizootics of the last quarter of the century and the decline in wool-based income that the 'pluk' outbreaks threatened are brought into the equation, peasant agriculturalists, lay and ecclesiastical estate-managers, elite households and merchants were confronted with a 'perfect storm' of environmental impacts against multiple facets of the economic regime that had developed across the MCA. It was a foretaste of what the fourteenth century would deliver but, in its intensity and longevity, it was a short series of occasionally traumatic dips amidst an otherwise continuing period of moderate economic and environmental conditions. No matter how positive a gloss is placed on this period, however, it was no 'Golden Age' and, though the cultivators, pastoralists and fishermen of late thirteenth-century Scotland may have sought to continue as they had done through the twelfth and earlier thirteenth centuries, the writing was very much on the wall.

[75] Serjeantson, *Farming and Fishing*, pp. 13, 53–57, 99; Oram, 'Innse Gall'; Oram and Adderley, 'Lordship, Land and Environmental Change', pp. 257–68; Oram and Adderley, 'Lordship and Environmental Change'.

[76] For general discussion of the livestock regimes in southern and eastern Scotland, see Oram, *Domination and Lordship*, pp. 242–46, 257–63.

Conclusion

But was this traditionally 'Golden Age' an era rather of mortality crises? Across the five decades examined above, both the historical records and climate proxy data, supported by palaeoenvironmental data, point to a deteriorating position in which conditions deemed adverse for the human populations of Scotland, northern England and western Ireland occurred in the majority of years. Historical records from across these geographical regions make increasing levels of comment on weather conditions (positive and negative) and the consequences of them in terms of agricultural productivity, human and livestock mortality, and material damage. Observations, too, are made on the impact of cold winter weather in respect of delayed spring ploughing and late sowing of grain-crops, but with the effects of that delayed start to the agricultural year come harvest time usually left open to implication rather than explicit comment. Unequivocal statements concerning crop-failure and resultant dearth commence in 1257 in Scotland and continue through to 1260, but from 1261 the accounts speak explicitly of famine and attendant human mortality. Famines are also reported for north-eastern Ireland in 1265 and 1268, for both Scotland and Ireland in 1270 and 1272, Ireland in 1281 and dearth in Scotland and northern England in 1293. All of these events occurred in the context of reported failures of the cereal harvest. In the absence of a robust set of yield figures or prices for this period and the lack of any quantitative data for mortality figures, all that can be said with confidence is that the period from the late 1250s to the early 1270s subjected the human populations of these regions to sustained food-supply related stress, with prices of grain, malt, and cereal-derived commodities likely subject to price-inflation. That position stood in sharp contrast to the experience of the 1240s and earlier 1250s.

All datasets also provide evidence, either explicitly stated or implicit from the scientific measures, for extended periods of pressure on other facets of the rural economy. Increased storminess and changes in oceanic surface- and bottom-water temperatures affected fishing opportunities as well as threatening coastal agriculture through sand-drift or marine inundation. The delayed springs that affected ploughing also affected pasture as the cold retarded grass growth. This retardation reduced the physical extent and altitudinal range of pastures available at critical times of the year for the maintenance of beef and dairy herds and wool-producing flocks; carrying capacities were slowly declining. These factors affected the physical condition of the livestock and may in part have been responsible for the spread of epizootic diseases through them that are also recorded often from the 1260s onwards. For the livestock- and fishing-dependent communities of the Atlantic west of Scotland, these impacts were probably as serious as the arable failures were for populations in the south and east of the country.

Our historical sources claim that there were at least six full-blown famine episodes in the later thirteenth century, five of which fell within the single decade 1261/62 to 1271/72. Crop yield data from England for the fourteenth century indicate that failures on this frequency would have hit hard on a rural regime that lacked the resilience to absorb recurrent shocks of this kind. Put bluntly but rather deterministically, the weather deteriorated, harvests failed, and people starved. Recognising the possibility of too deterministic a reading of peasant populations' responses to events and irrespective of our understanding of the social, political, and economic factors that contributed to the interplay of human and environmental agency that stimulated such responses, these headline episodes of crisis must also be set against a wider context of climate-driven environmental change that represented a downturn from the optimal conditions of the MCA. The 1260s apart, Scotland's rural populations would not have recognized immediately or felt the decisive shift in conditions affecting their agricultural or fishing practices which the year-on-year decline from the twelfth- and early thirteenth-century maximum entailed; that made it itself manifest in the fourteenth century and had its greatest impact in the fifteenth. All of the data currently available, however, points to systems under extreme pressure and populations experiencing distress, even if only intermittently. From the political perspective of post-Wars of Independence chroniclers, Alexander III's reign may well have seemed a Golden Age of peace, stability and prosperity when set against the violence and social dislocation, epidemic and epizootic disease, and manifestly poorer agricultural conditions of the centuries that followed. For the farmers and fishers of Alexander III's Scotland, however, any sense of living through a 'Golden Age' had probably been dispelled swiftly as they struggled to subsist in the grim years after 1256.

A Post-Plague Golden Age? Scotland, *Antiqua Taxatio*, and Thirteenth-Century Climate Change

Alasdair Ross*
Formerly of History, Heritage and Politics, University of Stirling

Introduction

For one small country sitting on the northern fringes of Western Christendom it is near impossible to discover the various impacts of climatic variation, war, and disease between the two plagues from written sources alone. In Scotland, documentary material really only comes to the fore in the twelfth century, and even then only as partial survivals with limited geographic coverage across the regnum. In contrast to England for the thirteenth century, for example, Scotland has no manorial accounts. Yet, this lack of written evidence has not hitherto stopped some historians from building a seemingly robust economic picture of medieval Scotland for the latter half of the thirteenth century and into the fourteenth after the second great plague. In these scenarios, volcanic eruptions, disease, and climatic downturn only feature as minor distractions, if at all. This offering will re-examine the data and present a different perspective on the thirteenth-century Scottish climate and economy post-plague.

* This paper was presented by Alasdair at the 'Mortality Crises before the Black Death' seminar at the University of Stirling in 2014 and originally intended for the volume of essays arising from that gathering. He finished this present version in 2016 and it represents his last completed work. Other than formatting changes to align it with the style of the present collection, it is as he submitted it for the 'Mortality Crises' volume.

'With Our Backs to the Ocean': Land, Lordship, Climate Change, and Environment in the North-West European Past. Essays in Memory of Alasdair Ross, ed. by Richard Oram (Turnhout: Brepols, 2023)
EHNAW 5, pp. 109–134 BREPOLS PUBLISHERS 10.1484/M.EHNAW-EB.5.131272

Between Old and New Extents

To date, the historiography of thirteenth-century Scotland has been almost wholly focused on fleshing out an idea given form in the fifteenth-century prose of Andrew de Wyntoun:

> Quhen Alexander our kynge was dede,
> Þat Scotland lede in lauche and le,
> Away was sons of alle and brede,
> Off wyne and wax, of gamyn and gle,
> Our golde was changit into lede.[1]

This is that the reign of King Alexander III (1249–1286) had been a golden or a silver age when the Scottish economy hugely expanded, the tax base increased, and the climate was benign,[2] despite the almost complete lack of quantitative data for population trends, crop yields, and commodity prices. Instead, this historical construct of a booming late-thirteenth-century Scottish economy is, apart from Wyntoun's rhyming, wholly dependent upon figures collected before 20 July 1366 when a new valuation, the *Verus Valor* ('true value/extent'), of Scotland's sheriffdom and ecclesiastic tax bases was made and entered into the parliamentary record (see Appendix 1). The calculations were made according to these methods:

> Item quod certi inquisitores nominatim ibidem ordinati per consilium inquirant diligenter et fideliter magno intervenient' sacramento per universas partes totius regni super omnibus et singulis redditibus tam ecclesiarum quam terrarum firmarum, bonis, bladis, catallis preter blada decima, et aliis possessionibus quibuscunque, exceptis albis ovibus, equis domitis et bobus, et domorum utensilibus; ita tamen quod qualibet ovis nigra et lachtana solvat ad contributionem; et quod nulla persona cuiuscunque status vel conditionis extiterit ab presenti contributione sit exempta. Et quod ipsi inquisitores presentent fideliter ad diem et locum iam per consilium ordinatur numerum omnium ovium cuiuscunque coloris et nomina ipsas possidentium una cum nominibus artificum et cuiusmodi artis fuerit et nominibus omnium aliorum queccunque bona possidentium, et quod inquirant diligenter de bobus assedatis ad firmam et quam summa et ipsam presentent ut supra.

> (Item, that certain inquisitors, ordained there by name by the council, shall take a great oath and diligently and faithfully enquire through all parts of the kingdom

[1] Andreew of Wyntoun, *The Original Chronice*, v, p. 145.

[2] Mayhew, 'Alexander III — A Silver Age?', p. 58 n. 38.

> upon rents, whether of churches or of fermes of land, goods, corn, chattels, excepting tiend corn, and other possessions whatsoever, excepting white sheep, broken-in horses and oxen, and household utensils; with the proviso that each black and dun-coloured sheep pay towards the taxation; and that nobody of whatever estate or condition he may be shall be exempt from the present taxation. And that the said inquisitors shall present faithfully at the day and place ordained by the council the numbers of all sheep, whatever their colour, and the names of those that possess them, along with the names of craftsmen of whatever craft they be, and the names of all others possessing whatsoever goods; and that they shall enquire diligently of oxen set to ferme, and for what sum; and they should present the same as above.)[3]

The calculations provide an invaluable snapshot of part of Scotland's taxation base across a *c.* hundred-year period because that parliamentary record of 1366 also compared *Verus Valor* to an earlier assessment of the Scottish economy known as *Antiqua Taxatio* ('old taxation/extent'). The latter is certainly thirteenth-century in date because a majority of early fourteenth-century references hark back to the Old Extent of the time of King Alexander [III] 'who died most recently'.[4] We should also take in good faith that since fourteenth-century Scots themselves contextualized *Verus Valor* with *Antiqua Taxatio*, the tax bases of both valuations must have been virtually identical. Otherwise, their 1366 comparison between the two sets of figures would have been utterly meaningless. This despite the reality that while the new extent was intended to help pay King David II's ransom (following his capture at the Battle of Neville's Cross), the old extent was in origin an ecclesiastic tax. It is glaringly obvious from these two sets of figures that there had been a massive decline in the value of both the Scottish ecclesiastic tax base and the tax base of sheriffdoms during the period between *Antiqua Taxatio* and *Verus Valor*. Deciding precisely when and why this decline began is another matter altogether.

The origins of *Antiqua Taxatio* have for long been a subject of interest for both historians and lawyers, partly because of its (relative) antiquity, and partly

[3] These conditions were set on 6 November 1358 but it was not until 20 July 1366 that a set of calculations first appeared in the parliamentary record: *The Records of the Parliaments of Scotland to 1707* (hereafter *RPS*), 1357/11/2.

[4] *RPS*, 1328/1. This distinction is key because the first *Antiqua Taxatio* may date to the latter part of the reign of King William I (1165–1214) and it was also used during the reign of his son, King Alexander II (1214–1249) — see Appendix 2. Accordingly, specific reference to the *Antiqua Taxatio* of Alexander III implies that it was different in value to the Old Extents in use during the reigns of his predecessors.

because it formed the long-term basis of land valuation in Scotland.[5] By the end of the twelfth century the Holy See found that its revenues were inadequate to meet the cost of its rapidly expanding activities, including Crusading in the Outremer. Imposed for the first time in 1199 by Pope Innocent III (1198–1216), these extents were a tax (usually a tenth of both temporal and spiritual sources) upon the income of the clergy and they were increasingly utilized during the thirteenth century. Virtually every thirteenth-century pope either demanded or was the recipient of money collected under such auspices in Scotland, even if the Scots were not always willing to pay (see Appendix 2).[6]

During the last two decades the topic of extents has been re-visited by a number of scholars, focusing almost exclusively on the figures for the ecclesiastic tax base. In 1996 Stevenson wrote extensively on this topic for the new edition of the *Atlas of Scottish History*. Here, he compared the values of the two 1366 extents and added further sets of figures to his analyses. Though these figures came from separate ecclesiastic registers, according to Stevenson they were copies of lost originals drawn up for a tax of 4d in the merk (a fortieth) demanded by Pope Innocent III in 1199 and levied in Britain in 1201, and the same extents were later used during the thirteenth century to collect subsequent ecclesiastic taxes of different values (see Appendix 3). All of which indicated to Stevenson that *Antiqua Taxatio* as recorded in these sources could be as old as 1201.[7] Tout too discussed these valuations and concluded that any minor discrepancies could easily be explained away by scribal error and they all derived from a common original.[8] There is no good reason to dispute this latter assertion but to these three sets of figures a fourth can be added from the records of Holm Cultram (see Appendix 4). The figures provided by Holm Cultram are also broadly similar with one exception, the total provided for the diocese of Argyll, which is perhaps harder to explain away as scribal error. One further set of undated figures exists for the Old Extent in the diocese of Moray, this time totalling £920,[9] which lies in between the two totals listed for the old and new extents of Moray in 1366.

[5] For example: Innes, *Scottish Legal Antiquities*; Mackie, ed., *Thompson's Memorial on the Old Extent*; McKechnie, 'Early Land Valuations'.

[6] Lunt, 'The Financial System of the Medieval Papacy', pp. 261, 280–82.

[7] Stevenson, 'Taxation in Medieval Scotland'.

[8] *The Register of John De Halton*, I, pp. vii–xxv.

[9] *Registrum Episcopatus Moraviensis*, pp. 361–66. The editor of this work thought these figures dated to the mid-fourteenth century but this is conjectural.

The second set of figures Stevenson utilized is dated to 1274 when Master Boemundo di Vezza from northern Italy (remembered in Scottish sources as Bagimond) was appointed as a papal official to collect the levy of a tenth of annual assessed value in Scotland, initially for six years. From the official records of the sums collected it is clear that the values of each assessed bishopric had risen greatly in relation to the Old Extents (see Appendix 5).[10] Importantly, the Bagimond tax collection was still incomplete at the time of the death of King Alexander III in 1286 so unless Pope Gregory X and his successors simultaneously imposed two ecclesiastic taxations upon Scotland between 1274 and 1286, it seems reasonable to assume that all of the figures relating to the collections of an Old Extent in Scotland must date to the period between *c.* 1200 and 1274, and one must specifically date to the reign of King Alexander between 1249 and the Bagimond demand post-1274.

The third set of figures employed by Stevenson dates to the period 1289–1291 when Pope Nicholas IV provisionally granted King Edward I of England a grant of a tenth of the British Isles in return for taking the cross. It was recognized at that point that England, Wales, Ireland and Scotland each possessed different methods of assessment so the pope granted Edward the right to make a common new extent for four countries.[11] The money was to be paid to Edward in two installments: the first on 24 June 1291 and the second one year later. In return, the king was to begin his crusade by 24 June 1293 and, if he did not, the money was to be repaid to the church.[12] In Scotland, Alan de St Edmund, bishop of Caithness, was appointed as co-collector with John de Halton, bishop of Carlisle, but he died before the operation began, leaving the latter in sole charge of the assessment process in Scotland. This award, and its subsequent quick implementation, must have been rather controversial in Scotland, firstly because of the new extent it introduced and secondly because King Edward I was acting as superior lord there during the Great Cause to choose the next king of Scots (1290–1292). The winner of that contest, John Balliol, was not inaugurated as king until November 1292.[13] Finally, Stevenson highlighted that since there was no discernable reduction in this taxation in Scotland in the 1320s, most of the big decline in value between *Antiqua*

[10] Lunt, 'A Papal Tenth'; Watt, 'Bagimond di Vezza and his "Roll"'.

[11] Graham, 'The Taxation of Pope Nicholas IV', p. 440.

[12] Lunt, 'Papal Taxation in England', p. 412.

[13] Penman, *Robert the Bruce*, pp. 29–37.

Taxatio and *Verus Valor* must have occurred later, most likely during the period of the Black Death (1349 in Scotland).[14]

Throughout his work, Stevenson made two key assumptions. First that all 'old extents' listed in ecclesiastic registers were identical to the old extent listed in 1366 alongside *Verus Valor*. Second, while he knew that the old extent taxed only spiritualities, that Bagimond taxed both spiritualities and temporalities, and that the Halton tax base was different to both, Stevenson made direct comparisons between them allowing him to claim 'average increases' in value of between thirty-eight per cent and sixty-nine per cent between the old extent and the Halton extent. He did this by assigning an undated assessment for the archdeaconry of Lothian found among the records of Coldingham priory to the period of the Halton tax collection.[15]

Mayhew has also written on this topic on a number of occasions. In his first offering he followed Stevenson and made direct comparisons between the values of the old extent, Bagimond's valuation, Halton's valuation, and the new extent. He accounted for the drop in value between Bagimond and Halton and the new extent by casting doubt (unfairly) on the veracity of the new extent,[16] by blaming high custom rates for depressing contributions, and taxpayer resistance.[17] Because of this he saw the relatively peaceful reign of Alexander III as crucial to the surge in Scottish economic growth between *c.* 1250 and *c.* 1300 before war damage and the Black Death combined to depress the economy and land values during the first half of the fourteenth century.[18]

Returning to this topic in 1995, Gemmill and Mayhew virtually abandoned any contextualization when comparing the different sets of figures. They made a direct comparison between the old extent and Halton's assessment, claiming that the rise in income was an indication of thirteenth-century economic growth — in fact, according to Gemmill and Mayhew, the figures represented a doubling in value of Scottish dioceses across the thirteenth century. Any vari-

[14] Stevenson, 'Taxation in Medieval Scotland', p. 303.

[15] Stevenson, 'Taxation in Medieval Scotland', p. 300.

[16] He noted that the new extent was completed in only seven weeks, before 20 July 1366. This, however, ignores the fact that the new extent assessment had originally been ordered to be prepared nine years previously in November 1357 (*RPS* 1357/11/3).

[17] Mayhew, 'Alexander III — A Silver Age?', p. 60.

[18] Mayhew, 'Alexander III — A Silver Age?', pp. 60–65.

ations in this increase between the individual diocesan totals were dismissed as an unreliability of the sources.[19]

The most recent comment on these extents appeared in 1999 when Donnelly introduced new evidence from Durham Priory's Coldingham estate in Scotland to the debate.[20] Using evidence from payments made by Durham Priory by 1331 for the churches it possessed in Scotland, 'according to the ancient taxation', he argued that it was at least the mid-1330s before there was any decline in the value of Scottish land assessments.[21] Therefore, according to him, warfare before 1329 had no measureable impact upon the Scottish economy. Like Tout in 1913,[22] Donnelly further tentatively dated the three complete surviving old extents known to him (see Appendix 5: Aberdeen, St Andrews and Arbroath) to 1254, making them contemporary with the Norwich taxation of England.[23] At the moment there is no evidence to support these choices of date other than wishful thinking.

Although Donnelly also argued that the rise in tax collected by Bagimond was due either to a failure of existing valuations to provide a true value or to (unspecified) changes in the Scottish economy,[24] Watt has since clarified that in fact Bagimond's collection introduced a radically different tax base to that of the old extent. While the Knights Templar, Hospitallers, hospitals, mendicants, and some secular clerks were exempt, clerical temporalities like revenues from forests and fisheries and spiritualities all became liable for tax. Cistercian establishments were also declared individually exempt but together made a special payment as an Order. It seems as though tax was also due on produce sold or revenues put out to farm. This new tax base also likely explains why the Scottish clergy asked Bagimond to return to the papal court after August 1275 with an alternative offer to pay the tax according to the old extent but for seven years, not six. They were unsuccessful.[25] This refusal did not end Scottish opposition to Bagimond's tenth and as late as 1285 Pope Honorius IV effec-

[19] Gemmill and Mayhew, *Changing Values in Medieval Scotland*, pp. 363–68.

[20] For some of the records of Coldingham see: Raine, ed., *The Correspondence, Inventories, Account Rolls, and Law Proceedings, of the Priory of Coldingham*.

[21] Donnelly, 'Skinned to the Bone'.

[22] *The Register of John De Halton*, I, p. xii.

[23] Donnelly, 'Skinned to the Bone', p. 7.

[24] Donnelly, 'Skinned to the Bone', p. 8.

[25] Watt, 'Bagimond di Vezza and his "Roll"', pp. 3–5.

tively accused King Alexander III of collusion to prevent papal representatives from uplifting collected bullion and taking it out of Scotland.[26]

In his discussion of the Halton assessment of 1291 Donnelly noted that it was by far the most detailed assessment of the church's wealth near the peak of medieval Scotland's economic prosperity. Crucially, he listed everything that was exempt from the tax but not what had been included for the first time.[27] Also problematic is the fact that Donnelly did not effectively contextualize his discussion of the Halton collection.

At this time, it was in King Edward I's best interests to raise money quickly. His favoured bankers, the Riccardi of Lucca, had failed and so he had nobody willing to lend him cash. This may help explain why ecclesiastic estates were also then first drawn into this general taxation. Both the Premonstratensian and the Cistercian orders were to be included in the assessment, for example.[28] These additions doubtless help explain the remarkable difference in values between the Bagimond and Halton valuations — both were working from radically different tax bases (see Appendix 6), and both of these valuations were also completely different to the only known tax base for the Old Extent in Scotland. This was recognized by Tout in 1913 but seems to have been forgotten in the intervening period.[29] In fact, anyone who has previously compared the three taxations (Old Extent, Bagimond, and Halton) with each other and used that evidence as a sign of economic growth is in effect comparing apples, oranges, and limes. In any event, the process of collecting the Halton valuation was still continuing as late as 1308, no doubt due at least in part to the various activities of William Wallace and King Robert Bruce (1306–1329). It was thereafter abandoned by the Scots but remained in use in England until the 1530s.[30]

To further complicate matters more generally, in September 1294 the Halton assessment was increased to a half on ecclesiastic estates in England.[31] There is no contemporary record of this being imposed on Scotland too but Abbot Bower's *Scotichronicon* does make a curious remark under the same year date when he claimed that: 'Hoc eciam anno rex Anglie a clericis cepit medium omnium bonorum ecclesiasticorum secundum estimacionem vere taxacionis ad

[26] Lunt, *Papal Revenues in the Middle Ages*, II, no. 275.

[27] Donnelly, 'Skinned to the Bone', p. 11.

[28] Prestwich, *Edward I*, pp. 403–12.

[29] *The Register of John De Halton*, I, pp. vii–xxv.

[30] Donnelly, 'Skinned to the Bone', p. 13.

[31] Prestwich, *Edward I*, p. 404.

debellandam Scociam' (in September 1294 Edward I took from the Scottish clergy a half of all their spiritual revenues according to an assessment of true valuation in order to subdue Scotland). Bower, of course, writing at the beginning of the fifteenth century, hated all things English but the date he provided is precise, much of *Scotichronicon* was sourced from earlier chroniclers, and the fact that this tax increment was also imposed in England makes his claim all the more interesting.[32] Though the editors of this volume of Bower's work argue that he was wrong to claim this tax of half all spiritual revenues was levied in Scotland too, the timing is interesting and there actually is no good reason to reject this claim as anti-English propaganda when we know that the English crown was desperate for bullion at that moment in time.[33] In addition, September 1294 was the period immediately following King John Balliol's two humiliating appearances at Westminster in September 1293 and June 1294 and also just after King Edward I had demanded military service from the Scots to fight for him in France.[34] In this context it is surely possible that Edward I might also have demanded that the Scottish church increase its contribution to his coffers to half of their annual income.

Reassessing the Old and New Extents in Scotland

Rejecting the Bagimond and Halton assessments as comparators against which the Old and New Extents in Scotland may be measured forces a new appraisal of the latter two. It is at this point that the first major problem is encountered. Unfortunately, none of the written data sets relating to the collection of an old extent in Scotland during the thirteenth century are precisely dated. Appendix 4 contains all the sets of ecclesiastic data relating to the Old Extent from four different sources. They appear to be closely identical but there are one, perhaps two, key differences in the amounts that cannot readily be explained away as scribal error. This suggests that at a stretch these figures could represent at least two different undated Old Extents of the period *c.* 1200–1274.

Further, if these four data sets are then compared to the old extent summarized in the document of 1366, while many of the numbers again are broadly compatible, there are five totals that are substantially different and it is unlikely they could be blamed on scribal error (see Appendix 7). Clearly, this evidence

[32] Bower, *Scotichronicon*, VI, pp. 58–59.

[33] Bower, *Scotichronicon*, VI, p. 220.

[34] Penman, *Robert the Bruce*, pp. 40–41.

strongly indicates there are definitely two, possibly as many as three, different old extents dating to the thirteenth century recorded in various Scottish and English sources. Whichever chronological way these figures are assembled, they either are evidence of a slightly declining thirteenth-century valuation before 1274 or a slightly rising thirteenth-century valuation before 1274. There is no way of telling which option is correct from the figures alone and it can again only be assumed that all of these old extents utilized an identical tax base.

The starting points in any re-analysis of this evidence are the valuations produced for King David II (1329–1372) in 1366. When this data is mapped there appear to be some noticeable regional differences in the amounts by which the *Antiqua Taxatio* had declined in comparison to *Verus Valor* with parts of the Highlands being particularly badly affected (see Appendices 8 and 9). Perhaps not too much should be made of his. Three years previously Scotland had experienced a major political rebellion as King David II (1329–1371) strove to defeat the political ambitions of the so-called 'Highland Party', led by the king's nephew and heir, Robert Stewart.[35] It is likely no coincidence that some of the areas that performed particularly poorly in the *Verus Valor* of 1366 also happened to be under the control of ex-'Highland Party' members, but lacking the original and other comparable accounts it is difficult to read too much into this, and it should not be forgotten that the king defeated the Highland Party rebellion.

Of these two sets of figures the greater discussion has focused on the ecclesiastic data, likely because of the greater amounts of available evidence for papal tax collections that, as we have seen, stretches back into the thirteenth century. Nevertheless, the two values provided for sheriffdoms in 1366 (see Appendix 1) are also worthy of more discussion because there is another data set to compare them against, this time dated to the period 1328–1330, and again collected according to the Old Extent. Unfortunately, thanks to broadly defined deferred payments in these latter sets of accounts, it is impossible to make absolutely detailed comparisons between these figures and the Old and New Extents of 1366.

As part of the Treaty of Edinburgh (27 March 1328) the Scots pledged to pay the English crown some £20,000 as a 'contribution to peace' (war reparations)[36] and the fuller Scottish exchequer accounts of the late-1320s contain some details about how this sum was raised across a period of three

[35] Penman, *David II*, pp. 283–84.

[36] Penman, *Robert the Bruce*, p. 283.

years.[37] The first year that the contribution to peace was raised in Scotland (1328) the sheriffdoms contributed, though some were allowed to defer payment until the following year. In total, the chamberlain received a total of approximately £6931 for the first two years' collections from sheriffdoms, giving an average of *c.* £3466 per annum for the tenth collected via the Old Extent in 1328 and 1329.[38] While this is only a rough approximation it compares unfavourably against the Old Extent tenth of £4484 listed in 1366, demonstrating a decline in the revenue of sheriffdoms of approximately twenty-three per cent between *c.* 1200 × 1274 and 1328. If these sums are subsequently compared to the values of sheriffdoms in *Verus Valor*, they tell us that the valuation of sheriffdoms further declined by another thirty-eight per cent between 1329 (a tenth of *c.* £3466) and 1366 (a tenth of *c.* £2172). Renewed warfare (both civil and international) after 1332 in combination with the Black Death likely account for much of this second decline in values. Using these same figures, the value of sheriffdoms in Scotland declined by forty-eight per cent between *c.* 1200 and 1366. It should be emphasized that these are at best approximate calculations but without more data it will be difficult to further improve upon their accuracy.

In contrast to sheriffdoms the Scottish church contributed nothing to the payment for peace in the first year (1328).[39] It was not until year two of the contribution that the first money was received from the church in two installments, amounting to just over £1863.[40] Since no other payments were recorded by the chamberlain across these two years there seems a strong likelihood that this was the church's contribution to peace for the first two years of the tax. Unfortunately, the records for the third year are missing so figures regarding possible arrears in 1328 and 1329 are possible but unknown. When this sum of £1863 is divided across the first two years of the contribution for peace the Old Extent tenth averages out at approximately £932 *per annum*. This is very close to the value of the ecclesiastic *Verus Valor* tenth calculated in 1366 of £939 but far below the value of the Old Extent listed in that same document (£1500); in fact, a decline in value of some thirty-two per cent.

All of which serves to demonstrate that the value of the Old Extent for both the church and sheriffdoms had already substantially declined from their thir-

[37] *Rotuli Scaccarii regum Scotorum*, I, pp. cvi–cx.

[38] *Rotuli Scaccarii regum Scotorum*, I, pp. 112–13, 181–82, 204.

[39] *Rotuli Scaccarii regum Scotorum*, I, p. 112.

[40] *Rotuli Scaccarii regum Scotorum*, I, p. 181.

teenth-century high points by the late 1320s, long before renewed warfare after 1332 and the arrival of plague in the late-1340s. In fact, a decline in the value of the Old Extent because of the vagaries of war in Scotland had already been recognized by the Papacy in 1308,[41] and again by the Scots themselves in 1328:

> Qui omnes et singuli comites, barones, burgenses et liberetenentes, tam infra libertates quam extra, de domino rege vel quibuscunque aliis dominis infra regnum mediate vel immediate tenentes cuiuscunque fuerint conditionis, considerantes et fatentes premissa domini regis motiva esse vera ac quam plura alia suis temporibus eis per eum comoda accrevisse suamque petitionem esse rationabilem atque justam, habito super premissis communi ac diligenti tractatu, unanimiter, gratanter et benevole concesserunt et dederunt domino suo regi supradicto annuatim ad terminos Sancti Martini et Pentecostes proportionaliter pro toto tempore vite dicti regis decimum denarium omnium firmarum et reddituum suorum tam de terris suis, dominicis et wardis quam de ceteris terris suis quibuscunque infra libertates et extra tam infra burgos quam extra iuxta antiquam extentam terrarum et reddituum tempore bone memorie domini Alexandri Dei gracia regis Scottorum illustris ultimo defuncti per ministros eius fideliter faciendam excepta tantummodo destructione guerre in quo casu fiet decidentia de decimo denario preconcesso secundum quantitatem firme que occasione predicta de terris et redditibus predictis levari non poterit prout per inquisitionem per vicecomitem loci fideliter faciendam poterit reperiri.[42]

> (Which things all and singular the earls, barons, burgesses and freeholders, both within liberties and without, whether tenants of the lord king or whatsoever other lords within the kingdom, with or without an intermediary, whatever their condition shall be, considering and acknowledging the afore going motives of the lord king to be true and, what is more, how many other benefits had accrued to them in his times through him, and his petition to be reasonable as well as just, having had common and diligent discussion upon the foregoing, unanimously, with joy and in a spirit of goodwill granted and gave to their above said lord king, annually, at the terms of Martinmas and Whitsun in proportion, for the entire time of the life of the said king, a tenth penny of all money from their fermes and rents, both from their lands, demesnes and wards and from whatsoever their other lands, within or out with liberties, both within and out with burghs, according to the old extent of lands and rents in the time of the lord Alexander by the grace of God illustrious king of Scots of good memory who died most recently; [to be collected] faithfully by his ministers, making exceptions only for the destruction of war, in which case there shall be a decrease of the tenth penny granted before according to the size of

[41] Theiner, ed., *Vetera Monumenta*, p. 300.

[42] *RPS*, 1328/1.

> the ferme which, by reason of the afore going, cannot be raised from the aforesaid lands and rents, as can be discovered by an inquest to be made faithfully by the sheriff of the place.)

War, naturally, is an obvious candidate that might be blamed for the collapse in the value of the Old Extent in Scotland between the reign of King Alexander III and 1328. The first phase of the wars of independence between Scotland and England (which also eventually drew Ireland, Wales, and France into the conflict) began in 1296 and was only officially terminated by the Treaty of Edinburgh in 1328, before resuming once again with renewed vigour in 1332 when Edward Balliol returned to the country. In between, these conflicts also included a civil war in Scotland between 1306 and *c.* 1310, which was accompanied by the deliberate destruction of economic assets, one of the most famous medieval invasions and pitched conflicts (Bannockburn) in 1314, the Scottish invasion of Ireland in 1315, various rebellions in north England against the rule of King Edward II (1307–1327), a rebellion by some Scots against King Robert I in 1320, and numerous expeditions by Scots forces into north England between 1314 and 1328. The Great European Famine of 1315–1322, the most damaging food shortage on record in north-western Europe, can be added into this mix, shortly followed in Scotland (1319) by a devastating bovine panzootic, and the recent works of Newfield and Slavin highlight the widespread social-economic impacts of these events.[43] Additionally, ovine populations suffered unusually often, it seems, from severe epizootic disease in the late thirteenth and early fourteenth centuries (between 1279 and 1330).[44] Anyone looking for reasons why there was a decline in the value of the Old Extent before 1328 would naturally be drawn to the cumulative effects of such events and the strains they must have placed upon society at large as climate deteriorated in the early fourteenth-century.

Even so, this story of Scottish medieval economic decline may be more nuanced; we may not be able to blame everything on post-1306 warfare, famine, and animal disease alone. Noticeably, when Scottish chronicle sources for the thirteenth century are examined they clearly demonstrate a change in weather patterns and extreme events after the ultra Plinian eruption of Gunung Rinjani in Indonesia in 1257, one of the two largest stratospheric-clouding eruptions of the last 2500 years, and an accompanying large increase in the recorded num-

[43] Newfield, 'A Cattle Panzootic'; Slavin, 'The Great Bovine Pestilence'; Stone, 'The Impact of Drought'.

[44] Slavin, 'Epizootic Landscapes'.

bers of livestock epizootics.[45] For example, Bower's *Scotichronicon* — fifteenth century in date but primarily based upon thirteenth-century sources — records five extreme climatic events before 1258: in 1205 (a bad winter that killed animals), in 1209 (severe flooding), in 1210 and 1211 (severe winters which killed livestock), and in 1223 (bad winter weather that killed livestock).[46] To these we can add occasional events noted in other sources like the *Chronicle of Lanercost* in 1256:

> [...] in this year there was so great a corruption of the air, and inundation of rain, throughout the whole of England and Scotland, that both crops and hay were nearly all lost. And some men's corn rotted in the fields from the day of harvest; some men's corn, shaken out by the wind, grew again under the straw; some men's harvest was so late that they did not reap it until about the festival of St Martin [11 November] or later.[47]

The fall-out from Gunung Rinjani and the accompanying decline from the relative stability and higher temperatures of the Medieval Climate Anomaly (MCA) are also noted in *Scotichronicon*:

> Ipso anno magna caristia fuit in partibus Scocie ita quod bolla farine ad iiiior solidos venderetur. Audita sunt tonitrua perhorrida et fulgara immania que homines in agris et pecora passim combusserunt ac messes et arbores pluribus locis urendo consumpserunt; pluviosus autumpnus messes suffocavit; post autumpnum tempestates ventorum domos subverterunt et strages plurimas fecerunt; et sic multa signa anno apparuerunt.
>
> (In that same year [1260] there was a great food shortage in parts of Scotland, so much so that a boll of flour was being sold for 4s. Dreadful claps of thunder were heard, and there were terrifying flashes of lightning which burned up men in the fields and animals far and wide, and destroyed crops and trees in several places by burning them up. A wet autumn choked the harvest with weeds. After the Autumn, gales of wind wrecked houses and caused a vast amount of destruction.)[48]

[45] Stothers, 'Climatic and Demographic Consequences'; Sigl and others, 'Timing and Climate Forcing of Volcanic Eruptions', p. 545 Figure 3. The identification of this eruption with Gunung Rinjani only occurred in 2013: Lavigne and others, 'Source of the Great A.D. 1257 Mystery Eruption'; Guillet and others, 'Climate Response'.

[46] Bower, *Scotichronicon*, IV, pp. 437, 457, 461, 467; V, p. 117.

[47] *Chronicon De Lanercost*, p. 64.

[48] Bower, *Scotichronicon*, V, p. 325.

It was noticed by the same chronicler that the bad weather continued without respite across 1261 and 1262 when failed harvests caused famine across Europe and the North Atlantic.[49] In fact, this weather pattern in Scotland continued across virtually all of the decade of the 1260s and it was accompanied by a major epizootic(s?) in 1268 that specifically killed red deer, fallow deer, forest ponies, and sheep across the country.[50] The 1270s started badly too: in 1272, for example, the crops failed, people starved, and this was accompanied by a cattle epizootic.[51] The 1280s and 1290s provided no respite. In fact, after 1258 chronicle notices of extreme weather, animal epizootics, and human disease in sources that relate Scottish events increase by over one hundred per cent from the pre-1258 period.

This is statistically significant and it doubtless was one aspect of the more general pronounced decline in northern hemisphere weather and temperatures that was also noted by the Irish Annals. Oram has commented upon this at some length and has posited that Scottish and Irish chronicle accounts of increasingly bad weather from this period all perhaps reflect a southwards displacement of the Jet Stream and the accompanying tracking of Summer low pressure cells across north Britain, bringing a succession of cold fronts (and rain) in from the Atlantic.[52] But perhaps the most important point to make here is that the Scottish climate clearly was on a downward trajectory in the thirteenth century as it approached the Wolf Minimum. A summer temperature series spanning 800 years from the Northwest Cairngorms seems to reflect just this.[53] But in each successive decade fewer good years are noted by chroniclers and, despite an almost complete lack of commodity price data, it seems logical to suggest that these cumulative weather and disease patterns must surely have had a long-term impact upon the relative value of the Old Extents. Indeed, it is possible that the Scots were already suffering economically by the mid-1270s. Then, they tried to reject Master Boemundo di Vezza's new six-year valuation and instead asked Pope Gregory X whether they might be allowed

[49] Bower, *Scotichronicon*, v, p. 335; Stothers, 'Climatic and Demographic Consequences', pp. 364–65.

[50] Bower, *Scotichronicon*, v, p. 369.

[51] Bower, *Scotichronicon*, v, p. 385.

[52] Oram, 'Between a Rock and a Hard Place'. See also: Campbell, 'Physical Shocks, Biological Hazards, and Human Impacts'.

[53] Rydval and others, 'Reconstructing 800 Years'. Although sample replication is worrisome before the sixteenth century, and more work is required, several of the coldest summers, and the coldest decadal run of summers, in this 800-year reconstruction fall in the thirteenth century.

to use the Old Extent valuations instead but for seven years.[54] Simply put, the Scots were offering to pay Pope Gregory X much less money with the payments spread over a longer period of time.

Summary

Both the Old and New Extents in Scotland are problematic sources to use yet they are all we have in the face of an almost complete lack of commodity prices and other financial data for the thirteenth century in that country. They are irregularly and conservatively used by contemporaries and can only provide evidence of a general downward trend in values when there must undoubtedly have been peaks and troughs in valuations at various times and along regional lines. More importantly, it has also been clearly demonstrated that they cannot be used in conjunction with either the Master Boemundo di Vezza or the Bishop Halton valuations because the latter two were predicated upon radically different tax assessments with much wider remits to extract cash. To do so merely provides an utterly misleading picture of a booming economy in late thirteenth-century Scotland.

Removing these latter two valuations and reassessing the remaining Old and New Extents in conjunction with chronicle evidence actually reveals nothing extraordinary in a European context. Instead of the alleged late thirteenth-century booming economy we are left with post-plague northern European society and economy that is clearly finding it increasingly difficult to stage any kind of long term recovery in the midst of dramatic climate change as some landscapes became increasingly marginal. Adding sustained warfare, famine, and epizootics to that mix after *c.* 1300 certainly worsened matters but the roots of the decline can be traced back to the sixth decade of the thirteenth century.

[54] Bower, *Scotichronicon*, v, p. 403.

Appendix 1[55]

Summe taxationum et verorum valorum ecclesiarum presentate in parlamento predicto	The amount of taxation and of the true value of churches presented in the aforesaid parliament
Episcopatus	Bishoprics
Candide Case: per antiquam taxationem, iij^{C} lxviij li. xv s. vj d. per verum valorem, C xliij li. xx d.	Galloway: by the old assessment, £368 15*s* 6*d* by the true value, £143 20*d*
Ergad': per antiquam taxationem, ij^{C} iiij^{xx} li. xxvj s. viij d.?per verum valorem, C xxxiij li. vj s. viij d.	Argyll: by the old assessment, £281 6*s* 8*d* by the true value, £133 6*s* 8*d*
Abredon': per antiquam taxationem, M iiij^{C} iiij^{xx} xij li. iiij s. iiij d. per verum valorem, M ij^{C} lviij li. xvij s. viij d.	Aberdeen: by the old assessment, £1492 4*s* 4*d* by the true value, £1358 17*s* 8*d*
Ross': per antiquam taxationem, iij^{C} xx li. vij s. xj d. ob. per verum valorem, ij^{C} xlvi li. xij s.	Ross: by the old assessment, £320 7*s* 11½*d* by the true value, £246 12*s*
Dunblan': per antiquam taxationem, vj^{C} vij li. xiij s. iiij d. per verum valorem, iij^{C} lxxvj li. xiij s. iiij d. Et de terris episcopi xxx li. xix s. iiij d.	Dunblane: by the old assessment, £607 13*s* 4*d* by the true value, £376 13*s* 4*d*. And £30 19*s* 4*d* from the bishop's lands.
Brechin': per antiquam taxationem, iiij^{C} xlj li. iij s. iiij d. per verum valorem, iij^{C} xxj li. xvj s. viij d.	Brechin: by the old assessment, £441 3*s* 4*d* by the true value, £321 16*s* 8*d*
Morau': per antiquam taxationem, M iiij^{C} xviij li. xj s. per verum valorem, v^{C} lix li. viij s. viij d.	Moray: by the old assessment, £1418 11*s* by the true value, £559 8*s* 8*d*
Cathan': per antiquam taxationem, ij^{C} iiij^{xx} vj li. xiiij s. x d. ob. per verum valorem, iiij^{xx} vj li. vj s. viij d.	Caithness: by the old assessment, £286 14*s* 10½*d* by the true value, £86 6*s* 8*d*
Dunkelden': per antiquam taxationem, M ij^{C} vj li. v s. viij d. per verum valorem, vj^{C} li. liij s. iiij d	Dunkeld: by the old assessment, £1206 5*s* 8*d* by the true value, £602 13*s* 4*d*
Glasguen': per antiquam taxationem, iiij^{M} iiij^{xx} li. xij s. ij d., set nunc est antiqua taxatio iij^{M} ij^{C} xxxix li. pro eo quod plures ecclesie decanatuum Tevydal et Eskdal' sunt ad fidem regis Anglie. per verum valorem, MM xxviij li. x s. vj d. ob. except' Vall' Anand' et pluribus ecclesiis dictorum decanatuum ad fidem ut supra.	Glasgow: by the old assessment, £4080 12*s* 2*d*, but now the old assessment is £3239 for that place because many of the churches of the deaneries of Teviotdale and Eskdale are at the fealty of the king of England. by the true value, £2028 10*s* 6½*d* excepting Annandale and the many churches of the said deaneries at the fealty as above.

55 *RPS*, 1366/7/18.

Sancti Andree: per antiquam taxationem, v^{M} iiij^{C} xiiij li., sed quia plures ecclessie infra le Mersk sunt ad pacem regis Anglie nunc est antiqua taxatio dicti episcopatus v^{M} iij^{C} xl li. xiij s. iiij d. per subtractionem lxxiij li. vj s. viii d. de ecclesia Sancte Trinitatis de Berwyc tantum. per verum valorem, iij^{M} v^{C} vij li. preter decanatum de Mersk qui se extendit ad Cxx li. estimative. Summa totalis taxationis ecclesiarum omnium episcopatuum Scotie preter episcopat' Sodoren' secundum antiquam taxationem, xv^{M} li. lvj s. Summa totalis verum valorem eorundem preter ut supra, ix^{M} iij^{C} iiij^{xx} xvj li. vj s. vj d.	St Andrews: by the old assessment, £5414, but because many of the churches within the Merse are at the peace of the English king, the old assessment of the said bishopric is now £5340 13*s* 4*d* by subtraction of so much as £73 6*s* 8*d* from the church of the Holy Trinity at Berwick. by the true value, £3507 except the deanery of Merse which extends to approximately £120. Sum total of the assessment of the churches of all the bishoprics of Scotland except the bishopric of the Isles according to the old assessment, £15,000 56*s*. Sum total of the true value of the same except as above, £9396 6*s* 6*d*
Terre et Redditus Vicecomitatuum	Lands and rents of the sheriffdoms
Clacman': per antiquam extentam, iij^{C} xxxj li. viij d. per verum valorem, ij^{C} xliij li. xiiij s. viij d.	Clackmannan: by the old extent, £331 8*d* by the true value, £243 14*s* 8*d*
Kynros: per antiquam extentam, lxv li. per verum valorem, xxxviij li. xiiij s. viij d.	Kinross: by the old extent, £65 by the true value, £38 14*s* 8*d*
Fiff': per antiquam extentam, iij^{M} iiij^{MC} lxv li. xiij s. iiij d. per verum valorem, ij^{MM} v^{C} lv li.	Fife: by the old extent, £3465 13*s* 4*d*: by the true value, £2555
Perth': per antiquam ext[entam], vj^{M} C iiij^{xx} xij li. ij s. vj d. per verum valorem, iij^{M} iiij^{xx} vij li. xix d.	Perth: by the old extent, £6,192 2*s* 6*d* by the true value, £3,087 19*d*
Forfar: per antiquam extentam, iij^{M} iij^{MC} lxx li. vj s. viij d. per verum valorem, ij^{M} ij^{C} xl li. vj s. viij d. de qua ad contributionem j^{C} xij li. et iiij d.	Forfar: by the old extent, £3370 6*s* 8*d* by the true value, £2240 6*s* 8*d*, of which £112 4*d* to the contribution.
Kyncardin: per antiquam extentam, M iiij^{xx} viij li. x s. viij d. per verum valorem, vij^{C} xxij li.	Kincardine: by the old extent, £1088 10*s* 8*d* by the true value, £722
Aberden': per antiquam extentam, iiij^{M} iiij^{C} xlviij li. vj s. per verum valorem, ij^{M} v^{C} iiij^{xx} viij li. v s. ij d.	Aberdeen: by the old extent, £4448 6*s* by the true value, £2588 5*s* 2*d*
Banff: per antiquam extentam, M C x li. vj s. per verum valorem, C xxviij li. xvj s. viij d.	Banff: by the old extent, £1110 6*s* by the true value, £128 16*s* 8*d*
Invernys: per antiquam extentam, iij^{M} C li. xiij li. xj s. viij d. per verum valorem, M iiij^{xx} li. x s. ix d.	Inverness: by the old extent, £3113 11*s* 8*d* by the true value, £1080 10*s* 9*d*

Ergad': per antiquam extentam per verum valorem Terre Johannis de Insulis infra vicecomitatum Ergadie se extendunt ad MCCCxx li. Terre Ghillaspic et aliorum baronum eiusdem vicecomitati secundum antiquam extentem ad vj^{C} li. Eedem terre per verum valorem: C xxxiij li. vj s. viij d. Terre Johannis de Loorn per antiquam extentem iiij^{C} xx li. de vero valore earundem terrarum ignoratur. Terre domini Senescalli Scotie de Bote, Cogwall, Knapdale, Arane et duabus Combrays se extendunt ad M li. per antiquam taxationem. Summa antiquarum taxationum dictarum summarum: xxvj^{M} v^{C} xxiiij li. xvij s. vj d.	Argyll: by the old extent by the true value John of the Isles' lands within the sheriffdom of Argyll amount to £1320 The lands of Gillespic and the other barons of the same sheriffdom according to the old extent [are] £600 The same lands by true value: £133 6*s* 8*d* John de Lorne's lands by the old extent [are] £420, concerning the true value of the same lands it is unknown. The lord steward of Scotland's lands of Bute, Cowall, Knapdale, Arran and the two Cumbraes extend to £1000 by the old assessment. Sum of the old assessments of the said amounts: £26,524 17*s* 6*d*.
Wygtona: per antiquam extentam, M ij^{C} xxxv li. iij s. iiij d. per verum valorem, C iiij^{xx} xv li. ij d.	Wigtown: By the old extent, £1235 3*s* 4*d* By the true value, £195 2*d*
Drumffres: per antiquam extentam, ij^{M} vj^{C} lxvj li. xiij s. iiij d. per verum valorem, viij^{C} iiij^{xx} li. lv s. iiij d.	Dumfries: By the old extent, £2666 13*s* 4*d* By the true value, £682 15*s* 4*d*
Dunbretane: per antiquam extentam, M iiij^{C} xlij li. ix s. vj d. per verum valorem, iiij^{xx} xvj li. ix s. vj d.	Dunbarton: By the old extent, £1442 9*s* 6*d* By the true value, £96 9*s* 6*d*
Peblys: per antiquam extentam, M ij^{C} lxxiiij li. xviij s. vj d. per verum valorem, viij^{C} lxiij li. xiij s. iiij d.	Peebles: By the old extent, £1274 18*s* 6*d* By the true value, £863 13*s* 4*d*
Selkirk: per antiquam extentam, iiij^{xx} xix li. ix s. x d. per verum valorem, iiij^{xx} li. xviij s. vj d. Terre abbatis de Melros non assedetur nunc sed secundum antiquam extentam liij li. xiij s. iiij d.	Selkirk: By the old extent, £99 9*s* 10*d* By the true value, £80 18*s* 6*d* The abbot of Melrose's lands are not assessed now, but according to the old extent [are] £53 13*s* 4*d*.
Roxburgh': per antiquam extentam, MC xxxiij li. xv s. per verum valorem, v^{C} xxiij li. xvij s.	Roxburgh: By the old extent, £1133 15*s* By the true value, £523 17*s*
Berwyk: per antiquam extentam, vj^{C} xxij li. ij s. iiij d. per verum valorem, iij^{C} iij^{xx} xij li. xvij s. iij d.	Berwick: By the old extent, £622 2*s* 4*d* By the true value, £372 17*s* 3*d*
Strivelyn: per antiquam extentam, M vij^{C} xlix li. xix s. iiij d. per verum valorem, vj^{C} iiijxx vij li. iij s. x d.	Stirling: By the old extent, £1749 19*s* 4*d* By the true value, £687 3*s* 10*d*
Are: per antiquam extentam, iij^{M} iij^{C} lviij li. xix s. x d. per verum valorem, M iij^{C} iiij^{xx} xvj li. xvj s. ij d.	Ayr: By the old extent, £3358 19*s* 10*d* By the true value, £1396 16*s* 2*d*
Lanark: per antiquam extentam, iiij^{M} lvij li. ix s. per verum valorem, M vij^{C} lv li. xix s. viij d.	Lanark: By the old extent, £4057 9*s* By the true value, £1755 19*s* 8*d*
Ramfrw: per verum valorem, v^{C} xxxv li. ix s. viij d.	Renfrew: By the true value, £535 9*s* 8*d*
Edynburgh' cum const[abulariato]: per antiquam extentam, iiij^{M} xxix li. xvj s. x d. per verum valorem, ij^{M} xxx li. xij s. ix d.	Edinburgh, with the constabulary: By the old extent, £4029 16*s* 10*d* By the true value, £3030 12*s* 9*d*

Appendix 2: Papal Extents Recorded in Thirteenth Century Scotland

Pope	Dates of tax	Amount	Basis collected and for what purpose	Recorded in Scotland
Innocent III	8/1/1198 to 16/7/1216	1/40th	Annually for 4th crusade	1199
Honorius III	18/7/1216 to 18/3/1227	1/20th	for 3 years for 5th crusade	1220
Gregory IX	19/3/1227 to 22/8/1241	1/10th	for 3 years for 6th crusade	(1229/30)†
Celestine IV	25/10/1241 to 10/11/1241			
Innocent IV	25/6/1243 to 7/12/1254	1/20th	for 3 years for 7th crusade	1247/48
Alexander IV	12/12/1254 to 25/5/1261	1/20th	for 6 years	(1254)†
Urban IV	29/8/1261 to 2/10/1264	1/100th	for 5 years	1263
Clement IV	5/2/1265 to 29/11/1268	1/10th	for 3 years	(1265/66)†
Gregory X	1/9/1271 to 10/1/1276	1/10th	for 3 years	(1267–1269)† × 2)
Ditto	1274–1288	1/10th	for 6 years	1274, Bagimond
Nicholas IV	22/2/1288 to 4/4/1292	1/10th	for 6 years	1291, Halton

† = Refusal to pay the tax by Scots.

Appendix 3

	Source		
Bishoprics	**Aberdeen**[56]	**St Andrews**[57]	**Arbroath**[58]
Caithness	287	287	387
Ross	352	352	352
Moray	1419	1419	1419
Aberdeen	1611	1611	1610
Argyll	281	281	281
Dunkeld	1206	1206	1206
Brechin	441	416	341
Dunblane	508	508	607
St Andrews	8023	8018	8018
Glasgow	4081	4081	4081
Galloway	359	358	359
Total	18,690	18,659	18,751[59]

[56] *Registrum Episcopatus Aberdonensis*, II, pp. 52–56.

[57] *Liber Cartarum*, pp. 29–38. An undated copy of the taxation for the diocese of St Andrews is also found in the Dunfermline records: *Registrum de Dunfermelyn*, pp. 203–11.

[58] *Liber S. Thome de Aberbrothoc*, I, pp. 231–47.

[59] What is listed here are approximate values, rounded down to the nearest pound.

Appendix 4

	Source			
Bishoprics	**Aberdeen**	**St Andrews**	**Arbroath**	**Holm Cultram**[60]
Caithness	287	287	387	286
Ross	352	352	352	353
Moray	1419	1419	1419	1409
Aberdeen	1611	1611	1610	1610
Argyll	281	281	281	300
Dunkeld	1206	1206	1206	1206
Brechin	441	416	341	410
Dunblane	508	508	607	606
St Andrews	8023	8018	8018	8018
Glasgow	4081	4081	4081	4086
Galloway	359	358	359	358
Total	18,690	18,659	18,751	18,641

[60] See 'The Register: Bishoprics of Scotland; Memorandum on writs; Caldbeck (continued)'. Again, what is listed here are approximate values, rounded down to the nearest pound.

Appendix 5

Bishoprics	Source				
	Aberdeen liber	**St Andrews liber**	**Arbroath liber**	**Holm Cultram**	**Master Boemundo di Vezza, 1276 × 87**
Caithness	287	287	387	286	317
Ross	352	352	352	353	628
Moray	1419	1419	1419	1409	2130
Aberdeen	1611	1611	1610	1610	2833*
Argyll	281	281	281	300	542*
Dunkeld	1206	1206	1206	1206	1730
Brechin	441	416	341	410	660
Dunblane	508	508	607	606	1095
St Andrews	8023	8018	8018	8018	10,833*
Glasgow	4081	4081	4081	4086	7625*
Galloway	359	358	359	358	1343*
Total	18,690	18,659	18,751	18,641	29,736*

* Indicates that arrears were due so these figures are not precise.

Appendix 6

	Source						
Bishoprics	**Aberdeen**	**St Andrews**	**Arbroath**	**Holm Cultram**	***Antiqua taxation* (as listed in 1366)**	**Master Boemundo di Vezza, 1276 × 87**	**Bishop Halton, 1291 × 1308**
Caithness	287	287	387	286	286	317	464
Ross	352	352	352	353	320	628	681
Moray	1419	1419	1419	1409	1418	2130	2496
Aberdeen	1611	1611	1610	1610	1492	2833*	3439
Argyll	281	281	281	300	281	542*	661
Dunkeld	1206	1206	1206	1206	1206	1730	2525
Brechin	441	416	341	410	441	660	1008
Dunblane	508	508	607	606	607	1095	1376
St Andrews	8023	8018	8018	8018	5414	10,833*	13,723
Glasgow	4081	4081	4081	4086	4080	7625*	11,143
Galloway	359	358	359	358	368	1343*	1322
Total	18,690	18,659	18,751	18,641	15,913	29,736*	38,838

Appendix 7

	Source				
Bishoprics	**Aberdeen**	**St Andrews**	**Arbroath**	**Holm Cultram**	***Antiqua taxatio* (as listed in 1366)**
Caithness	287	287	387	286	286
Ross	352	352	352	353	320
Moray	1419	1419	1419	1409	1418
Aberdeen	1611	1611	1610	1610	1492
Argyll	281	281	281	300	281
Dunkeld	1206	1206	1206	1206	1206
Brechin	441	416	341	410	441
Dunblane	508	508	607	606	607
St Andrews	8023	8018	8018	8018	5414
Glasgow	4081	4081	4081	4086	4080
Galloway	359	358	359	358	368
Total	18,690	18,659	18,751	18,641	15,913

Appendix 8: Sheriffdoms

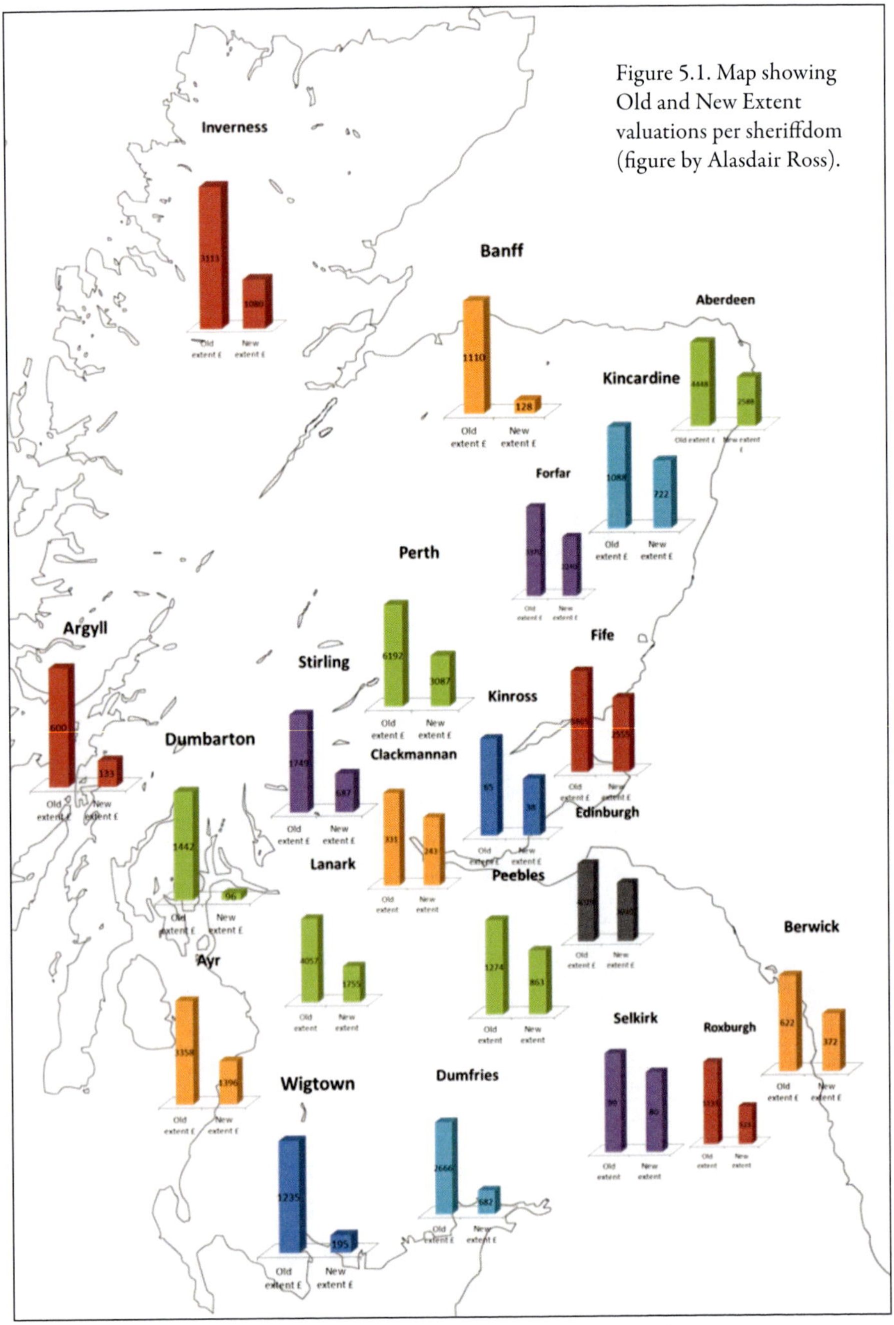

Figure 5.1. Map showing Old and New Extent valuations per sheriffdom (figure by Alasdair Ross).

'To the abbottis profeit': The Cistercian Abbey of Coupar Angus and the Scottish Export Economy

Victoria Hodgson
Department of History, University of Bristol

Introduction

Medieval monasteries were economic powerhouses and their impact was felt throughout Europe. Much research is still to be done, however, to determine the precise character of the Scottish monastic economy within that broader European context and its responsiveness to fluctuations in resource availability, trade routes, and market demand. The Cistercian abbey of Coupar Angus makes for a particularly good case study as the house was a major landowner and possesses an uncommonly (in a Scottish context) good set of surviving records. Founded by King Malcolm IV in 1164, the abbey came to control a vast and varied portfolio of landed property, spanning a distance of over 200 miles and comprising arable, pasture, woodlands, and waterways.

These resources were strictly managed and fulfilled specialized functions, both during the period of direct monastic control and the following episode of large-scale leasing of the fifteenth and sixteenth centuries.[1] As with all Cistercian houses, Coupar was involved in the trade of its surpluses, facilitated by a network of urban properties in neighbouring east coast burghs. The available evidence for the discussion of this commercial activity is still, however, fragmentary. While the extant Coupar material is comparatively extensive, documentation relating to trade is limited. More broadly, Scottish customs rolls are unavailable until the fourteenth century and extant burgh records are generally later still. Nevertheless, when the abbey's records are placed within the context of what can be determined of local, national and European trends,

[1] For detailed discussion of some of these activities, see Hodgson, 'The Landholding and Landscape Exploitation'.

'With Our Backs to the Ocean': Land, Lordship, Climate Change, and Environment in the North-West European Past. Essays in Memory of Alasdair Ross, ed. by Richard Oram (Turnhout: Brepols, 2023)
EHNAW 5, pp. 135–154 BREPOLS PUBLISHERS 10.1484/M.EHNAW-EB.5.131273

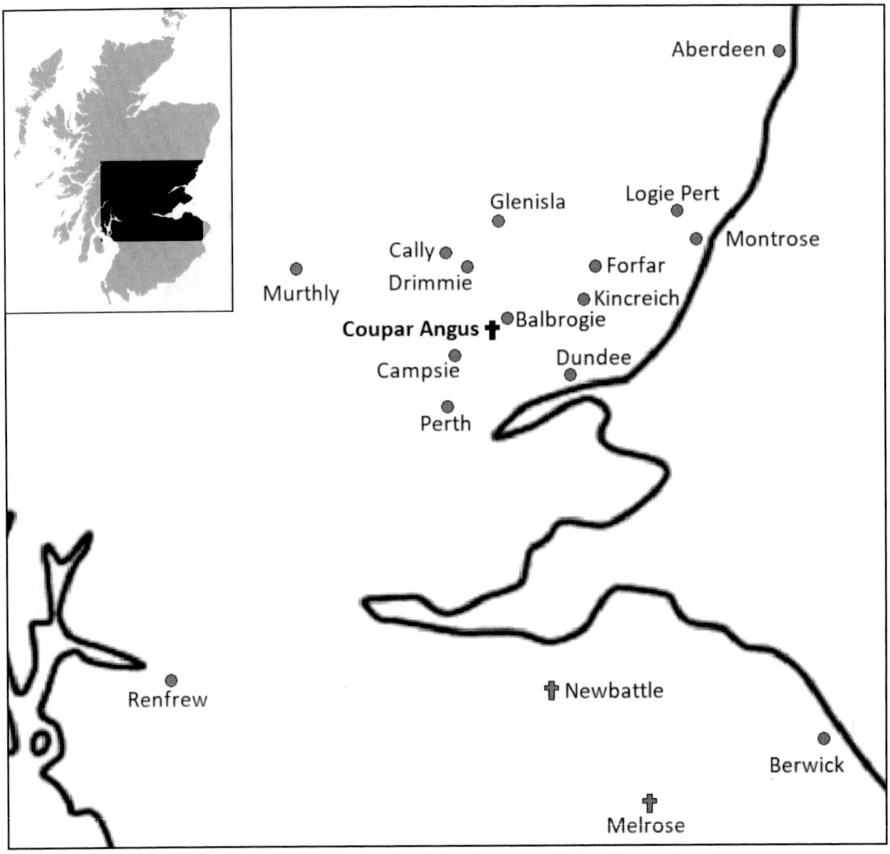

Figure 6.1. Location map of the main Coupar Angus properties discussed in the essay (image by the author).

a clearer picture begins to emerge. Two commodities may be singled out here as of particular importance, to Coupar and also the Scottish economy: wool and salmon.

The Wool Trade

The predominantly pastoral nature of medieval Scotland was reflected in its commercial interests. The national economy was highly dependent on the export of wool, the demand for which was found in the Flemish cloth industry. Cistercian houses are well known for their role in the medieval wool trade, particularly in Britain, and Coupar was no exception. The notebook of Italian merchant Francesco Balducci Pegolotti, probably dating from around the turn of the fourteenth century, records that Coupar produced thirty sacks of wool annually. This would place the flock at around 7500 sheep, working on the basis that a thousand sheep would produce four or five sacks of wool, though these figures can be considered a minimum as Pegolotti has been shown to have underestimated levels of production. We may consider the evidence for Melrose Abbey for comparison: Pegolotti records production at 50 sacks annually which would suggest that the flock numbered around 12,000, but based on later export figures it has been argued that the total was closer to 17,000 in the 1390s, or 20,000 in the late 1420s.[2]

Pegolotti's figures include prices for fifteen Scottish monasteries, graded into three levels of quality and priced separately. The highest value is assigned to Coupar's good wool; indeed, it exceeds the majority of prices listed in this class by the merchant for English and Welsh houses. Another price schedule compiled in Douai in *c.* 1270 values Coupar's wool, along with that of Melrose and Glenluce, the highest of the five Scottish monasteries on the list and this price also compares favourably with many of the English houses.[3] The textile products of Flanders during this period can be divided into two fundamental categories, the first of which was the distinct and superior class of fine coloured cloths. The other comprised a range of cheaper varieties, including special types such as says, stanforts, *biffes*, and rays, never referred to as coloured.[4] During the thirteenth century, Douai specialized in 'coloureds' and, indeed, generally commanded the highest prices amongst the elite manufacturing towns. These cloths were largely made from English wool. The category of coloureds, however, spanned a wide range of qualities and prices; while unable to compete with the finest English wool, it is possible that Coupar's best product was utilized in the lower grades of fine cloth manufactured in Douai.[5]

According to Pegolotti, Coupar's poor wool was also priced highest in its category, while its middle wool came a very close second to Glenluce Abbey. Such wool was obviously unsuitable for use in fine coloureds, yet there is no

[2] It is very difficult to account for relative price fluctuations since the precise dating of Pegolotti's notebook is unknown and could be anywhere between *c.* 1275 and *c.* 1320. Pegolotti, *La Practica della Mercurata*, p. 259. Figures discussed in Duncan, *Scotland*, p. 430; Oram, 'A Fit and Ample Endowment?'; McNeill and MacQueen, *Atlas of Scottish History*, p. 251. There were approximately 24 stones of wool per sack (Gemmill and Mayhew, *Changing Values*, p. 403).

[3] Figures quoted in Munro, 'Wool-Price Schedules', p. 122.

[4] Chorley, 'Cloth Exports', pp. 350, 359.

[5] Chorley, 'Cloth Exports', pp. 359–60, 362, 373.

doubt that it too found a ready market in Flanders. The miscellaneous range of cheaper cloths represented a very substantial part of the trade, with many major centres concentrating on these types. One example was Bruges, where products such as rays were graded on quality based on whether English, Scottish, Irish or Flemish wool was used. St Omer, another of the 'grandes draperies urbaine', specialized in says while also marketing other lines such as stanforts; here, the guild regulations refer far more frequently to Scottish wool than English.[6] In the later thirteenth century, Scottish wool arriving at St Omer from Perth, Berwick, Aberdeen and Montrose was being differentiated by port of origin to be woven separately, Perth's being the most highly valued.[7]

The issue of wool quality is a topic of debate. Cistercian wool in general fetched very high market prices and appears to have been considered an inherently better product than that of their competitors. Breed, climate, and quality of pasture all affected the standard of wool produced. Superior breeding techniques asserted to have been conducted by Cistercian houses are often cited, but the extant monastic records are silent on this topic and the tracing of breeds back to the medieval period is problematic. It is possible that Coupar imported breeds from south of the border, perhaps from Rievaulx Abbey in Yorkshire through their filiation link. However, while England exported wool considered to be of the finest quality in Europe, much inferior wool was also produced, and it is not clear if breed was the all-important determinant.[8]

In terms of good quality pasture, the expansive estates of a monastery like Coupar would have allowed the operation of transhumance regimes and ensured access to both summer pasture and winter grazing.[9] Another factor may have been colour; white wool dominated the export market.[10] Not all Scottish wool was white, something evident in 1357 when legislation issued by King David II pronounced that black and dun-coloured sheep would be subject to taxation while white sheep were exempt.[11] Indeed, the majority of medieval wool remains which have been excavated in Perth and Aberdeen are

[6] Chorley, 'Cloth Exports', pp. 362–63, 373 n. 49.

[7] Stevenson, 'Trade Between Scotland and the Low Countries', pp. 19–20.

[8] Munro, 'Wool-Price Schedules', p. 118; Ryder, 'Medieval Sheep and Wool Types'. The introduction of new stock into areas by monasteries is noted by Ryder at p. 24.

[9] Transhumance regimes, or the seasonal movement of livestock, in Scotland are discussed in Ross, *Assessing the Impact of Past Grazing Regimes*.

[10] Ditchburn, *Scotland and Europe*, p. 163.

[11] *RPS*, 1357/11/2.

coloured wools, chiefly grey, leading M. L. Ryder to suggest that these types predominated among Scottish flocks. Perhaps, then, monasteries like Coupar held a controlling interest in the trade of white wool, though it is unlikely that their stock was exclusively so. Even at an abbey as specialized in wool production as Fountains, a percentage of the clip was black, grey, and brown.[12]

But there may have been another, more significant, factor in raising the value of Cistercian wool. Unconvinced by arguments which cast the Cistercians as expert breeders, R. A. Donkin highlighted their pre-eminence in the preparation of wool for sale, something also emphasized by others.[13] The highly-skilled, costly and time-consuming processes of cleansing, sorting, grading and packing were carried out in-house, meaning that buyers could have confidence in what they were purchasing and the condition it would arrive in.[14] Moreover, Cistercian houses regularly exported wool produced by other local flocks, known as *collecta*, alongside their own, acting as middlemen between smaller lay producers and merchants. While this was priced lower than the best Cistercian wool, the English evidence reveals that *collecta* was often valued at higher rates than middle-grade abbey wool. It has been argued convincingly that it was the specialist dressing of this wool that gave the monks the ability to broker these amounts; merchants were 'brand aware' and had faith in the quality of lay product supplied under a Cistercian house's name.[15]

Routes to Market

By the early thirteenth century, Coupar held property in the nearby burghs of Perth, Forfar and Dundee, and further afield in Renfrew.[16] A further property in Berwick had also been granted to the abbey, but was transferred shortly afterwards into the possession of the more conveniently located Newbattle Abbey.[17] We may suggest that Coupar's thirteenth-century export interests operated out

[12] Ryder, 'Medieval Sheep and Wool Types', pp. 24–26.

[13] Donkin, 'Cistercian Sheep-Farming'; Bell, Brooks, and Dryburgh, *The English Wool Market*, pp. 134–35; Lloyd, *The Movement of Wool Prices in Medieval England*, p. 10.

[14] The process is described in Eckenode, 'The English Cistercians', pp. 258–59.

[15] Bell, Brookes, and Dryburgh, *The English Wool Market*, pp. 51–52, 148.

[16] Somerville, ed., *Scotia Pontificia*, no. 163; *Rental Book of the Cistercian Abbey of Cupar-Angus*, I, pp. 336, 350; Turnbull, ed., *The Chartularies of Balmerino and Lindores*, no. 31.

[17] *Registrum Sancte Marie de Neubotle*, no. 190. Property in Montrose was leased in 1304 × 1305 but the date of its acquisition is unknown (*Charters of the Abbey of Coupar Angus*, I, no. LXXV).

of Perth based on the only surviving reference to independent shipment of goods by the abbey. On 11 April 1225, Coupar was granted a licence by King Henry III of England to send a vessel to Flanders laden with wool and other merchandise.[18] The ship was stated to be in the charge of Robert of Perth and a monk, Gilbert Faber. It seems likely that we can identify the former as Robert Faber, burgess of Perth, who appears as a charter witness in 1219, and the two thus appear to have been related.[19] The high status of wool arriving from Perth at St Omer perhaps suggests that a monopoly had been established by a producer of consistently high quality wool, at least in a Scottish context. A. A. M. Duncan has made the very plausible suggestion that Coupar was in control of a collection centre at Perth for their own and other Cistercian houses' wool.[20] Furthermore, if Coupar was engaging extensively in the purchase and sale of *collecta* then it may be that the vast majority of wool being exported through Perth during the thirteenth century was being subjected to expert preparation and marketed under Coupar's 'brand'. As a result, wool produced in this locality may have enjoyed some repute abroad; contracts for *collecta* wool frequently specified the area from which the product must be supplied and places like Yorkshire, for example, commanded distinctly high prices.[21]

It has also been argued, however, that the lack of evidence for Coupar trading on its own behalf after 1225 may indicate that Melrose took charge of exporting Coupar's wool along with its own.[22] There is little reason, however, to suppose that this was the case. Alexander Stevenson has argued that, up until the end of the thirteenth century, the Scots primarily favoured a policy of allowing foreign merchants to come to them and take the risks involved in the seaborne transport of goods.[23] That Coupar was actively involved in 1225 was part of a wider response to a situation caused by Anglo-French hostilities after the Truce of Chinon expired on 14 April 1224. Talks to renew it were broken off on 5 May, and Flemish goods were seized in England that September. It was not, however,

[18] *Calendar of Documents Relating to Scotland*, I, no. 904.

[19] *Liber Ecclesie de Scon*, no. 82.

[20] Duncan, *Scotland*, p. 513.

[21] Lloyd, *The English Wool Trade*, pp. 294–95; Bell, Brookes, and Dryburgh, *The English Wool Market*, p. 52.

[22] Stevenson, 'The Monastic Presence', pp. 110–11. Stevenson also cites the lease of Coupar's only property in Berwick, which she argues was the primary port used by Scottish monastic exporters.

[23] Stevenson, 'Trade Between Scotland and the Low Countries', pp. 16–18.

the intention of the English to block the wool trade, and so the problem faced by the merchants and monasteries was thus one of access. Coupar along with Melrose were part of a long list of monasteries, which includes thirteen English Cistercian houses, who received licences to ship wool between June 1224 and July 1225.[24] Disruption to the usual shipping patterns, therefore, had resulted in the monks of Coupar taking charge of this themselves, but there is no reason why they would not revert once normal service could resume and the export of wool by foreign merchants certainly continued. When the goods of French merchants were seized at Dunwich and Yarmouth in 1242, they included forty-one and twenty-eight sacks of Scottish wool, respectively.[25]

In order to facilitate the sale of its wool, Coupar regularly sent representatives to Troyes to attend the annual summer trade event held there. This information is revealed indirectly by the arrangement which was put in place for Coupar to convey pension payments owed from the parish church of Airlie to the abbey of Cîteaux in support of the hosting of the annual General Chapter. In January 1220, it was agreed that the annual payment would be made on the feast of the Apostles Peter and Paul (29 June) at the fair of Troyes.[26] This procedure was repeated in 1246 when a dispute which had occurred over ownership of the church was settled.[27] The 'hot' or 'warm' fair of Troyes was one of the six Champagne fairs, beginning on the first Tuesday after the fortnight of St John's Day (24 June) and typically lasting for fifty-two days.[28] The Champagne fairs were at the peak of their importance during the thirteenth century, functioning as the key trading hubs for merchants and merchandise from both northern and southern Europe; they were particularly important centres for the cloth trade, ten days of each fair being officially devoted to it. Indeed, R. D. Face argues that the entire schedule of business conducted at the fairs was geared towards simplifying operations for northern cloth merchants.[29] Represented at Troyes, then, was Coupar's core market for the disposal of raw wool. Attendance pro-

[24] Lloyd, *The English Wool Trade*, p. 16.

[25] Duncan, *Scotland*, p. 514.

[26] Wilson, 'Charter of the Abbot', p. 173; Wilson, 'Original Charters of the Abbey', p. 273.

[27] *Charters of Coupar Angus*, I, no. LI.

[28] Face, 'Techniques of Business', p. 427 n. 2; Armstrong, Elbl, and Elbl, eds, *Money, Markets and Trade*, p. 332.

[29] Face, 'Techniques of Business in Trade'; Hunt and Murray, *A History of Business*, pp. 28–29; Pirenne, *Economic and Social History*, pp. 100–03.

vided the opportunity to secure buyers, draw up contracts, arrange shipments and conduct the associated monetary transactions.

The argument could be made that such an arrangement reflects considerations beyond Coupar's own: during the thirteenth century, Order representatives were sent to the Champagne fairs during periods of Cistercian taxation to collect contributions.[30] But, while not extraordinary, it certainly seems that the remission of this type of payment to a location other than the General Chapter meeting itself was certainly atypical; this was the directive issued to other houses who were responsible for transmitting such sums.[31] It was also the case in thirteenth-century charters which record various grants made by Irish kings which stipulated that the money was to be transmitted by Irish abbots to Cîteaux at the time of the General Chapter.[32] This is particularly significant considering that, like Scottish houses, Irish abbots were only required to attend the meeting every four years, yet no alternative arrangement was made for the delivery of what were stated to be annual payments.[33] The stipulation that Coupar make payment at Troyes, then, cannot be considered standard protocol and must reflect the realities of Coupar's activities; there was evidently an expectation that members of the abbey, or their procurators at least, would consistently be present at the summer fair at Troyes, and presumably more regularly than they were expected to be in attendance at the General Chapter.

Coupar continued to attend the Troyes fair into the late thirteenth century, as indicated by a receipt dated July 1287 which records transmission of the pension payment.[34] At this time, Coupar's, and indeed Scotland's, commercial prospects were bright: Bruce Campbell estimates that *c.* 1290, on a per capita basis, Scotland's export trade in wool was on par with or outperforming England's.[35] The fourteenth century and beyond would see significant changes for the abbey and the nature of the European market. The outbreak of war with

[30] King, 'Coupar Angus and Cîteaux', p. 54; King, *The Finances of the Cistercian Order*, pp. 90–91.

[31] For example, see *Statuta Capitulorum Generalium*, I, *s.a.* 1206 no. 76, II, *s.a.* 1247 no. 27. Aside from Coupar's arrangement, I have been unable to find any examples which stipulate payment at a location other than the General Chapter.

[32] Flanagan, 'Irish Royal Charters and the Cistercian Order', nos 12–15.

[33] *Statuta Capitulorum Generalium*, I, *s.a.* 1157 no. 62; Waddell., ed., *Twelfth-Century Statutes*, *s.a.* 1190 no. 1.

[34] King, 'Coupar Angus and Cîteaux', pp. 54, 58 no. 5.

[35] Campbell, 'Benchmarking Medieval Economic Development', p. 933.

England in 1296 had serious implications for Coupar's agricultural and economic pursuits, not least in terms of diminished production levels. The effects of war were also felt in the impact on Scottish trade. In addition to the obvious obstruction of military occupation, since medieval sea-travel was largely coastal, ships sailing between Scotland and the continent followed the English coastline and customarily put into English ports.[36] The monks of Coupar were now dependent on the good favour of the English king for foreign travel and their export activities. On 16 July 1297, Brother John of Coupar was issued with a safe conduct to go beyond seas on the abbot's business, no doubt on his way to Troyes.[37] And the hostility between Scotland and England was far from the only obstruction to Coupar's continental trade in the fourteenth century. The abbey's available export avenues were also greatly restricted by concurrent Anglo-French and Franco-Flemish wars which massively disrupted established trade routes.[38]

The monks of Coupar initially responded by turning to the commercial networks utilized by English Cistercian houses, likely facilitated through Melrose Abbey which had pre-existing trade links in England, and perhaps Rievaulx Abbey. By the later thirteenth century, Italian merchants had come to dominate the English monastic wool market, and it is to one such merchant house that a debt of Coupar of 180 marks (£120) is recorded in 1306, when an order to arrest the goods of the merchants of the Pulci-Rembertini of Florence was issued by Edward I.[39] By the end of the thirteenth century, English Cistercian houses were heavily involved in a cycle of advance wool contracts and indebtedness to Italian societies: in 1282 × 1283, Kirkstall Abbey is recorded as owing 670 marks to this particular merchant house. Coupar had clearly become involved in such transactions, as had Melrose which is recorded with a lesser debt of 130 marks in 1306. It may be that the sums represent loans taken out by the houses, with wool used as surety, or even advance payments made for contracted wool. It is significant that the arrest order was directed to the sheriffs of Lincoln, York and Northumberland; the principal collection centres for wool were located at Boston, in Lincolnshire, and at York, while, further north, Cistercian abbeys such as Holm Cultram and Newminster delivered to Newcastle. Boston was the pre-eminent centre, attracting wool producers from

[36] Stevenson, 'Trade Between Scotland and the Low Countries', pp. 165–66.

[37] *Calendar of Documents Relating to Scotland*, I, no. 961.

[38] Munro, 'Medieval Woollens', pp. 235–36.

[39] *Calendar of Close Rolls*, V, no. 426.

great distances due to St Bodulph's Fair, an internationally important trading event at which Melrose Abbey was active in the early thirteenth century.[40]

If the Scottish abbeys were represented at the Boston fair in the early fourteenth century, however, it is unlikely that this continued for long. By then, the great provincial English fairs had already entered a decline which would only accelerate; in the 1320s and 1330s, average wool exports from Boston were half what they had been in the later thirteenth century.[41] There is one documented instance where Coupar was present at the fair of Troyes during this period: in July 1320, 100 *livres tournois* were paid as the equivalent of £20 for the Airlie pension.[42] The use of French currency would seem to indicate that the abbey was still conducting business transactions at the fair at this date. On this occasion, the monks of Coupar present are named as John de Breneciro, John 'Clonkerdim' (Cloquhat), and William de Pilmore. It is noteworthy that the latter was almost certainly a member of a Dundee burgess family, as another contemporary monk of Coupar who shared this surname was.[43] Moreover, the Pilmore family appear to have had links to the wool trade: in 1292, Roger de Pilmore appeared as cautioner for a debt of seventeen sacks of wool.[44]

The involvement of William de Pilmore in the abbey's trading affairs mirrors the involvement of Gilbert Faber of Perth a century previously. By 1320, Dundee had superseded Perth in terms of trade. The earliest available customs figures record that, between February 1327 and June 1330, 2658 sacks of wool were customed at Dundee, while the figure at Perth was 794. The figures for woolfells at Dundee are incomplete for this period, but even the available numbers are over double those for Perth.[45] Coupar's mercantile base seems to have shifted accordingly. Of only six monks whose personal names appear on

[40] Bell, Brooks, and Dryburgh, *The English Wool Market*, pp. 21–24, 29 n. 79, 54–58, 146, 250–51; Fawcett and Oram, *Melrose Abbey*, pp. 250–51.

[41] Quinton, 'The Drapers and the Drapery Trade', pp. 103–04.

[42] King, 'Coupar Angus and Cîteaux', p. 58 no. 6.

[43] John de Pilmore, monk of Coupar, was the uncle of another John de Pilmore, bishop of Moray from 1326–1362 (*Registrum Episcopatus Moraviensis*, no. 117; Watt and Murray, *Fasti Ecclesiae Scoticanae*, p. 279). Robert Keith (d. 1757) refers to a charter in the possession of the antiquarian Walter Macfarlane (d. 1767) which records that Bishop John was the son of Adam de Pilmore, burgess of Dundee (Keith, *An Historical Catalogue of the Scottish Bishops*, p. 140).

[44] Maule, *Registrum de Panmure*, II, pp. 151–52.

[45] Stevenson, 'Trade Between Scotland the Low Countries', p. 249; Rorke, 'English and Scottish Overseas', pp. 110, 381, 463. The shift of export activity from Perth to Dundee was part of a long-term trend which continued from the fourteenth to sixteenth centuries.

record in the first half of the fourteenth century, four of these men hailed from Dundee.[46] The far more extensive available evidence for the later period indicates that recruits were being drawn in sizeable numbers from both Dundee and Perth.[47] Such men provided direct links to urban communities, affording intimate access to networks which could be utilized to the benefit of the abbey's commercial interests. Moreover, as a product of their background, the monks themselves likely had valuable skills and experience; such credentials may have been what caused men like Gilbert Faber and William de Pilmore to be selected to represent the abbey in business matters far beyond the precinct.

1320 would appear to be the final year the abbey attended the Troyes fair. A major impact of the chronic, economically debilitating European warfare of the fourteenth century was the decline in the importance of the Champagne fairs.[48] For Coupar, the absence of Flemish merchants meant that their principal market was no longer represented. In 1324, payment to the abbot of Cîteaux on Coupar's behalf was made by Guidonis de Alacrimonte, monk of the French abbey of Preuilly.[49] The money was delivered to the Cistercian college of St Bernard in Paris, which was often used by the Order as a financial collection centre in the later Middle Ages.[50] The location of the payment made in 1327 by Richard, who was a monk of Coupar, is not recorded but given that it is dated at the octave of the Purification of the Blessed Virgin Mary (February) it was not made at Troyes. In 1329, it was stipulated that payment was to be made in that year to the abbot of Stratford Langthorn in England, and thereafter at the General Chapter, despite the fact that, in the same charter, the abbot of Coupar was excused from the Chapter for the next six years.[51]

In 1350, however, it was pronounced that payment of the pension was now to be made to the abbot of Ter Doest, itself a producer of wool, near Bruges.[52]

[46] Apart from John and William Pilmore, Walter de Dundee appears in 1326 and Nicholas of Dundee in 1344 (*Charters of the Abbey of Coupar Angus*, II, no. CIX; King, 'Coupar Angus and Cîteaux', p. 60 no. 10).

[47] Hodgson, 'Cistercian Abbey', pp. 160–68.

[48] Munro, 'Medieval Woollens', pp. 235–36.

[49] King, 'Coupar Angus and Cîteaux', pp. 58–59 no. 7.

[50] King, 'Coupar Angus and Cîteaux', p. 54; King, *The Finances of the Cistercian Order*, p. 111.

[51] King, 'Coupar Angus and Cîteaux', pp. 59–60 nos 8 and 9.

[52] King, 'Coupar Angus and Cîteaux', pp. 61–62 no. 12. A 1315 account of Ter Doest records an income of 877 *livres* in local currency from wool sales and a further 257 *livres* from sheep sales. It has been estimated that the abbey flock at this date numbered at least 3000 (Williams, *The Cistercians in the Early Middle Ages*, p. 359; Verhulst, 'De inlandse wol', p. 13).

This would suggest that, by this date, representatives of Coupar were considered more likely to be present in Bruges on a regular basis than anywhere else, including at the General Chapter. It has been argued that Ter Doest, along with its motherhouse, Ten Duinen, had a close personal and economic relationship with Melrose Abbey, and it seems that this was also the case for Coupar.[53] The Scottish commercial presence was well-established at Bruges by the later thirteenth century, and by the mid fourteenth century at least, and perhaps earlier, it had become the staple port for Scottish wool.[54] The revised arrangement for the Airlie pension, then, seems to confirm that the monks of Coupar were now conducting their commercial transactions directly through Bruges. There is every reason to believe that the monastic trade in wool continued to thrive in the following few decades. Between 1362 and 1375, Scottish wool exports reached their highest-ever recorded levels, peaking in the early 1370s when nearly 9500 sacks were being exported annually.[55]

The Decline of the Scottish Wool Trade

This situation was not to last, however, and shortly afterwards the medieval Scottish wool trade entered a long and sustained decline from which it would not recover. From 1375 through to the mid-fifteenth century, Scottish wool exports fell dramatically, by over two-thirds by the 1460s. This was a period of general contraction in the European economy, characterized by severe trading slumps and monetary crises. The Low Countries' textile industry underwent a restructuring to focus on high quality cloth; at the same time it appears that Scottish wool was increasingly considered a much inferior product, evidenced by a great slump in value as compared to Flemish and Zealand wools and various bans on its use. For the remaining century up to the Scottish Reformation in 1560 this sharp decline was halted, largely due to great expansion in the export of woolfells, but an overall gradual descent continued.[56]

[53] Coomans, 'From Flanders to Scotland', p. 249.

[54] Stevenson, 'Trade Between Scotland the Low Countries', pp. 11, 17–18, 158–59, 253.

[55] Rorke, 'Scottish Overseas Trade', pp. 303–04. It is important to note that Scottish customs records do not survive until the early 1300s so a comparison with the previous century is not possible.

[56] Rorke, 'English and Scottish Overseas Trade', pp. 270–71; Rorke, 'Scottish Overseas Trade', pp. 304–10, 621; Stevenson, 'Trade Between Scotland and the Low Countries', p. 21.

Alexander Stevenson has argued that the scale of decline cannot be satisfactorily explained simply by a shrinking cloth market. Instead, he points to environmental factors which may have caused the general deterioration of the quality of Scottish wool. The climatic downturn which followed the mid thirteenth-century waning of the Medieval Climate Anomaly became established by the fourteenth century and continued for the rest of the medieval period in what is referred to as the 'little ice age'. This episode was marked by the cooling and wettening of the climate, which it is suggested would have increased the number of faults in the wool such as matting, discolouration, and even fungal infection. A further contributing factor may have been the resultant reduction in biomass which increased pressure on grass resources, affecting both summer pasture and supplies of winter fodder, and placed winter grazing under strain as it came under use for longer periods each year.[57] By the later medieval period, Coupar's estates may no longer have been capable of sustaining the numbers of animals it had done previously. Access to common pasture was also likely reduced due to local reassessment of souming levels.[58] Indeed, the fifteenth and sixteenth century evidence reveals that Coupar was involved in disputes with neighbouring landowners regarding rights to pasture.[59]

Regardless, however, flocks were still kept on Coupar's lands. In the mid-fifteenth century, the abbey's tenant at Pitlochrie in Glenisla was required to provide pasture for twenty score (400) sheep, while sixteenth-century leases refer to the *schyphird* (shepherd) land of nearby Dalvanie which supported fourteen score (280) sheep. Here, the tenants were instructed to provide hay and fodder, and given charge of the 'hyrdin and keping of thame to the said abbottis profeit'.[60] And there was still a foreign market for Scottish wool during this period. Coarse wool and fells supplied the light cloth industries of western Flanders at Diksmuide, Poperinge and Tourcoing; from 1497–1515, Scottish and Newcastle wool dominated this market. Moreover, some fine Scottish wool was still supplied to Bruges.[61] Another market was found in Holland, at Gouda

[57] Stevenson, 'Trade Between Scotland and the Low Countries', pp. 21–25; Oram and Adderley, 'Lordship and Environmental Change'; Oram, ed., *The Lordship of the Isles*, pp. 55–59.

[58] Souming was the division of pasture into units which supported a finite number of livestock. See Ross, 'Scottish Environmental History'.

[59] *Charters of the Abbey of Coupar Angus*, II, nos CLVI, CXXXII, CXXXIII; *Acts of the Lords of Council, II*, pp. 424, 438; *Acts of the Lords of Council, III*, pp. 5, 13.

[60] *Rental Book of the Cistercian Abbey of Cupar-Angus*, I, p. 119, II, pp. 236, 260–62, 267–68.

[61] Blanchard, 'Northern Wools and Netherlands Markets'.

and Haarlem.[62] It seems likely that Coupar's wool did still make up some percentage of the much-reduced Scottish wool trade.

The Cloth Market

Other of Coupar's wool seems to have gone to the local production of cloth. Scotland was never a major exporter of cloth, nor did it become one in the way that England did after the collapse in the raw wool trade.[63] Nevertheless, there may have been a reasonable level of domestic trade in such products. Scottish burgh legislation demonstrates that coarse cloth was being manufactured in the countryside by the twelfth century onwards.[64] As Alexander Grant has pointed out, in the 1370s Scotland was able to increase exports by 2000 sacks almost overnight in response to demand created by the restricted availability of English wool; in normal circumstances, this excess must have been absorbed by the domestic cloth industry.[65] To an extent, the cloth manufactured by Coupar will have been utilized internally rather than traded. Based on the account of the keeper of the wardrobe at Beaulieu Abbey in 1269–1270, John Oldland has calculated that annual cloth requirements for clothing were 12,565 yards (at two yards wide) for monks, and 7535 yards for lay brothers.[66] Beaulieu's fulling mill and cloth store was situated just beyond the abbey complex; a very similar setup was found at Coupar where a fulling mill stood at the home grange of Keithick which bordered the precinct.[67] The abbey also possessed another fulling mill, however, at Kincreich grange, erected by the mid-thirteenth century and still in operation in the fifteenth and sixteenth centuries.[68] Kincreich was Coupar's most distant grange and it seems unlikely that it also manufactured products intended for internal use. Instead, low quality wool may have been woven into cheap cloth here to be sold at the nearby burgh of Forfar.

[62] Ditchburn, *Scotland and Europe*, p. 173.

[63] Rorke, 'English and Scottish Overseas Trade', pp. 275–76; Ditchburn, *Scotland and Europe*, pp. 182–83.

[64] Gemmill and Mayhew, *Changing Values in Medieval Scotland*, p. 350.

[65] Grant, *Independence and Nationhood*, p. 79.

[66] Oldland, 'Cistercian Clothing and Its Production', pp. 75, 79–80.

[67] *Rental Book of the Cistercian Abbey of Cupar-Angus*, I, pp. 126, 161, 187–88, 204, 214.

[68] *Charters of the Abbey of Coupar Angus*, I, no. LX; *Rental Book of the Cistercian Abbey of Cupar-Angus*, I, pp. 141, 145, 148; II, 81–82, 82, 201.

Nevertheless, the volume of wool involved must have been a meagre amount as compared to the abbey's exports at their peak.

The Salmon Trade

Noting that nothing compensated for the decline in wool, Martin Rorke has concluded that the 'story of Scottish overseas trade from 1300–1600, therefore, is one of a failure to diversify successfully'.[69] While this may be broadly true, use of the customs records to chart export patterns is problematic for many other types of goods. One such commodity, and one which appears to have been of great importance to Coupar, was salmon. Total figures for salmon exports at any burgh are not available until the fifteenth century, and national figures not until the 1470s, preventing comparison with the earlier period. Moreover, the figures are incomplete for the period 1482–1537 when the burgesses of Aberdeen, Scotland's most important salmon exporting burgh, held a customs exemption, and continue to be recorded inconsistently after this date.[70] What is apparent from the patchy data is that between the 1470s/80s and the 1530s/40s there was a great increase in salmon exports. Yearly averages for Aberdeen rose from 1200 barrels to over 2000. During this period, Aberdeen's pre-eminence continued but was reduced.[71] Of interest here is that average numbers of barrels of salmon customed at Perth rose from 275 to 430, and at Dundee from 159 to 537.[72] These burghs continued to be the hubs of Coupar's commercial activity in the later medieval period. The fifteenth- and sixteenth-century rental records reveal that significant investment in property had been made by the abbey in these burghs. Coupar maintained a permanent *hospitium* in both, complete with resident tenants under strict instructions to be ready to receive the monks at all times, cellars available for the storage of goods, both within the residences themselves and elsewhere in the burgh, and a further 'geir lodging' in Perth.[73]

[69] Rorke, 'Scottish Overseas Trade', p. 317.

[70] Rorke, 'Scottish Overseas Trade', pp. 182–89, 194–95.

[71] Rorke, 'Scottish Overseas Trade', p. 195. These figures may be compared to the collective totals for all burghs south of Aberdeen: 630 barrels exported per year *c.* 1475 to 1482, increasing to 1700 barrels per year in 1537 to 1542.

[72] Rorke, 'Scottish Overseas Trade', pp. 383, 386, 465–66, 468. Based on the figures for 10 July 1472 to 4 September 1482 and 21 August 1537 to 12 July 1542 at Perth, and 30 June 1473 to 24 November 1482 and 27 August 1537 to 13 July 1542 at Dundee. It should be noted that goods customed at Perth were frequently then shipped out of Dundee.

[73] *Rental Book of the Cistercian Abbey of Cupar-Angus*, I, pp. 145, 147, II, pp. 64–65,

There is no direct documentation for the monks of Coupar trading in salmon, though that means little considering the sparsity of evidence generally. There is no doubt that the abbey had the resources to do so. The initial twelfth-century endowment of the abbey included the right to fish in both the Rivers Ericht and Isla and salmon fisheries were established on both rivers on the bounds of Coupargrange.[74] In the twelfth and early thirteenth centuries, the abbey established a net, two cruives and a yair on the Tay, courtesy of grants from the Hay family.[75] The monks added fishing in the North Esk to their portfolio around the turn of the fourteenth century when they acquired the land of Logie Pert.[76] From 1326, the abbey also benefitted from the right to catch salmon in the close season in all four of these major rivers thanks to a grant of King Robert I.[77] By the fifteenth century, Coupar had established fisheries at Campsie, Balbrogie, Drimmie and Cally, on the Tay, Isla, Ericht and Ardle (a tributary of the Ericht) respectively.[78] At this time, the abbey was also actively promoting the establishment of new fisheries: tenants of Murthly in Atholl were instructed to establish a *schot* (a place from which nets are shot) for salmon fishing in the Tay, while the fishings of Campsie were let 'as well new as old, and those that may yet be found'.[79]

Fish was a key part of the monastic diet and all houses sought access to an adequate supply. Coupar's extensive portfolio of fisheries, however, must have generated catches far exceeding internal requirements, and it seems clear that they were utilized for commercial purposes. The evidence would indicate that Coupar was taking the protection of these fisheries very seriously indeed. Particularly stringent control was maintained at Campsie. Two fifteenth-cen-

205–06; Morgan, 'Economic Administration', II, p. 234; National Records of Scotland, CH 6/2/4.

[74] Barrow, ed., *The Acts of Malcolm IV*, no. 226; *Rotuli Scaccarii regum Scotorum*, XXIII, p. 464. The estate which became Coupargrange, situated within the confluence of the Rivers Ericht and Isla, had been intended by King Malcolm IV as the site of the monastic precinct. Instead, the abbey came to be constructed to the south of the Isla where the modern village of Coupar Angus now stands.

[75] Stuart, 'The Erroll Papers', no. 7; *Rental Book of the Cistercian Abbey of Cupar-Angus*, I, pp. 340, 342; *Charters of the Abbey of Coupar Angus*, I, no. LXXXII.

[76] *Charters of the Abbey of Coupar Angus*, I, nos LXVII, LXVIII.

[77] Duncan, ed., *The Acts of Robert I*, no. 298.

[78] *Rental Book of the Cistercian Abbey of Cupar-Angus*, I, pp. 124–25, 222, 303–04, II, 166–67, 187.

[79] *Rental Book of the Cistercian Abbey of Cupar-Angus*, I, pp. 169–70, 222, 231–32, 243.

tury leases stipulated an annual quota of thirty dozen salmon per tenant, of which four were named in each instance. The teind sheaves ('garbal' or grain tithes) of the land belonged to the tenants in return for working the fishings, but the abbey appointed its own fishermen who were stated to be at the command of the abbot. Detailed instructions were issued regarding the net and strict penalties applied should any fault be found with it; the nearby tenant of Blair was instructed to superintend the fishings at Campsie and 'warne ws lawtefully quhen that he knawis any falt with the fissaris' (warn us faithfully when he knows of any offence with the fishers).[80]

Furthermore, from the 1440s Coupar began to become embroiled in disputes, usually as the aggressive party, with neighbouring lay landholders regarding fishing rights. These included a clash over fishing in the Ericht at Drimmie, and another in the North Esk, where the abbey successfully pressed its right to fish in the close season while also being warned to remain within the bounds of Logie Pert at all other times.[81] Another dispute began in 1518, this time over fishing in the Isla where, following the death of Alexander Cuming of Couttie at Flodden, the abbey had taken the opportunity to seize control of the Couttie fishery. When the abbot, cellarer, and prior were summoned to the sheriff court in Perth to answer for their actions, Coupar's defence was to cite a twelfth-century charter of King Malcolm IV, apparently referring to the abbey's foundation charter and the rights to fish in the Isla that it conferred. The issue rumbled on until 1535.[82] This situation was mirrored elsewhere in Scotland; the later fourteenth century saw the commencement of a long and bitter dispute between the Augustinians of Cambuskenneth Abbey and the inhabitants of the burgh of Stirling over salmon fisheries on the River Forth. Cambuskenneth was similarly guilty of the illegal occupation of burgh fisheries and even encouraged poaching by others.[83]

These conflicts are indicative of just how valuable a commodity these fish had become. Freshwater fish were a luxury item afforded by only the privileged few and were therefore used in demonstrations of aristocratic status, served at feasts and presented as gifts. Salmon were highly prized due to their cultural

[80] *Rental Book of the Cistercian Abbey of Cupar-Angus*, I, pp. 121, 127, 220–22, 227, 237, 274–75, II, 181.

[81] *Charters of the Abbey of Coupar Angus*, II, nos CXXXII, CXXXIII, CXXXIV.

[82] *Charters of the Abbey of Coupar Angus*, II, nos CLXII, CLXIII; Ramsay, ed., *Bamff Charters*, no. 46.

[83] Hoffmann and Ross, 'This Belongs to Us!'.

and social significance, generating elite demand and fetching high prices.[84] Records attest to the medieval domestic trade in Scotland with prices per fresh fish dictated by both size and availability.[85] Across Europe, however, the impact of human activity had taken its toll on riverine fish populations. Analysis of fish bone evidence indicates that from *c.* 1000 fishing catches went from being overwhelmingly comprised of freshwater and migratory species to being dominated by marine fish such as gadids, a fundamental change which has been labelled the 'fish event horizon'. This dramatic shift has been attributed to the damage caused to river systems by rising populations, land clearance, agricultural production and, in particular, milling.[86] Migratory species went into rapid decline; the deterioration of salmon stocks by the thirteenth century on both the European mainland and in England is well-documented.[87] Further research has suggested that the event horizon was a two-stage process, initially based upon the intensification of local marine fisheries but later, as demand outstripped local supply, upon long distance trade; thus, commercial fishing took over from subsistence fishing.[88]

But in sharp contrast to the fate of riverine fish populations elsewhere in Europe, Richard Hoffmann has identified that the extant Scottish evidence shows high economic return but no reduction in yields from Scotland's salmon rivers.[89] This 'sustainable abundance' can be attributed to a combination of factors. Royal legislation which placed private fisheries under public regulation was established by around 1200, perhaps the earliest of its type in Europe, putting measures in place which protected salmon stocks. Other developments also inadvertently aided their sustainability, for example, Scottish mills tended to be erected upon artificial streams rather than upon major rivers, leaving the main stream open to migrant fish species.[90] It is likely that it was the density

[84] Dyer, 'The Consumption of Fresh-Water Fish'; Van Dam, 'Fish for Feast and Fast'.

[85] Gemmill and Mayhew, *Changing Values in Medieval Scotland*, pp. 303–04; Hoffmann, '*Salmo salar*', p. 361.

[86] For example, see: Enghoff, 'Fishing in the Southern North Sea Region'; Barrett, Locker, and Roberts, '"Dark Age Economics" Revisited'; Hoffmann, 'Economic Development'.

[87] Hoffmann, '*Salmo salar*', p. 356.

[88] Barrett and others, 'Detecting the Medieval Cod Trade'; Barrett and others, 'Interpreting the Expansion of Sea Fishing'; Taylor, 'Problems and Possibilities'.

[89] The initial basis for much of the following discussion was research undertaken in 2011 by Richard Hoffmann and Alasdair Ross as part of the project *Inland Fishings in Medieval Scotland*.

[90] Hoffmann, '*Salmo salar*'. See also Hoffmann's paper in this volume for discussion of

of Scottish river systems which rendered the development of complex systems of fishponds and breeding techniques, evident at English Cistercian houses, unnecessary.[91]

Thus, the destruction of other European riverine fish populations, coupled with the commercialization of fishing and the growth of an international market, saw Scotland emerge as a leading exporter of salmon from the late fourteenth century well into the sixteenth century. The demand was almost limitless and unfailing. During this period, it became the norm to set prices by the barrel, rather than per fish, and salmon became as acceptable as ready money in some circles. Salmon packaged in large Hamburg barrels were exported in their thousands through Scottish burghs to England, the Low Countries (particularly Bruges and Veere), and, slightly later, France. Markets were also found in Amsterdam, Zeeland, Hamburg, Copenhagen, and Gdańsk.[92] In the fifteenth and sixteenth centuries, the monks of Coupar were perfectly poised to take full advantage of these opportunities.

Conclusion

Rich and powerful monasteries like Coupar Angus had the resources and networks to respond to commercial opportunities and change. The evidence demonstrates that markets for trade were actively sought out. In the thirteenth and early fourteenth centuries, abbey representatives travelled to Troyes to conduct the sale of wool to European merchants. During this period, Coupar's product supplied Flemish cloth manufacture in such elite towns as Douai. The course of the fourteenth century brought change and challenges in the form of endemic warfare, at home and abroad, and its impact on export outlets. Coupar responded by identifying other avenues through which to reach its core market, turning first to Italian merchants already active in England, and then shifting focus to Bruges as the city came to dominate Scottish commercial life. Crucial

salmon stocks in early modern Europe. While the detrimental effects of the colder temperatures of the little ice age are demonstrated here, Hoffman notes that late medieval Scottish stocks were far healthier than elsewhere.

[91] Bond, 'Monastic Fisheries'; McDonnell, *Inland Fisheries*; Currie, 'The Role of Fishponds'. In Scotland, the evidence indicates that neither monks nor the laity engaged in this type of aquaculture. Singular fishponds do appear to have existed at Sweetheart and Balmerino Abbeys; these were holding ponds used to store live fish in the manner of 'live larders'.

[92] Hoffmann, '*Salmo salar*'; Gemmill and Mayhew, *Changing Values in Medieval Scotland*, pp. 303–17; Ditchburn, *Scotland and Europe*, pp. 146–47.

here was the cohesive organizational structure and transnational identity of the Cistercian Order. Strong inter-house ties meant that the monks of Coupar could call upon their English and Flemish counterparts to capitalize on trading networks elsewhere as traditional channels disintegrated. Of equal importance were the abbey's close ties to the burghs of Perth and Dundee, Coupar's domestic commercial hubs, and the merchant families who inhabited them, members of whom were to be found amongst the ranks of the monks.

The Scottish wool trade was hit hard in the fifteenth and sixteenth centuries, perhaps partly, as Stevenson argues, due to the impact of climatic deterioration. Whatever the cause, it seems clear that wool lost its position as Coupar's preeminent commercial commodity. The decline must have provided the impetus to seek out alternative export markets; the question is, what, if anything, filled the void? When we look at access to resources and available market opportunities, the obvious answer is salmon. The evidently abundant supply from Scottish rivers contrasts sharply with the decline of salmon stocks elsewhere and the consequent levels of demand on the European market. That the monks too recognized this opportunity is strongly suggested by a seemingly growing emphasis on the operation of fisheries. This is evidenced by the strict management of these sites, but more particularly by the emergence of an aggressive policy towards fishing rights. Legal disputes regarding land and access to common resources are a consistent feature throughout the abbey's history, but those relating to fishing are unique to the later period. It is certainly not proposed that Coupar underwent a straightforward transition from wool to salmon supplier, nor that salmon dominated the abbey's economic interests in the way that wool had likely done. Instead, the shift helped to soften the blow and is thus a further example of Coupar's ability to respond to market developments and, in this case, successfully diversify.

A New Theoretical Approach to the Study of Medieval Scottish Park Emergence and Resource Management

Kevin Malloy
University of Wisconsin

Introduction

The emergence and significance of Scotland's medieval park enclosures at the end of the twelfth century has remained a poorly understood and under-researched form of medieval land-management. Often considered carbon copies of contemporary English sites, Scottish parks are frequently reduced in historical discourse to simple hunting reserves owned and used by the elite members of society for that most aristocratic leisure pursuit, the hunting of deer. Notably, the processes behind their initial emergence and the functions that they performed peripheral to hunting have been largely ignored within the broader historical context of climate change, deforestation, and population expansion. This essay seeks to place the study of parks within a multi-disciplinary framework that weaves together the archaeological, historical, and environmental contexts of medieval Scotland.

To do so, in this paper I will present the wider historic, social, climatic, and environmental contexts in which medieval Scottish park enclosures emerged. From this basis, I will then propose a new theoretical model that examines the necessary conditions that made *imparking*, the process of enclosing a tract of land in order to create a park landscape, beneficial as both a social and economic process. The model will consider each of the variables that may have directly or indirectly influenced the motivations for imparking, incorporating paleoenvironmental and paleoclimatic data pertaining to the medieval period in Scotland and the North Atlantic after CE 1000 with existing archaeological and historical data to form a coherent whole.

'With Our Backs to the Ocean': Land, Lordship, Climate Change, and Environment in the North-West European Past. Essays in Memory of Alasdair Ross, ed. by Richard Oram (Turnhout: Brepols, 2023)
EHNAW 5, pp. 155–182 BREPOLS PUBLISHERS 10.1484/M.EHNAW-EB.5.131274

Background

Figure 7.1. The author standing on top of the earthen bank at Buzzart Dykes. The ditch is located on the interior (right) side of the embankment (photo by Derek Hall and courtesy of the author).

Medieval parks were enclosed tracts of land surrounded by an earthen embankment, stone wall, or combination of the two, with a ditch running along the interior face of the barrier (Fig. 7.1). Amongst contemporary English sites, a wooden fence structure or palisade lining the top of this barrier referred to as a 'pale' and meant to serve as a 'deer-proof' fence, was common.[1] The added height provided by the pale and ditch combination prevented animals from easily escaping the enclosure, while also serving as a deterrent to trespassers. Parks ranged in size from less than a hundred to thousands of acres. The similarities between Scotland and England's park construction methods have been addressed elsewhere but is a question that continues to require further archaeological investigation.[2]

Existing documentary records suggest that parks first appeared in Scotland in the late twelfth century. The first concrete textual evidence of park construction dates the reign of King William I (reigned 1165–1214) with his creation

[1] Rackham, *The History of the Countryside*, p. 122.

[2] Malloy and Hall, 'Archaeological Excavations'; Malloy and Hall, 'Medieval Hunting and Wood Management'.

of what became known as the King's Park at Stirling.[3] However, the foundation of such hunting reserves in Scotland is arguably rooted in the introduction of a formal 'forest' system during the reign of his grandfather and predecessor, King David I (reigned 1124–1153).[4] In this medieval context, the term 'forest' refers to an area of land reserved for royal hunting and placed under a specific legal regulatory regime to manage and protect that status as hunting land, rather than a wooded tract of land as the term is usually understood today.[5] While other regulated activities, such as grazing for example, were practiced in forest lands, the legal constraints revolved around hunting.[6] Many Scottish parks were created out of *forest* lands, and they have long maintained a reputation as serving as little more than exclusive hunting reserves for the aristocracy, used for sport and leisure and the entertainment of fellow noblemen, and little else.

Although it is well established that Scottish kings devoted considerable energy to the hunt, and issued numerous laws regarding the practice, medieval park landscapes typically garner little more than passing attention in wider historical research.[7] Yet, simple characterizations of medieval parks as mere aristocratic hunting enclosures are reductive, and ignore the complexity of their sophisticated plans and designs, expressions of wealth and prestige, intricate roles within the broader human-designed landscape, and most importantly, the diverse array of functions each structure had often served. Furthermore, parks came into fashion during a period when Scotland was in the midst of widespread socio-cultural and environmental changes that included but was not limited to: profound cultural reorientations, population growth, climate change, and deforestation.[8]

[3] Barrow, ed., *The Acts of William I*, no. 130.

[4] Gilbert, *Hunting*, pp. 6, 13, 82. Gilbert is one of the few scholars in the last fifty years to have undertaken detailed research into hunting, the process of afforestation and parks in medieval Scotland.

[5] Alternatively, and quite confusingly, *forest* was also used in Scotland in the French tradition to mean a wooded area of land (Gilbert, *Hunting*, p. 19). For the purposes of this study however, *forest* will refer only to a medieval hunting reserve.

[6] Neville, *Land, Law and People*, p. 46.

[7] Neville and Simpson, eds, *The Acts of Alexander III*, p. 36.

[8] See Rotherham, 'Reinterpreting Wooded Landscapes'; Mileson, *Parks in Medieval England*; Liddiard, ed., *The Medieval Park*; James and Gerard, *Clarendon*.

Other Functions of a Medieval Park

Robert Liddiard has argued that the physical act of hunting within the confines of a park, many of which were less than a hundred acres, may have been difficult.[9] Instead, parks may be viewed more accurately as venison farms or live-larders, thereby providing their owners with regular access to venison, an elite delicacy, for lavish feasts. Nevertheless, documentary and archaeological evidence suggests hunting did occur in parks.[10] While few textual references to Scottish park hunting exist, Gilbert has identified evidence for James VI hunting in Falkland Park, before 1603, likely a tradition carried on from the earlier medieval period.[11] Archaeologically, several features suggest hunting was practiced in parks, as remnants of a handful of so-called 'deer traps' have been identified throughout Scotland, and recent work by me and my colleague Derek Hall has found archaeological evidence for hunting at multiple parks in the zone along the south-eastern highland/eastern lowland interface.[12]

John Fletcher argues 'it was the deer and the association with hunting deer, that conveyed the prestige that motivated the park builder', regardless of any of the alternative functions they may have served.[13] Yet, Aleksander Pluskowski has argued that some parks were simply too small to maintain a herd of deer or even livestock and suggests that some parks were never intended to house animals.[14] Fiona Beglane's comprehensive investigation of Irish parks lends credence to Pluskowski's assertion, arguing that the management of wood was a far more significant function than any role deer played in their utilization in Ireland.[15]

In his discussion of the effects of grazing, Ian Rotherham has described the ecologies of medieval parks as reflections of the pre-existing landscapes from which they were created and 'driven by multifunctional systems of economic

[9] Liddiard, 'Introduction', p. 4.

[10] Liddiard, 'Introduction', pp. 4–5; Mileson 'The Sociology of Park Creation', p. 11; Malloy and Hall, 'Medieval Hunting and Wood Management'.

[11] Gilbert, 'Falkland Park', p. 21.

[12] Malloy and Hall, 'Medieval Hunting and Wood Management'; Malloy and Hall 'Archaeological Excavations'.

[13] Fletcher, *Gardens of Earthly Delight*, p. 9.

[14] Pluskowski, 'The Social Construction of Medieval Park Ecosystems', p. 66.

[15] Beglane, *Anglo-Norman Parks*.

utilization'.[16] He considers them to be a 'form of "pasture-woodland", related to forests, heaths, moors and some commons, with grazing animals and variable tree cover'.[17] The perspective of Melvyn Jones in his work in South Yorkshire, England, further supports this view, stating:

> The deer in medieval parks were carefully farmed. Besides their status-symbol role, the main functions of parks were to provide for their owners a reliable source of food for the table, supplies of wood and timber, and in some cases quarried stone, coal and ironstone. They were, therefore, an integral part of the local economy.[18]

In Scotland, indirect evidence from surviving textual accounts and, as we are beginning to find, the archaeological remains of these landscapes indicate this was also the case there. Functions will have varied between parks, determined by the differing goals of individual landowners, and included: secure grazing locations for livestock; wood-pasture, timber production and woodland management; fuel production; *pannage* (the autumn grazing of pigs); species management; winter fodder production; and arable agriculture. This present study is interested principally in the role of wood production, as many of these functions are related to or revolved around more general wood management activities. Nevertheless, each mode of park utilization was important in contributing to the shape of the wider landscape and is worthy of further analysis in its own right.

Wood-Management

Wood was typically managed within parks across the British Isles in one of two ways: wood-pasture or woodland.[19] In its simplest description, wood-pasture is a form of management in which trees grow intermittently and animals, including deer and cattle, graze on the 'lawns' among them.[20] The defining attribute of wood-pasture is the level of tree cover, possessing discontinuous tree spacing and a relatively sparse, open canopy.[21] A common element of English parks, wood-pasture was an important component of medieval park economies and

[16] Rotherham, 'Reinterpreting Wooded Landscapes', p. 73.

[17] Rotherham, 'Reinterpreting Wooded Landscapes', pp. 73–74.

[18] Jones, 'Woods, Trees, and Animals', p. 28.

[19] Mileson, *Parks in Medieval England*.

[20] Rackham, *Woodlands*, pp. 25–27.

[21] Rackham, 'Woodland and Wood-Pasture', p. 16.

their function.[22] Conversely, the term woodland refers specifically to islands of trees, perhaps on a small scale, and into which access was possibly restricted to prevent grazing.[23] Wood-pasture and woodland differ primarily in the level of tree density. While wood-pasture is an area with a canopy open enough to encourage pasture growth, woodlands possess too much shade for pastures to thrive.[24] Woodland canopy is closed and results in a less vegetated understory consisting of plants accustomed to shady environments, as opposed to the well-developed understories and lawns associated with wood-pastures.[25]

Although it remains unclear how wood-pasture and woodland figured within the medieval Scottish model of parks, it may be useful to understand the goals of park ownership, construction methods, and the regional historical contexts of individual parks to determine how they may have been used. It is difficult to overstate how valued wood was as a natural resource. Properly managed oakwood sources, for example, allowed for economically lucrative activities like yearly pannage to bring in extra revenue, the production of supplementary fodder for wintering one's animals to reduce annual loss of livestock, and for the production of fuelwood.[26] However, a park's wood source may have been most important in its role as a construction material. Wood was integral to medieval Scottish society as the primary building material of the time, used to manufacture the woven hurdles for wattle and daub walled houses, fences and palisades, tools, ladders, and carts, farming implements, and larger structural timbers used for the roofing and floor members of major secular and ecclesiastical buildings.[27] Despite international fascination with Scotland's stone castles, stone was employed rarely in Scottish secular elite architecture before the early thirteenth century, and even then this transition from largely timber to largely stone construction was not wholesale.[28] Some early stone castles comprised outer stone curtain walls but still possessed a combination

[22] Rackham, *Woodlands*, pp. 25–26; Rotherham, 'Reinterpreting Wooded Landscapes'; Rackham, 'Woodland and Wood-Pasture', pp. 11–14.

[23] Rackham, *Woodlands*, p. 21; Rackham, 'Woodland and Wood-Pasture', pp. 11–14.

[24] Rackham, 'Woodland and Wood-Pasture', pp. 11–14.

[25] Rackham, *Woodlands*, p. 22.

[26] Rotherham, 'Reinterpreting Wooded Landscapes', pp. 76–78; Jones, 'Woods, Trees, and Animals', p. 31.

[27] Crone and Watson, 'Sufficiency to Scarcity'.

[28] Oram, *Alexander II*, pp. 250–55. See also Oram, 'Royal and Lordly Residence', for a broader discussion.

Figure 7.2. Large coppiced oak from Dalkeith Park that most likely dates to the medieval period (photo by author).

of both internal stone and wood structures, and other aristocratic residential structures, like hunt halls, continued to utilize wood throughout the medieval period.[29] Thus, timber persisted as an important construction material in elite architecture that was never totally abandoned.

To maintain the productivity of woodland or wood-pasture *coppicing* and *pollarding* were often employed. Coppicing refers to a management practice whereby trees are cut down to near ground level and multiple new shoots grow from the stumps or stools (Fig. 7.2).[30] While not all trees (such as most conifer species) do this, many British tree species like oak, birch, ash, and elm respond well to this type of management.[31] Pollarding is a similar management practice but differs from coppicing in that trees are cut at a height of two or three metres instead of near ground level (Fig. 7.3).[32] A common feature of wood-pasture, pollarding discourages grazing animals from eating the new shoots,

[29] Oram, *Alexander II*, pp. 245–50.

[30] Rackham, *Woodlands*, pp. 16–17; Quelch, 'Ancient Trees', p. 27.

[31] Quelch, 'Ancient Trees', p. 27.

[32] Quelch, 'Ancient Trees', p. 27.

Figure 7.3. Example of a large, old, oak pollard from Dalkeith Park that may date to the medieval period (photo by author).

while ensuring regular supplies of fodder for animals in winter when grass does not grow dependably, and shade for animals to keep cool in the summer.[33] Both coppicing and pollarding allow for timber sources to be replenished over time and if managed properly can be a sustainable source of timber that lasts for centuries when done in regular intervals. Through coppicing and pollarding, a landowner could successfully extend the life and productivity of a tree by considerable margins.[34]

Several examples of formerly pollarded and coppiced trees still exist within the remnants of medieval Scottish parks. Perhaps the best-known example of pollarding being practiced within a park can be found among the Cadzow oaks of the Hamilton High Park in Lanarkshire, where local tradition maintains (incorrectly) that the oak trees date to as early as the reign of David I in the twelfth century.[35]

[33] Rotherham, 'Historical Ecology', p. 87.

[34] Rotherham, 'Historical Ecology', p. 87.

[35] Dougal and Dickson, 'Old Managed Oaks', pp. 77–79. While it is possible these trees date to the twelfth century, they were most likely planted in the fifteenth century (Mills and Crone, 'Dendrochronological Evidence'; Crone and Watson, 'Sufficiency to Scarcity', p. 74).

Other examples of such managed trees can be seen in the woodland around Darnaway Castle in Moray, in Dalkeith Park in Midlothian and the park at Lochwood Castle in Dumfries and Galloway. While pollarding is well attested to in English Parks, the above examples provide useful insight into at least late medieval period Scottish parks that possessed distinct wood management regimes.

As will be discussed below, parks emerge at a time when timber supplies were growing increasingly scarce, and incorporating wood into parks provided an economically lucrative and socially prestigious opportunity for park owners.[36] As the medieval period progressed and accessible wood supplies dwindled further, adequate timber sources could only be found in the more remote and inaccessible parts of Scotland (i.e. uplands where logging was difficult and costly) or imported from the continent.[37] Realistically, it was easier to manage and produce timber locally, much nearer to the edge of a settlement in private supplies.[38] With such resource control came added prestige. By maintaining and controlling private supplies of wood, lords demonstrated their social standing and economic stature by sending an unmistakable signal regarding who wielded power and who did not within society. Unfortunately, in Scotland, these roles are poorly understood due to the sparse documentary record from before the fifteenth century and the limited archaeological work conducted to date. Nevertheless, understanding the complex, multifunctional nature of parks is invaluable for informing understanding of the larger cultural and environmental history of Scotland during the medieval period.

Considering a Different View of Medieval Scottish Park Utilization and the Foundations of the Model

Estimates suggest that as many as 3200 parks had been created in England by the start of the fourteenth century. By way of contrast, perhaps between only 80 and 110 such structures were ever constructed in Scotland, a contrast that may account for the comparative lack of research into Scottish parks.[39] In other words, there were between 30 and 40 English parks for every one Scottish park. The disparity in park numbers between the two countries could be attributed

[36] Fletcher, *Gardens*, p. 9; Mills and Crone, 'Dendrochronological Evidence'.

[37] Mills and Crone, 'Dendrochronological Evidence'.

[38] Fenton, 'A Postulated Natural Origin', p. 121.

[39] Gilbert, *Hunting Reserves*, p. 356; Rackham, *The History of the Countryside*, p. 123.

to a number of factors, but the simplest explanation may be that the lower population density of Scotland made land restriction through imparking less economically lucrative than in the more populous and resource-pressured south. Nevertheless, the climatic/environmental and socioeconomic/cultural conditions prevalent during this period in Scotland may still have made it beneficial for some to impark their lands.

Given the limited body of archaeological research and sparse contemporary documentary evidence to draw upon, a new approach, one that considers the more thoroughly studied aspects of medieval Scottish history and places park enclosures within the broader historical context, is needed to provide insight into the enigmatic nature of these sites. With this in mind, the following sections discuss an informal model that examines the wider socioeconomic/cultural and climatic/environmental circumstances of the period to further our understanding of how parks were used and why they emerged. The goal is to place park construction and ownership within a broader historical framework, in which multiple variables interconnect and human influence on the landscape is seen as part of a larger system. There are two lines of thought that have been integral to the construction of this model, the first being Gunderson and Holling's theory of *Panarchy*.[40] Panarchy, or resilience theory, strives to organize economic, ecological, and institutional systems and how they interact to achieve different outcomes.[41] These ideas are employed in this study to explore the wider economic, ecological, and institutional systems of medieval Scotland, mapping the interactions that led to park creation.

The second influential line of thought is the method advocated by Schreg called the 'village ecosystem' approach, a perspective that encourages archaeologists to approach the past by considering humans as part of a complex ecosystem. Schreg argues that this is a particularly useful approach for archaeological studies of the medieval period, as it is a period often regarded as the realm of historians.[42] In Scotland, this approach is even more relevant considering the dearth of surviving historical sources prior to the fifteenth and sixteenth centuries. The true advantage of this approach is that it allows us to move away from the confines of localized and narrow spatial analyses, focused on singular landscape aspects, towards an examination of the past that is ultimately broader, balanced, and inter-disciplinary. The goal is not to present a definitive descrip-

[40] Gunderson and Holling, *Panarchy*.

[41] Gunderson and Holling, *Panarchy*.

[42] Schreg, 'Ecological Approaches', p. 87.

tion of the past but to present a fluid model of a complex system of interplaying climatic, environmental, social, economic, and political factors.[43]

Both of these perspectives seek to be comprehensive in their interdisciplinary perspectives, but the most significant commonality is that in both approaches humans are not divorced from the nature with which they are interacting. Instead, humans are active participants in a complex system. Separating humankind from the larger picture promotes a perception of a reactionary relationship with the world that considers humans to be alien outsiders that have invaded what would otherwise be a pristine, unspoiled ecosystem rather than influential contributors in its formation. In truth, people are just as enmeshed in that ecosystem, and if we are to understand the actions and activities of the people of the past we must view said actions and activities in an all-encompassing context.

In the case of medieval Scottish parks, these theories allow us as researchers to move beyond narrowly defined perceptions that focus on the parks' significance as hunting grounds, and recognize them as intricately designed human landscapes that were a reflection of a period of significant climatic, technological, and social change. It is the human response to this dynamic period that is reflected in how these parks were designed and utilized. Parks are a physical manifestation of human-land interactions. Although a village or densely populated urban centre differs significantly from a lordship, particularly within a landscape that was inherently exclusionary by definition, utilizing the ideas of the village ecosystem approach and the tenets of Panarchy allows us to look outside, to the broader context of the medieval period. It allows us to examine what was occurring within the socioeconomic/cultural and environmental/climatic spheres of society and how the interaction of those spheres led to an outcome of parks. Humans designed these landscapes to fit their needs and their dynamic world, shaped by their society, their culture, the environment, and the climate. Thereby, the following discussion aims to emphasize the human role and the interconnected system within which humans operated to the study of medieval Scottish parks.

[43] Schreg, 'Ecological Approaches', p. 100.

The Model Variables

The Historical Context of Climate, Environment, and a Human Shaped Ecosystem

While Scotland's modern climate is generally characterized as temperate it is by no means homogenous, and instead highly variable, often even between short distances and elevations.[44] Scotland's geology and topography varies considerably from region to region and because of this the same is true of its climates and ecosystems. This has a dramatic effect on different aspects of Scottish ecology and soils, agricultural capabilities and growing seasons, and vegetation, and simple changes to the climate and environment have wide-ranging effects.[45] Given this variety, it is important to consider the influence of medieval environmental and climatic conditions on decisions to enclose lands within parks. During this period, there were considerable climatic changes globally that resulted in the so-called 'Medieval Climate Anomaly' (MCA) and the subsequent 'Little Ice Age' (LIA), with the lands bordering the North Atlantic Ocean arguably experiencing some of the most profound effects of these changes.

The MCA, lasting from roughly CE 850 to 1250, was a period of more benign climatic conditions, particularly in the British Isles and Northern Europe. The fierce winter storms of later centuries were, for a period, less pronounced.[46] Data based on the analysis of tree-rings, ice cores, corals, sediments, speleothems, and other environmental proxies, demonstrate a defined period of warmer temperatures for the North Atlantic region running from approximately the mid-tenth century through to the mid-thirteenth century. This was followed by a period of cooler temperatures that commenced later in the thirteenth century and lasted into the mid-nineteenth century, with one of the coldest periods being from CE 1400–1700.[47]

Although conditions were never as balmy in Scotland during the MCA as they were in supposedly vineyard filled England, and tough winters were by no means a thing of legend, conditions were generally more favourable for crop production.[48] Lowland agricultural emphasis on cereals seems to have proliferated during this period with a transition to an increased focus on arable agricul-

[44] Mather, 'Geology, Soils, Climate', p. 68.

[45] Mather, 'Geology, Soils, Climate', pp. 68, 73–75.

[46] Dawson, *So Fair and Foul a Day*, p. 88.

[47] Mann and others, 'Global Signatures', p. 1257; Proctor and others, 'A Thousand Year Speleothem Proxy Record'.

[48] Dawson, *So Fair and Foul a Day*, pp. 96–99.

ture occurring during the MCA. Gains in crop yields were bolstered by climate change during the period between the eleventh and thirteenth centuries, spanning the height of the MCA. Evidence from paleoclimatic proxy records reflect three particularly warm intervals that lasted from CE 1010–1040, 1070–1105, and 1155–1190.[49] While estimates suggest that on average the mean temperature was only about 0.2°C warmer than the temperatures of the LIA, peak and trough episodes of warming and cooling throughout the MCA period could vary by as much as ± 1.0°C.[50]

By the late thirteenth century and throughout the fourteenth century, changes in atmospheric circulation, mainly stronger low-pressure systems over Iceland and high pressure cells over Russia and Canada, led Scotland to experience the bite of strong, cold easterly winds from Europe with a pronounced increase in severe winters, violent storms, and expanding sea ice.[51] With this climatic downturn of the LIA came devastating episodes of famine and longer-lasting food shortages, recurrent and prolonged episodes of epidemic and epizootic diseases, and the consequent collapse of population levels across much of Europe.[52] For Lowland Scots, cereals had become a staple by this time, making poor harvest years brought about by unpredictable LIA weather particularly stressful.[53] Yet, despite greater emphasis on non-arable lifestyles, Highland economies were no more insulated from the effects of climate change, as grazing animals still required biomass to sustain them, yet the growing season was equally affected by adverse climate.[54]

That parks first appear during the height of the MCA is probably not a coincidence but actually related to the more favourable climatic conditions. How their appearance in Scotland was influenced by the milder climate is open to debate, but when examined with other variables from the same time period, climate does appear to have been a factor. With the cooler conditions of the LIA came new challenges related to resource management and availability, but parks were likely an important part of how the landscape was shaped and used by landowners during this time.

[49] Crowley and Lowery, 'How Warm Was the Medieval Warm Period?', pp. 52–54.

[50] Crowley and Lowery, 'How Warm Was the Medieval Warm Period?', p. 53; see Mann and others, 'Global Signatures'.

[51] Dawson, *So Fair and Foul a Day*, pp. 99–105.

[52] Oram and Adderley, 'Lordship and Environmental Change', pp. 11–12.

[53] Dawson, *So Fair and Foul a Day*, p. 10.

[54] Oram and Adderley, 'Lordship and Environmental Change', pp. 7–8.

Introduction of New Technologies: the Heavy Plough and Horse Collar

Scotland possesses limited *land capability for agriculture* (LCA), with much of the country's highest quality land, designated as Class 1 and 2, or land with the greatest potential to produce consistently high yields of a wide range of crops, confined to scattered patches near the east coast and in river valleys.[55] These conditions are based on modern assessments, but are unlikely to have varied drastically during the medieval period. The greatest difference is likely to be found in the elevation in which fruitful cultivation and grazing was possible during the MCA.[56] Most of southern and eastern Scotland is classed as 3, or of moderate capability, while the central Highland and western portion of the country is classed as 4, or non-prime, moderate to severe in capability.[57] Mather describes the quality of Scottish agricultural land as such:

> Little more than a quarter of the land area is classed as capable of crop production (this does not, of course, mean that it is used for crop production), and less than 6 per cent is regarded as prime land. Potential cropland in general, and prime land in particular, are mostly in the east while most of the north and west lie in the lowest classes. Slightly over half of the land area is capable of use only as rough grazing, often supporting only very low densities of livestock, and then only seasonally.[58]

With the limited availability of useful agricultural land and the poor quality of that which was available, the combination of improved climate between the ninth and thirteenth centuries and technological and methodological advances bolstered soil quality and had significant societal impacts. One such technological advance was the introduction of a new type of wheeled, heavy plough, known as the *carruca*, which was capable of drastically increasing agricultural yields within the dense clayey soils underlying valley floors or the tough rocky upland soils at higher elevations, and reducing the amount of time required to till a field.[59] The *mould-board*, or part of the plough that overturns the soil and assists in releasing vital nutrients, was crucial in making otherwise intractable Scottish soils more amenable to agriculture by helping to reduce weed growth.[60]

[55] Brown and others, 'Influence of Climate Change', p. 51; Bibby and others, *Land Capability*.

[56] Bibby and others, *Land Capability*; Parry, 'Secular Climatic Change'.

[57] Bibby and others, *Land Capability*, p. 25; Brown and others, 'Influence of Climate Change', p. 25.

[58] Mather, 'Geology', p. 76.

[59] Oram, *Domination and Lordship*, p. 234; Duncan, *Scotland*, pp. 310–11.

[60] Fenton, 'Cultivating Tools and Tillage', pp. 655–57; Ross, *The Kings of Alba*, pp. 10–11.

While physical remains of such ploughs have been recovered in Denmark, the appearance of such an implement in Scotland is known primarily from the recovery of 'plough pebbles', small quartzite stones that were once embedded in the wooden mould-board to prevent the tool from wearing too rapidly, and depictions of the tool in stone carvings from the fourteenth century.[61] Today examples of *rig* or distinct ridges created by the heavy plough can still be identified across Scotland's landscapes, and Piers Dixon argues these ridges can be effectively distinguished from both earlier and later types of ploughed fields.[62] The constant turning over of the soil over time builds the rig up and creates a corresponding deepened 'valley' between known as the *furrow*.[63] Rig and furrow landscapes are useful features for establishing an area's land use history and for understanding individual park utilizations. The difficulty, however, is in dating accurately the use of the land for cultivation as opposed to other park functions like hunting, and subsequently piecing together the chronology of landscape use.

The advent of new collars, particularly ones that fit around a horse's neck, also played an important role. Oxen, the preferred draught animal up until this time, were notoriously difficult to manage. Unfortunately, the ox-bow, which was used to harness oxen to the plough, when placed on the more easily employed horse, would crush the animal's windpipe. This problem was rectified by the introduction of the horse collar.[64] Horses did not replace oxen outright but instead were incorporated into draught teams along with oxen and their use may indicate an expansion of arable agriculture into lands of poorer quality.[65]

The plough and horse collar were not the only advances in agriculture to emerge during this period. Different systems of land management and utilization were not insignificant changes, that contributed to increased crop production and agricultural expansion. The new technologies provided a level of efficiency and insight that had not been witnessed in earlier periods in Scotland. However, with efficiency came an increased pressure on lands, as improved

[61] Fenton, 'Cultivating Tools and Tillage', pp. 655–57; Duncan, *Scotland*, pp. 310–11. Examples have been recovered from East Lothian, Midlothian, Roxburghshire, and as far north as Shetland.

[62] Dixon, 'Crops and Livestock', p. 231; Dixon, 'Hunting, Summer Grazing and Settlement', pp. 35–45.

[63] Dixon, 'Crops and Livestock', p. 231; Dixon, 'Hunting, Summer Grazing and Settlement', pp. 35–45.

[64] Ross, *Kings of Alba*, pp. 10–11.

[65] Duncan, *Scotland*, pp. 310–11.

crop yields helped sustain higher population levels and ultimately led to still greater demand.

Health and Population

The assertion that population went up in Scotland during the MCA is generally agreed upon but inherently problematic. Of all the variables of the model, population is arguably the most difficult to discuss with any certainty. No source comparable to the late-eleventh-century English *Domesday Book* and the later fourteenth-century Poll Tax accounts exists for Scotland. Yet, the overwhelming consensus is that population was most certainly growing, possibly even doubling between CE 1000 and CE 1300.[66] As Ian Whyte explains, this argument is based on 'indirect evidence, such as the creation of new settlements and an expansion of the area of land under arable cultivation, that suggests Scotland shared in the general growth of population in Europe during the twelfth and thirteenth centuries'.[67] Information from contemporary charters indicates a transition in agricultural practices from subsistence cultivation to the creation of much larger agricultural surpluses and the rise of more urbanized, market-oriented centres.[68] There is also evidence to suggest that the lands of many regions were under considerable stress. The emergence of *assarts* in the twelfth century, or areas of land that had previously not been used for agriculture or deemed *waste* but were now cleared and used for arable or grazing, paints a picture of a landscape that was increasingly under anthropogenic pressure from a rapidly growing population.[69]

Despite the population increase, the physical health of the people was not necessarily improving, with diseases like leprosy and tuberculosis rife and still exceptionally real threats between the eleventh and thirteenth centuries. Yet, these pale into insignificance against the catastrophic impact of the Black Death in the mid-fourteenth century and its recurrence in waves of epidemic across the remainder of the Middle Ages.[70] Our evidence is, of course, largely limited to either chronicle accounts or, where there is physical record, to diseases that had an impact on bones or teeth rather than the rapidly decaying soft

[66] Whyte, 'Rural Society and Economy', p. 159; Oram, *Domination and Lordship*, pp. 234–35.

[67] Whyte, *Rural Society*, p. 159.

[68] Whyte, *Rural Society*, p. 159; Oram, *Domination and Lordship*, pp. 233, 235.

[69] Duncan, *Scotland*, pp. 365–67; Oram, *Domination and Lordship*, pp. 234–38.

[70] Oram, 'Disease, Death and the Hereafter'.

tissues of the human body. Such physical evidence is illustrated by the excavation of upper-class burials at Glasgow Cathedral, which found that 17% of the eighty-four individuals examined displayed enamel hypoplasia and Harris lines, pathologies which fingerprint significant bouts of metabolic stress in childhood and adolescence related to malnourishment or severe metabolic insult, and are indicative of prolonged episodes of illness.[71] Excavations from along Perth's High Street and Kirk Close have also shown that local populations were afflicted by a prevalence of intestinal parasites like whipworm (*Trichuris*) and roundworm (*Ascaris*).[72] Malnourishment was also likely a problem, with remains from Aberdeen and Linlithgow displaying *cribra orbitalia* or orbital pitting, once thought of as a condition of elderly females but now understood as usually caused by childhood anaemia.[73] Such a condition is typically brought about by unhygienic living conditions and iron deficient diets, which Catherine Smith argues were probably common problems in medieval Perth.[74]

Thus, evidence suggests that although population still continued to increase, human health was not improving. Instead, the steady rise in medieval population may have been related to improved fecundity, with a greater percentage of children surviving the perils of infancy and early childhood rather than a dramatic improvement in general health. A greater amount of food available from an increase in crop yields and the expansion of agriculture likely improved the nutritional state of the female population and led to higher reproductive success rates, or lower mortality rates for infants, young children, and women.[75] This, coupled with more efficient agricultural practices may have increased the amount of time and attention both parents were able to give their children. As Robert Kelly has found in hunter-gatherer societies, fecundity improves when there is a greater degree of childcare from both the parents.[76] In medieval Scotland, food surpluses and new agricultural technologies and practices likely improved fecundity, playing a critical role in population growth. Yet, with the growing population came greater demands for food and land.

[71] Oram, 'Disease, Death and the Hereafter', p. 198.

[72] Smith, 'Conclusions', p. 90.

[73] Smith, 'Conclusions', p. 92.

[74] Smith, 'Conclusions', p. 92.

[75] Kelly, *Lifeways of Hunter-Gatherers*, pp. 192–93.

[76] Kelly, *Lifeways of Hunter-Gatherers*, pp. 192–93.

Pressures on the Land

With the improved climatic conditions of the MCA, the technological advances of this period, and expanding populations, a circular pattern took shape, in which more food meant steady population growth, requiring continuous expansion into new arable and grazing lands. It was during this period between the late eleventh and the mid-thirteenth century that many regions of Scotland were exhausted of their natural timber and wood resources. Primarily, it was the clearance of woodland in favour of arable land and grazing pasture that led to such challenges. For instance, research near Oban in Argyll found that medieval wood clearance wiped out a system of agricultural woodland management 4000 years old within the span of just two-hundred years, resulting in the district's permanent deforestation by the twelfth century.[77] This rapid period of clearance was likely the result of one or several stressors that included: population increase and greater demand for arable land as well as wood for fuel and construction, the effects of climate change on the soil compounded by human mismanagement, or the export of timber to distant regions.[78] Similarly, pollen evidence recovered from Dogden Moss in Southern Scotland, illustrates considerable deforestation during the eleventh and twelfth centuries in the region, followed by brief regenerations during the thirteenth and fourteenth centuries, before clearance resumed in the mid-fourteenth century.[79] Yet, perhaps the most detailed investigation of medieval woodland clearance in Scotland undertaken to date is Richard Tipping's work on the Bowmont Valley, which demonstrates extensive deforestation in several locations across the Cheviot Hills along the Anglo-Scottish border in the interest of agricultural production during this time.[80]

Of course, the intensity of such clearance activities was not uniform across the country and varied between regions. For instance, while many highland zones may have experienced little change in resource availability, partially due to the difficulty of accessing said resources and a lack of population expansion in those areas, other regions, such as the Mearns in eastern Scotland, saw available woodland restricted to preserved areas of land.[81] Unfortunately, the earliest

[77] Macklin and others, 'Human-Environment Interactions', p. 118.

[78] Macklin and others, 'Human-Environment Interactions', p. 118.

[79] Dumayne-Peaty, 'Late Holocene Human Impact'.

[80] Tipping, *Bowmont*.

[81] Oram, *Domination and Lordship*, p. 246.

information available for the extent of Scottish woods only dates to the 1590s with the work of illustrious cartographer Timothy Pont.[82] Thus, understanding the state of wood cover in medieval Scotland is not straightforward, and much of the discussion is based on indirect evidence. One such line of evidence is the appearance of references to the aforementioned *assarts* in twelfth-century documents, which indicates an increased occurrence of agricultural expansion in then-wooded areas during this time. The unrelenting clearance of woods on lands once considered unusable or *waste* would have undoubtedly led to the shortage of valuable timber and fuel sources.[83] Assarts, however, were not the only manner in which land clearance manifested itself on the landscape. With the help of new technologies capable of tilling the difficult upland soils and the milder climate, agriculture activities were expanded into higher elevations than either before or after.[84] Evidence of this expansion is still visible in various rig and furrow features across the landscape. Simultaneously, the twelfth century also witnessed the draining of marshlands in the interest of agricultural production.[85] Overall, the expansion of agricultural regimes was intensive and thorough.

Theses dramatic changes in available wood supplies are borne out by the dendrochronological record as well. In their review of Scotland's existing dendrochronological evidence, Mills and Crone highlight important trends in the history of native Scottish timber use.[86] Most significantly, they identify a sharp transition made around CE 1450 from timber derived from native oak, to oak imported from Scandinavia and the Baltic States.[87] After the mid-fifteenth century there is little data to analyse, indicating the poor quality of native-grown timber during this time.[88] Young oak was not commonly used for construction purposes, leaving little surviving material, and making it difficult to

[82] For discussion of woodland cover in Ponts maps, see Smout, 'Woodland in the Maps of Pont'; Rackham, *Woodlands*, p. 57.

[83] Crone and Watson, 'Sufficiency to Scarcity', pp. 61–70; Oram, *Domination and Lordship*, p. 248.

[84] Parry, 'Secular Climatic Change'; Tipping, 'Climatic Variability'; Oram, *Domination and Lordship*, p. 234.

[85] Oram, *Domination and Lordship*, p. 251.

[86] Mills and Crone, 'Dendrochronological Evidence'.

[87] Mills and Crone, 'Dendrochronological Evidence'; Crone and Mills, 'Seeing the Wood', p. 792.

[88] Mills and Crone, 'Dendrochronological Evidence', p. 23.

employ dendrochronology to identify pollarding and coppicing as timber production practices. Several acts of parliament that date to the fifteenth century were partially intended to address the damage and destruction of wood due to the limited supply, emphasising the scarcity of wood during this time.[89] For example, King James II (reigned CE 1437–1460) ordered the first enclosures specifically for trees to protect particular areas of woods.[90]

Unfortunately, of the few palaeoecological studies that touch upon the medieval Scottish period many are of temporally low resolution and provide little clarity into minute but significant changes that took place over a short period. Despite the relatively small sampling of regions sporadically scattered across Scotland, available data indicate that wood as a resource was stressed. As demand grew, resources dwindled, leading to the need for foreign supplies and restricting access to remaining sources.

Deer

Given the poor quality of land and the stress of population growth discussed above, it should be noted that people and trees were likely not the only organisms to feel the strain of land pressures. Deer, particularly roe deer (*Capreolus capreolus*) populations but red deer (*Cervus elaphus*) as well, rely on woodland habitats. Roe deer prefer to subsist on hazel and colonize tracts of woodland.[91] As habitats disappeared in medieval England, deer herds sharply declined.[92] This is hardly surprising when considering the fierce territoriality of roe deer and their predilection for living in small groups.[93] These trends were most pronounced in the English Midlands and the south where, due to a combination of disappearing woodland and other anthropogenic pressures, some estimates suggest the roe deer were nearly extinct by the 1400s.[94] The introduction of the exotic fallow deer (*Dama dama*) in the early twelfth century may have been a means of rectifying this problem.[95]

[89] Gilbert, *Hunting*, pp. 104, 234–35.

[90] Gilbert, *Hunting*, pp. 235–37.

[91] Mileson, *Parks in Medieval England*, p. 29; Geist, *Deer of the World*, p. 313.

[92] Pluskowksi, 'Social Construction', pp. 75–76.

[93] Geist, *Deer*, p. 313.

[94] Mileson, *Parks in Medieval England*, p. 28.

[95] Mileson, *Parks in Medieval England*, p. 28.

One method of bolstering disappearing deer populations was to enclose herds within parks.[96] Pluskowski argues that south of the Lake District in north-west England, red deer could only be found in parks.[97] Despite some growth in deer numbers after the Black Death, English deer populations again declined drastically in the fifteenth century. Parks subsequently became vital to the survival of the roe and red deer populations of England. This was no altruistic medieval naturalist movement, however, and herds were not enclosed in parks for the noble task of preserving a disappearing species for posterity, but, rather, in the interest of preserving the sport of hunting. As Milseon states 'deer-hunting [in England] would not have been feasible in most areas by the later-thirteenth century (or even earlier) without parks to protect deer stocks and, presumably, to provide a space for undisturbed hunting in an increasingly busy landscape'.[98]

Of course, this was the state of affairs in England. In Scotland, we are once again wanting evidence, hampered by the few surviving historical sources and an equally limited degree of excavated material upon which to base our analysis. What can be said is that Scotland faced many of the same land pressures. Large areas of land were cleared of trees in the interest of agriculture, and human populations were expanding in number and geographic range. Some species like the wild boar, another popular hunting quarry, already faced near total extinction in Scotland, virtually disappearing by the thirteenth century.[99] According to the early-fifteenth-century chronicler Walter Bower, abbot of Inchcolm, fallow deer populations were present in Scotland by the mid-thirteenth century.[100] Keen to prevent the loss of another hunting animal, as in England the appearance of fallow deer here may have been a response to disappearing native species of deer. In this way, parks functioned similarly as animal 'preserves' and the gregarious nature, delicious meat, and ornate appearance of fallow deer endeared them to park owners across Europe.[101] While it is entirely possible that the presence of fallow deer was born out of fashionable emulation as opposed to 'necessity' and considering that Scottish deer populations had more area to roam than in England, with a greater abundance of remote

[96] Mileson, *Parks in Medieval England*, p. 28.

[97] Pluskowski, 'Social Construction', p. 75.

[98] Mileson, *Parks in Medieval England*, p. 28.

[99] Fenton, *Scottish Life and Society*, pp. 34–36.

[100] Bower, *Scotichronicon*, v, p. 369.

[101] Sykes, 'Animal Bones and Animal Parks'.

regions that remained underdeveloped, many of their preferred habitats were still under anthropogenic pressure.

Additional faunal material is needed from the medieval period from a variety of sites to help piece together the history of deer in Scotland. Since primarily the elite ate deer, their presence in the archaeological record is far less than other species typically recovered from both elite and non-elite settlement sites. This, coupled with poor preservation, has left us with little faunal material relating to deer in the medieval period and prevents a deeper understanding of human/cervid interactions in Scotland.

The Anglo-Norman Influence in Scotland

It is particularly telling that parks emerged during a period when Scotland was experiencing dramatic cultural shifts that witnessed the emulation of far more English and continental styles of elite behaviours.[102] Arguably, the period between roughly the mid-eleventh century and the early to mid-thirteenth century was one in which Scotland's rulers aimed to establish their kingdom as a contender on par with England and the powers of continental Europe. This cultural reorientation included the adoption of luxury goods, fashion styles, fine foods, styles of land management like the forest system and pannage, and even child naming conventions.[103]

King Malcolm III's youngest son, David I (reigned 1124–1153), introduced the Anglo-Norman style forest system during the first half of the twelfth century that is often perceived as the precursor to medieval Scottish parks, considering many parks were created out of pre-existing forest-lands.[104] John Gilbert has suggested that royal parks were a secondary response to the administrative difficulties of maintaining a forest, as it was easier to control and protect enclosed lands than difficult-to-define tracts of open land.[105] Undoubtedly, as resources became more strained, effective control would have become a priority.

[102] Oram, *Domination and Lordship*, pp. 28–29; Stephen Boardman and Alasdair Ross's editorial introduction to *The Exercise of Power in Medieval Scotland*, pp. 15–16 usefully discusses this subject.

[103] Oram, *Domination and Lordship*, pp. 16, 28–29; Ross, *Kings of Alba*, pp. 159–63.

[104] Gilbert, *Hunting Reserves*, pp. 12–13, 20–22.

[105] Gilbert, *Hunting Reserves*, p. 22.

Mileson argues that park building was likely often the result of the emulation of other aristocrats.[106] While he is referring to a more localized process within England itself, this idea can be appropriately expanded to include 'international emulation', where members of the Scottish aristocracy emulated their English and continental counterparts. If park construction in Scotland was a result of cultural transmission between England and Scotland, as is often argued, such emulation was undoubtedly meant as a message of power and status. Given the historical context of a period that witnessed the frequent assertion of English political superiority over Scotland, it is understandable that aristocratic Scots may have wished to convey their social parity as one means of asserting their independence and garnering respect on the international stage. An effective means of accomplishing this task would have been through the demonstration of equivalent lifestyles and luxuries. Hunting was just as fashionable among the elites of England and continental Europe as it was in Scotland, and access to private resources like wood when deforestation was plaguing not only northern parts of mainland Britain but other regions as well, would have been telling about one's social stature.

Park Emergence as a Response

The following model (as well as Fig. 7.4) illustrates the interactions of the above climatic/environmental and socioeconomic/cultural variables to present a likely scenario where it was socially beneficial and/or economically profitable for some enterprising (and elite) individuals to enclose tracts of land in parks.

The Model

- In medieval Scotland warmer temperatures during the climatic optimum known as the 'Medieval Climate Anomaly' along with improvements in agricultural technology, including the introduction of the heavy plough and new collars for horses in the eleventh century, made it possible to cultivate dense, heavy soils and lands previously deemed *waste* more effectively and efficiently, leading to higher crop yields.
- Higher crop yields improved human reproductive success or fecundity, thereby leading to population growth in medieval Scotland.

[106] Mileson, 'Sociology of Park Creation', pp. 24–25.

- The increase in population, the new agricultural technologies and practices, and higher crop yields increased environmental stress, creating greater demand for arable land and pasture obtained through the extensive clearance of woodlands.
- The increased clearance of woodlands led to depleted supplies of wood and timber, key resources for construction, fuel, animal fodder and many other common uses.
- Continued clearance of the land and the depletion of habitats strained deer populations, leading to the need to preserve deer herds for hunting populations and venison production by restricting access to them.
- Simultaneously, this period witnessed an increased dissemination and incorporation of English and continental cultural practices and land management systems that included the forest system, pannage, and possibly parks, beginning in the late eleventh century and continuing throughout the twelfth and thirteenth centuries.
- This culminated in the restriction of small quantities of land by the upper classes through the process of imparking, effectively privatising portions of highly sought-after lands and resources for economic and social purposes. In doing this park owners could utilize the enclosure for multiple purposes that included: private management of wood supplies for timber production, wood-pasture, fuel, and grazing; the creation of secure/restricted areas for grazing and crop production; and manage deer populations in the interest of hunting and/or maintaining a live-larder for ready access to venison.

Faced with significant land and resource pressures, restricting access to certain areas of land through imparking allowed for a variety of potential utilizations and bestowed both economic and social benefits upon the owners. It is possible to view the medieval parks of Scotland as multifaceted entities that operated as an effort at conservation by avoiding a 'tragedy of the commons' scenario, as a vehicle for expressing one's social status through conspicuous consumption, and as an example of a market response.

The 'tragedy of the commons', as made popular by Garrett Hardin's 1968 *Science* article of the same name, is the idea that resources held in common will inevitably be overexploited to the point of complete destruction.[107] He used the

[107] Hardin, 'The Tragedy of the Commons'.

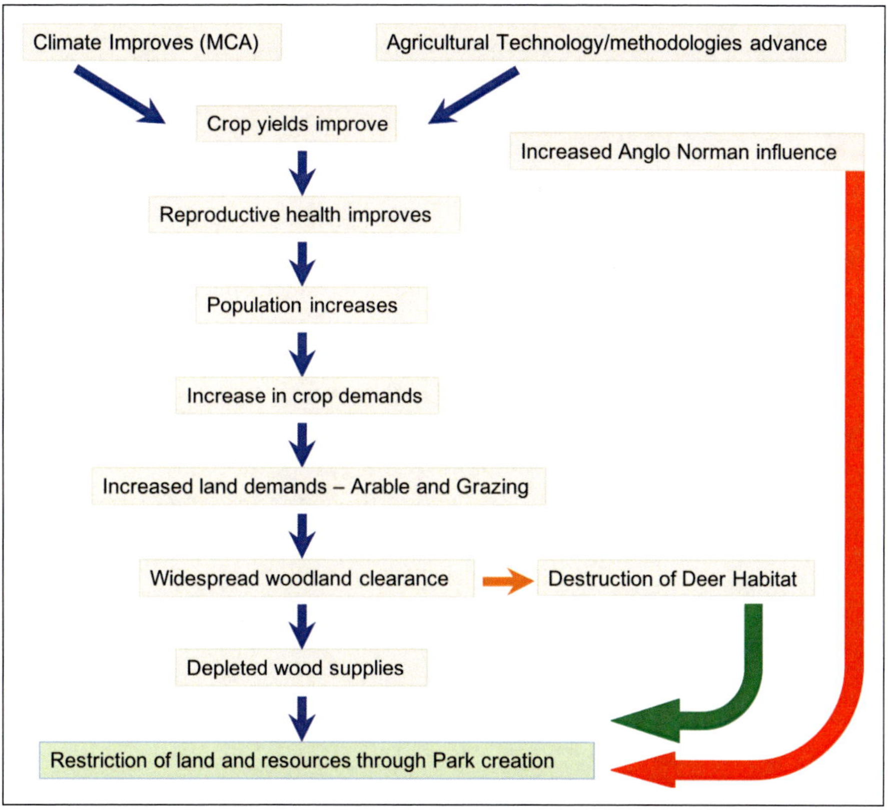

Figure 7.4. Informal model examining the interaction between the environmental, climatic, and socioeconomic variables that may have created a scenario in which it was beneficial to enclose tracts of land within parks (figure by author).

example of a common pasture to illustrate how social stability will ultimately encourage each herdsman to increase their herd size so as to maximize profits, ultimately leading to the destruction of the common land through complete degradation.[108] Despite the superficial strength of Hardin's argument, it fails to account for the many mechanisms of regulation societies employ to prevent such disasters. In this way, parks were just such a mechanism for regulating land and resources by restricting access and privatising portions of it, thereby removing it from the commons. What seems clear is that the pressures on the land, the people, and even perhaps the animals were significant enough to motivate

[108] Hardin, 'The Tragedy of the Commons', p. 1244; Feeny and others, 'The Tragedy of the Commons: Twenty-Two Years Later', p. 2.

the wealthiest people to rethink their approaches to land use and management in a way that was socially and economically worthwhile.

There is an important distinction to be made about the assertion of parks being an economically viable entity, however. Parks were not necessarily financially profitable endeavours, requiring the substantial organization of resources and labour. Building and maintaining a park, particularly one that housed deer, was incredibly costly. Rather, the economic value of parks may have been in the costs they ultimately offset for the owner. Requirements for landowners and their tenants for resources like timber, fodder, and fuel necessitated importing supplies from often distant locations if no local source was readily available.[109] The immediacy and immensity of these requirements is demonstrated in Alasdair Ross's analysis of wood use as detailed in two sixteenth-century documents, in which he revealed the massive quantities of wood consumption necessary simply to maintain built structures within one lordship unit.[110] This could be expensive and inconvenient, especially if there was an immediate need for a given resource. Therefore, instead of being a financially profitable enterprise to impark an area of land it may have been the least-costly option for wealthy individuals, ensuring available supplies of certain resources in the immediate vicinity. Mileson argues that parks were not particularly lucrative in England, providing relatively little income.[111] Instead they functioned in a way that balanced preservation of woodland, arable and pasture, as well as other resources with an environment that was ideal for deer habitats in the interest of the hunt and venison production.[112] Much of a park's resources were likely dedicated to the mere maintenance of fences or pales and animal herds, as well as some of the other parts of a lord's lands and buildings. Park fencing alone required staggering amounts of wood.[113] As resources became increasingly stressed by population and agricultural expansion, parks became important to the aristocracy.

As discussed above, many parks were born out of the medieval forest system. While these can be most easily understood as hunting reserves, it is important to emphasize that it was not simply animals and hunting activities that were regulated within these lands. Gilbert illustrates how 'vert' or the different varieties of vegetation present within forestlands were protected, effectively regu-

109 Mileson, *Parks in Medieval England*, pp. 59, 66.

110 Ross, 'Two 1585 × 1612 Surveys'.

111 Ross, 'Two 1585 × 1612 Surveys', p. 71.

112 Mileson, *Parks in Medieval England*, p. 71.

113 Malloy and Hall, 'Medieval Hunting and Wood Management', p. 11.

lating activities like agriculture, grazing, wood-cutting and all other activities that might involve the utilization, taking, or clearing of wood.[114] It is likely that many of these regulations were extended to park enclosures, if not expanded even further. Wood significance as a natural resource did not diminish during the course of the Middle Ages, even as local wood sources were depleted.

The above informal model lays out a scenario based on the historical, archaeological, and palaeoenvironmental evidence, however limited that evidence may at times be, to argue that the motivations behind park building in medieval Scotland were not simply consequences of the desire to display one's wealth or to possess a private hunting arena, though these were partial motivations. It was instead driven by an array of factors that included a 'conservation' effort. While the word conservation may imply a medieval environmentalist or 'green' movement, this is not what is being suggested. With increased clearance of woodland came an opportunity for enterprising individuals to save or even make money by privatising, preserving, and managing their own wood supplies, ensuring continued availability of the resources which they controlled. These efforts also included managing and 'preserving' dwindling deer herds that had been stressed and reduced as a result of habitat destruction. With the management of park wood came the ideal habitats to maintain deer herds. Whether this consisted of native roe or red deer, or the ornamental fallow deer imported from abroad requires much deeper analysis. Regardless of species, however, successful deer management within parks ensured that the favoured hunting quarry and prized meat of the elites persisted for the enjoyment of the aristocracy.

Conclusions

Approaching the history of imparking in Scotland and why it appeared is inherently complicated, requiring multiple avenues of investigation. Medieval parks cannot be explained simplistically with sweeping generalities that negate the veritable complexity of such carefully designed anthropogenic landscapes. The goal of this paper was to contextualize parks and their initial appearance and subsequent utilizations, within the socioeconomic/cultural and environmental/climatic framework of the Middle Ages. Scottish parks arose from the interaction of many complex variables that created a situation in which owning an enclosed park was both socially and economically beneficial. While the

[114] Gilbert, *Hunting Reserves*, pp. 103–04.

data is limited, and the evidence at times lacking, what is undeniably certain is that the context was multifaceted. Parks were far more than mere hunting reserves and were intricately linked to the broader anthropogenic activities and climatic changes that shaped the landscapes around them. Although the scope of this study is merely a snapshot of a short, yet transformational period in time, roughly 300 years, it is meant to examine what motivated individuals to initially create parks. Why they continued to impark lands is another question worthy of investigation, particularly after the climatic downturn of the LIA and the sweeping demographic changes brought about by the famines and plague of the fourteenth century. The evolution of park landscapes between the fourteenth and sixteenth centuries is just as complex but is undoubtedly just as intimately connected to the environmental and climatic conditions of the period, as well as the socioeconomic and cultural context of the time. Parks are not isolated pieces of niche historical esoterica, valuable only to the landscape historian or archaeologist. They are an important part of Scottish cultural heritage and environmental history with much to tell us about historic climate change and human interactions with the land. Ultimately, Scottish parks should be considered a unique vessel that could be studied through many avenues of medieval history. Although not the intended aim of this paper, economic theory in particular may provide significant insights into the history and significance of parks within the broader context of the medieval period. Nevertheless, to understand the history of parks and their place in Scotland's landscape history, it will be essential for future research to consider the broader historical and environmental contexts of the medieval period.

From Kestrels to Foul Marshes: Light on a Parish in the Merse *c.* 1200

Simon Taylor
University of Glasgow

This chapter will explore both the physical and the tenurial landscape of the parish of Langton in Berwickshire around the year 1200, remarkably illuminated by a charter which has survived in the cartulary of Kelso Abbey. Issued by Sir William de Vieuxpont (Vipont), it describes in minute detail four units of land which Sir William granted or confirmed to the abbey along with the church of Langton. All in all, it contains 12 place-names and 15 references to other local features such as land-holdings, ditches, faulds, shielings and stones. The following essay will use these names and features to attempt to reconstruct the medieval landscape of Langton, relating as many of them as possible to the modern landscape. It will also analyse the place-names as an important source for the environmental and linguistic history of the area, drawing on research conducted by the Leverhulme-funded project on the place-names of Berwickshire, 'Recovering the Earliest English Language in Scotland: Evidence from place-names', University of Glasgow.

Although this contribution will look in depth at a little piece of the kingdom of the Scots well outwith Alasdair's davoch-country, its detailed engagement with Scotland's medieval landscape and namescape, as well as its careful re-evaluation of the sources for that engagement, are subjects which were dear to his heart and scholarship. It is in that spirit that I offer this study in his memory.

Introduction

At some point between the birth of the future King Alexander II in August 1198 and the death of his father, King William I (the Lion) in December 1214, William de Vieuxpont (III) confirmed to Kelso Abbey earlier grants made by

'With Our Backs to the Ocean': Land, Lordship, Climate Change, and Environment in the North-West European Past. Essays in Memory of Alasdair Ross, ed. by Richard Oram (Turnhout: Brepols, 2023)
EHNAW 5, pp. 183–204 BREPOLS PUBLISHERS 10.1484/M.EHNAW-EB.5.131275

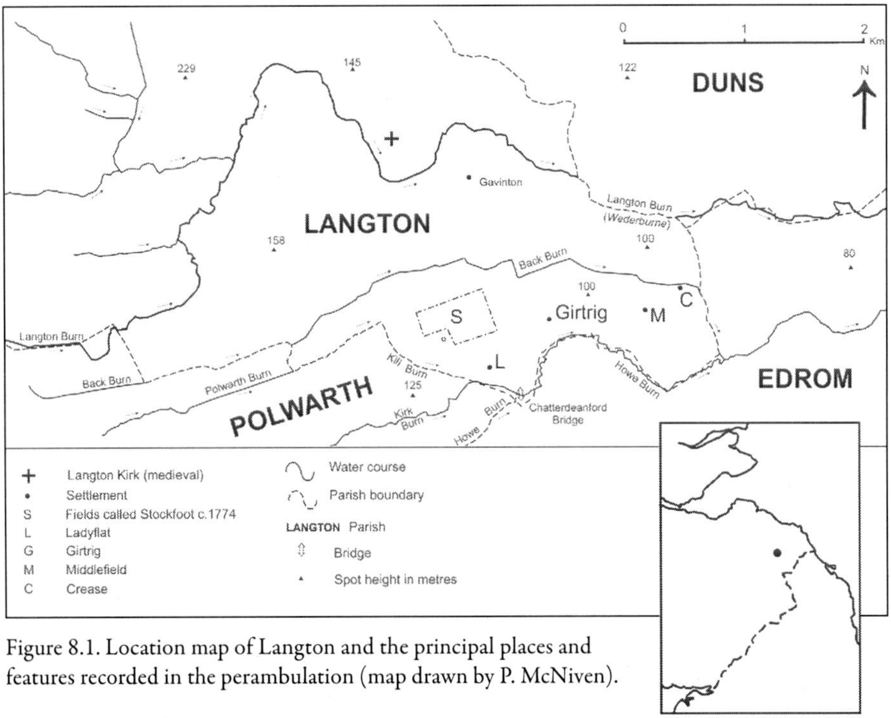

Figure 8.1. Location map of Langton and the principal places and features recorded in the perambulation (map drawn by P. McNiven).

his father, William de Vieuxpont (II) and grandfather, William de Vieuxpont (I), as well as making some new grants of his own.[1] Our only source of this information is the abbey's manuscript cartulary, an edition of which was printed for the Bannatyne Club in two volumes in 1846, which will be referred to below as *Kelso Liber*.[2] The most extensive expression of these grants is the charter printed as *Kelso Liber* i no. 140. That document chiefly deals with land in and around the Berwickshire parish of Langton, which lies to the west of the town of Duns, although it also confirms to Kelso Abbey the church of Horndean (now part of Ladykirk parish, Berwickshire) and some shielings in the Lammermuirs. What makes this charter so special is that it contains an unusually high number of place-names (12) and descriptions of minor features (15) which bring the medieval landscape in and around Langton alive in a way that is rare for a Scottish parish 800 years ago. A new, annotated edition of this charter with translation forms the core of this chapter. I will begin, however, with a closer look at the Kelso cartulary itself.

[1] For the suggestion that this charter is likely to have been issued after May 1203, see Appendix, below.

[2] National Library of Scotland, Adv. MS 34.5 (Cartulary of Kelso Abbey); *Liber S Marie de Calchou* [hereafter *Kelso Liber*].

The Kelso Abbey Cartulary

The cartulary of the Tironensian abbey of Kelso has survived as NLS Adv. MS 34.5 (henceforth abbreviated simply as MS), a small bound volume, the bulk of which dates from 1321 or shortly thereafter. This collection of charters was made soon after the destruction in the Anglo-Scottish wars of many of the earlier documents recording the abbey's rights and possessions. We are fortunate that this cartulary not only survived but was edited by Cosmo Innes and printed as *Kelso Liber* for the Bannatyne Club. Alasdair Ross, in his important article on Bannatyne Club cartularies,[3] was critical of many of the Club's editorial practices, using the edition of the episcopal register of Moray, printed as *Registrum Episcopatus Moraviensis* in 1837, both as a case study and an object lesson in editorial bad practice. Each cartulary brings with it its own problems, but *Kelso Liber*, while not perfect, has been said recently to be 'generally regarded as one of the most valuable works that the Bannatyne Club produced'.[4] As the bulk of it is taken from one manuscript, there has been no silent conflation of various sources, and any deviation from the original arrangement of the material has been clearly signalled.[5] Furthermore, the Kelso cartulary has been the subject of a detailed analysis by Andrew Smith in his 2011 University of Glasgow PhD thesis.[6]

The cartulary forms part of a 220-folio codex organized in a total of nineteen quires, with the original collation or lay-out of the codex probably slightly different from its present state, a reconstruction of which is given by Smith.[7] He concludes that the original cartulary (his 'phase one') began on fol. 8^{r} and ended on fol. 164^{v}.[8] Thus, all the Langton material which forms the basis of this chapter is within the original cartulary, produced by a small group of scribes between 1321 and 1326.[9]

The earliest charter in the cartulary dates from *c.* 1120.[10] However, it must always be borne in mind that the cartulary itself is a product of the 1320s, and

[3] Ross, 'The Bannatyne Club'.

[4] Smith, 'The Kelso Abbey Cartulary', p. 38.

[5] *Kelso Liber*, I, p. xviii.

[6] Smith, 'The Kelso Abbey Cartulary', p. 38.

[7] Smith, 'The Kelso Abbey Cartulary'.

[8] Smith, 'The Kelso Abbey Cartulary', p. 98.

[9] Smith, 'The Kelso Abbey Cartulary', pp. 99, 118–22.

[10] This is even before the abbey was moved from Selkirk to Kelso in 1128 (Chronicle of

was therefore prone to manipulation, with some spurious material added. This whole issue of forgery is explored at length by Smith, above all in Chapter 3, 'The Authenticity of the Charters in the Kelso Abbey Cartulary'.[11] He would argue that one of the Langton de Vieuxpont charters, namely no. 142, can be regarded as a forgery, but the evidence for this is tenuous, and even if it had been slightly 'massaged', it does not affect the authenticity of the Langton material as a whole.[12]

The de Vieuxponts

The name, which means literally 'of the old bridge', is a toponymic surname probably deriving from the commune Vieuxpont-en-Auge, Dép. Calvados, Normandy, roughly halfway between Falaise and Lisieux.[13] It has developed into the modern Scottish surname Vipont. Here is not the place to go into the complexities of the de Vieuxpont family in the late twelfth and early thirteenth centuries. The first detailed treatment of the family was by R. C. Reid in his 1955 article on the Veteriponts, as he calls them, with a family tree included in that paper.[14] The family line which concerns us here is what may be called the William-line, for reasons which will become immediately apparent. Reid's family tree was tentatively modified by Geoffrey Barrow,[15] with some more detail in editorial notes to two charters printed in the second volume in the series Regesta Regum Scottorum.[16] Both Reid and Barrow agreed, however, that there was a direct line from a William de Vieuxpont termed (I), to his son with the same name (II), to William (II)'s son, also William, termed William (III), and that William (III) is the William of the charter at the core of this

Melrose translation, from Anderson, ed., *Early Sources of Scottish History AD 500 to 1286*, II, p. 171). This earliest charter is MS fols 9^{r}–10^{r}, printed as *Kelso Liber*, I, no. 1, and more recently, *The Charters of David I*, no. 14.

[11] Smith, 'The Kelso Abbey Cartulary', pp. 125–67.

[12] *Kelso Liber*, I, no. 142; Smith, 'The Kelso Abbey Cartulary', pp. 267–70. See endnote [v], following the translation of the charter, below.

[13] Barrow, *The Anglo-Norman Era*, p. 94.

[14] Reid, 'De Veteripont', p. 106.

[15] Barrow, *The Anglo-Norman Era*, pp. 73 n. 73, 76 n. 88.

[16] Barrow, ed., *The Acts of William I*, nos 84 and 146. Stringer, *Earl David of Huntingdon*, p. 302 n. 12, gives a further reference, namely to Powicke, *The Loss of Normandy*, p. 357.

chapter.[17] What we know for certain is that William (II) married Emma de St Hilaire, and by her had another two sons also called William, namely 'middle' William (*medius*) and William 'the younger' (*iunior*). In a charter of *c.* 1200 by which William (III) granted to Holyrood Abbey the teind of his 'coal-mine' or 'coal-workings' (*carbonarium*) of Carriden, West Lothian, as well as the tenth penny 'de omnibus nauibus et battellis' (from all the ships and boats) loaded and unloaded at nearby Blackness, he describes himself as the first-born of the three sons of Lady Emma de St Hilaire. It is witnessed by his brothers, the above-mentioned middle William and William the younger ('Willelmo Medio et Willelmo Juniore fratribus meis'), as well as by Roger de la Léqueraye (de la crai), who also witnesses nos 140 and 141.[18] It is unclear who exactly Emma de St Hilaire was,[19] but she was clearly important in her own right. Not only does William (III) frequently describe himself as 'William the first born of the sons of William de Vieuxpont and Emma de St Hilaire', he also sometimes terms himself simply 'William son of Emma'.[20]

William de Vieuxpont (III)'s wife would appear to be Matilda of St Andrew, who witnesses no. 141, and whose name may have been carelessly omitted by the copyist from no. 140.[21] However, in William de Vieuxpont (III)'s grant (or rather, confirmation) of the church of Langton to Kelso Abbey, 1198 × 1214, she appears in the witness-list immediately after William de Vieuxpont junior as 'his mother' (*matre eius*).[22] As it is assumed that William de Vieuxpont junior is the son of the granter, William (III), it is perhaps strange that she is designated William the younger's mother rather than William (III)'s wife.

Not only do we have a proliferation of Williams, there are even two contemporary de Vieuxpont men designated William the younger. There is William the younger, brother of William de Vieuxpont (III), and William the younger, son of William de Vieuxpont (III). Both these men witness no. 140. They also

[17] *Kelso Liber*, I, no. 140.

[18] *Kelso Liber*, I, nos 140, 141; the charter is printed as *Liber Cartarum Sancte Crucis*, no. 41.

[19] Her name may derive from Saint-Hilaire-du-Harcouët, Normandy, *c.* 20 kilometres east of Mont St Michel.

[20] For example, in Stringer, *Earl David of Huntingdon*, no. 27, an original document (first printed in *Calendar of Writs*, no. 4), dated 1173 × 1174, possibly 1185; Stringer, *Earl David of Huntingdon*, no. 28, an original document, dated 1172; also no. 35 (first printed as *Kelso Liber*, I, no. 226), dated 1174.

[21] *Kelso Liber*, I, nos 140, 141.

[22] *Kelso Liber*, I, no. 139.

both witness the following charter, but owing to a relatively rare editorial error in the printed edition, William the younger, William (III)'s brother, has been omitted: both are in the original cartulary, both witnessing as William the younger (MS fol. 55^{r}).[23]

The De Vieuxponts formed part of the wider circle of David, King William's younger brother, who from 1185 was earl of Huntingdon. William de Vieuxpont (III) attended David during the eventful first half of 1174, first in Scotland then on campaign in the English Midlands.[24] And he is found as witness to four of David's charters, placing him in what Keith Stringer describes as David's 'outer circle'.[25]

Langton

The parish of Langton lay in the bishopric of St Andrews, deanery of Merse. The name probably dates from the Old English period (*c.* 600–1000), when what later became Berwickshire formed part of Bernicia, the northern part of the kingdom of Northumbria. It contains the Old English elements *lang* and *tūn*, 'long toun', 'long village',[26] probably referring to a settlement strung out along a road. If this is the case, then the road in question would be the one which ran from Duns westwards to Lauder and to the old Roman road known since the Middle Ages as Dere Street, passing much nearer to the site of the medieval kirk and village than the main road does today. Our earliest reference to Langton is from the mid-twelfth century, when it was already important enough to be the centre of a lordship with its own church. In a charter which can be dated *c.* 1150 × 1162 Roger de Eu gave to Kelso Abbey, 'the church of my toun of Langton' (*ecclesiam ville mee de Langtune*) with all its pertinents or facilities (*pertinenciis*) in free and perpetual alms just as Henry, parson (*persona*) of the church (of Langton), had held it, for the salvation of the soul of Earl Henry, Roger's lord,[27] and for the salvation of his own soul and the souls

[23] *Kelso Liber*, I, nos 140, 141.

[24] Stringer, *Earl David of Huntingdon*, p. 26.

[25] Stringer, *Earl David of Huntingdon*, pp. 155–58.

[26] For a discussion, with early forms, of this and all other Berwickshire place-names which appear on the Ordnance Survey Landranger series (1:50,000), see the online database *The Berwickshire Place-Name Resource* (hereafter BPNR Database).

[27] This is Earl Henry, son of King David I, who died in 1152, a year before his father. It is not entirely clear whether he was already dead by the time of this charter, as the wording is ambiguous.

of his ancestors and successors. Before this, as the charter tells us, the church had been held by its parson (*persona*) Henry, who was also one of the witnesses.[28] While the earliest possible date for this charter is approximate, it cannot be later than 1162, since this grant is also mentioned in a confirmation charter issued by Bishop Arnald of St Andrews, who died in that year.[29]

The de Eu family took its name from Eu on the north-east boundary of Normandy, Dép. Seine-Maritime, on the coast west-north-west of Amiens. The family's origins, its landholdings and networks in Scotland are discussed by Barrow.[30] By the 1170s Langton had passed to the de Vieuxponts, who were clearly connected to the de Eus, perhaps by marriage. The earliest document linking the de Vieuxponts with Langton appears to be a charter of 1173 × 1178 in which King William I confirmed to William de Vieuxpont[31] his (de Vieuxpont's) whole land of Bolton (East Lothian), Carriden (West Lothian) and Langton in feu and heritage, to be held 'with rights of warren' (*in warrennam*), for which reason no one may hunt in the said lands without his (de Vieuxpont's) licence on pain of the king's full forfeiture.[32] Then in 1195 King William confirmed William (III) de Vieuxpont's (re-)grant of the church of Langton to Kelso Abbey.[33]

The site of the medieval church of Langton (NT762525) is still marked by the remains of a burial vault and a graveyard within a high wall in mature woodland a short distance south of the now demolished Langton House.[34] It lay in the old village of Langton, which was demolished shortly after David Gavin bought the estate in 1758, the inhabitants being moved to the newly built village of Gavinton *c.* 700 metres to the south-east. The church survived for a while longer, but was probably demolished when the new church at

[28] *Kelso Liber*, I, no. 138. Also printed Lawrie, ed., *Early Scottish Charters*, no. CXCII.

[29] *Kelso Liber*, II, no. 451.

[30] Barrow, *The Anglo-Norman Era*, p. 179. A member of this family, Philip de Eu, who was alive in 1165, gave his name to Philpingstone in Bo'ness WLO (*c.* NT007813) (*Philpdawystoun Ew* 1165, *Registrum Domus de Soltre*, no. 3; baroniam de *Phillipston de Eu* 1327, *RMS*, I, app. 1 no. 25). It later became known as Grange, the name surviving in Philpingstone Road, Bo'ness. Forms and references from MacDonald, *Place-Names of West Lothian*, p. 32.

[31] Probably William (II).

[32] Barrow, ed., *Acts of William I*, no. 182. This survives as an *inspeximus* of David II dated 1366. It was first printed as *RMS*, I, no. 233.

[33] *Kelso Liber*, I, no. 144; Barrow, ed., *Acts of William I*, no. 381.

[34] See *Corpus of Scottish Medieval Parish Churches*: *s.n.* Langton.

Gavinton was built in 1798,[35] although the churchyard continued to be used until at least 1868 (NMRS Site Number NT75SE 6; Canmore ID 58685).[36] It was probably dedicated to St Cuthbert, since according to an almanac printed in 1578 a fair was held there on St Cuthbert's day (20 March).[37]

The Charters

The bulk of the Kelso cartulary is organized by place.[38] The Langton section forms nos 138–44.[39] While the de Vieuxpont charters anent Langton are not arranged chronologically, the first charter of this section (no. 138) is clearly the oldest, being Roger de Eu's original grant of the church of his vill of Langton to Kelso Abbey *c.* 1150 × 1162 (see above). The key charter on which I will focus in this chapter is no. 140 (MS fols 53^{v}–54^{v}), with details also of the closely related no. 141 (MS fols 54^{v}–55^{v}). Another charter purporting to have been issued by William de Vieuxpont (II) records the original grant of land attached to Langton church and contains the same detailed marches,[40] which the charter states were perambulated by William (II) himself on the day the charter was made. This charter is discussed in more detail in endnote [v] following the translation of the charter, below.

Summary

Dating from 1198 × 1214, or perhaps more precisely 1203 × 1214,[41] *Kelso Liber* no. 140 can be briefly summarized as follows: by it, William de Vieuxpont (III), son of William de Vieuxpont (II) and Emma of St Hilaire, in favour of Kelso Abbey (1) confirms the church of Langton and land, described in detail, as his father had given it; (2) adds enough land to make up his father's donated land to one full ploughgate, the added land being Girtrig, again described in

[35] This church was in turn replaced by the present church in 1872.

[36] For details of the medieval and early modern parish, with full references and a good selection of images, see *Corpus of Scottish Medieval Parish Churches*: *s.v.* Langton.

[37] 'St Cudbert in Langtoun in the Mers ane fair', Wedderburn, *Ane compendious buik*, p. 2 [5].

[38] Smith, 'The Kelso Abbey Cartulary', p. 99.

[39] *Kelso Liber*, I, nos 138–44.

[40] *Kelso Liber*, I, no. 142.

[41] See Appendix.

detail; (3) adds another piece of land called Colemannesflat; (4) grants the church of Horndean (Berwickshire); and (5) grants some shielings in the Lammermuirs called Dirringtons, as confirmed by the charter of his grandfather and his father.

Editorial conventions

Abbreviations have been silently expanded except in cases in which more than one expansion is possible. In such cases the supplied letters are put in angled brackets. Editorial additions, including footnotes, are put in square brackets. Capital and small letters have been retained, in as far as they can be clearly identified, as has original punctuation, including the forward slash, which the scribe or his exemplar used as a form of prosodic marking. I have indicated line breaks by double forward slashes. The rubric is printed in red. An almost identical charter covering the first two grants (nos (1) and (2) in the summary above), and with an almost identical witness-list, follows as no. 141 (MS fols 54^{v}–55^{r}). Significant differences between these two charters in terms of text and witness lists are given in the footnotes, as are differences between the MS text and that of the printed version (*Kelso Liber*).

> Item Will<el>mus de veteri ponte primog<enitus> . super dicta ecclesia et terr<is> de Langt'.[42]

> Vniuersis sancte matris ecclesie filiis et fidelibus / Will<elmu>s de veteri Ponte primogenitus // filiorum Will<elm>i de veteri Ponte / eorum scilicet quos habuit de Domina[43] Emme [*sic*] de sancto Hil//lario eternam in domino salutem / Nouerint vniuersi fideles ad quorum noticiam scriptum // hoc peruenerit . me de consensu coniugis mee . dedisse[44] et concess//isse . et hac presenti carta mea confirmasse . Deo et Ecclesie[45] sancte Marie de Kelchou // et monachis ibidem deo seruientibus Ecclesiam de Langtone cum terris et decimis /fol. 54^{r}/ et omnibus rectitudinibus Ecclesiasticis[46] ad eam pertinentibus sicut pater meus ante me fecerat cum // eadem terra quam ipse eidem ecclesie assignauit / et per eosdem

42 Rubric.

43 *d'na, Kelso Liber.*

44 followed by a space of *c.* 10 characters.

45 *eccl'ie, Kelso Liber.*

46 *eccl'iasticis, Kelso Liber.*

terminos . Videlicet[47] . sicut // via iacet ab orientali parte Ecclesie usque[48] in Wederburne[49] / Et ita per Wedyrburne[50] usque // humpulles[51] / Et in langelandes sicut diuisa uadit inter terram domini et terram ecclesie / // Et per Langelandes[52] uersus occidentem usque in Wedyrburne[53] [54] uersus aquilonem usque ad // Toftum Radulphi / et Toftum Gilberti[55] / Et ita usque ad diuisas Tofti Henr<ici> quondam // persone / Et ante Toftum eiusdem Henr<ici> iuxta quandam veterem fossam ubi ipse Henr<icus> // quondam posuit faldum suum / Preterea assignaui eis et addidi in territorio eiusdem // ville ad illam terram quantum illi defuit terre assignate ut sit una carucata plena//ria cum illa quam pater meus dederat . scilicet Gretryg'[56] / per has diuisas sicut fosse fac// te sunt de Holeburne usque in stocfotcluh[57] / Et sicut sica inde tendit usque in fulstro//thir / Et inde uersus orientem sicut terra culta et pratum de fulstrother [58] se diuidunt / // Et inde per fossas et lapides usque ad sicam que descendit inter Gretrig [59] et inter Stain//kilchestre [60]/ Et sicut sica descendit per tres frutices usque ad semitam que vadit in [61] hole//burne / ab occidentali parte de Chimbelawe [62]/ et sic sursum per holeburne usque ad fos//sas prenominatas que tendunt usque in Stocfotclouh[63][64] / Preterea dedi eis et concessi quan//dam particulam terre in territorio de Langtune que vocatur Colemannesflat per // suas rectas diuisas unde ecclesia de Langtone quondam saisiata fuit[65] / Conces//si

47 *terminos videlicet . sicut, Kelso Liber.*

48 *eccl'ie vsq', Kelso Liber.*

49 *Wedirburne, Kelso Liber* no. 141.

50 *Wederburne, Kelso Liber* no. 141.

51 *Hum pulles, Kelso Liber.*

52 *langelandes, Kelso Liber.*

53 *Wedirburne, Kelso Liber* no. 141.

54 *et ita per Wedirburne,* added *Kelso Liber* no. 141.

55 *Gileberti, Kelso Liber* no. 141.

56 *Gretrig', Kelso Liber* no. 141.

57 *stocfocclich, Kelso Liber; stocfocclou, Kelso Liber* no. 141; *stocfotclou,* MS fol. 54^{v}.

58 *fulstrothir, Kelso Liber* no. 141.

59 *Gretryg', Kelso Liber* no. 141.

60 *stainkylchestre, Kelso Liber* no. 141.

61 *que vadit usque in, Kelso Liber* no. 141.

62 *Chimbelaue, Kelso Liber* no. 141.

63 *Stocfocclouh, Kelso Liber; stocfocclou, Kelso Liber* no. 141; *stocfotclou,* MS fol. 54^{v}.

64 Here the close correspondence between *Kelso Liber* nos 140 and 141 ends.

65 The section from *Preterea* to *saisiata fuit* is found only in this charter.

etiam eis et hac presenti carta mea confirmaui Ecclesiam de Horuerdene et quasdem // scalingas in Lamb<er>more que vocantur Diueringdounes per suas rectas diui//sas tam plenarie sicut eas possident / et carta[66] aui mei et patris testantur et con//firmant /[67] Quare volo ut prefata ecclesia et prefati monachi de Kelchou predictas // ecclesias et predictas terras teneant et possideant inperpetuum in liberam et puram . et // perpetuam elemosinam ecclesiis collatis ita libere quiete honorifice et plenarie // sicut aliquas[68] alias ecclesias in Regno Scoc<ie> [69] liberius quiecius honori//ficencius et plenius tenent uel possident / Has autem donaciones et confirmaciones[70] feci // eis pro salute Dominorum[71] meorum Regis Will<elm>i et Regine et eorum filii Alexandri // et ceterorum liberorum eorum et pro salute mea et coniugis mee et heredum meorum / et pro // animabus Regis Dauid et Malcholom'[72] et comitis Henr<ici> / et pro animabus patris // mei et matris mee et omnium antecessorum et successorum meorum / Hiis testibus /fol. 54^{v}/ Johanne de Maccuswel vicecomite de Rokesburg'[73] / Johanne Decano[74] // de fogghou / Hugone ca//pellano domini Regis / IngelR<amo> clerico domini Regis / AndR<ea> Mansel / Rogero de lalescrie / // Will<elm>o de veteri ponte iuniore fratre Domini[75] / Will<elm>o iuniore filio Domini[76] [77] et multis aliis .

66 This should probably read *carte* (plural of *carta* 'a charter'), as the verbs which it is the subject of are plural.

67 The section from *Concessi etiam* to *confirmant* is omitted from *Kelso Liber*, no. 141. See Notes, below.

68 followed by *elemosinas* underscored for deletion; omitted *Kelso Liber*.

69 While this is almost certainly the correct expansion of *Scoc'*, *Kelso Liber* no. 140, the abbreviation underlying *Scoc'* in *Kelso Liber* no. 141 in MS fol. 55^{r}, is something like *scoc't'*. It may stand for *scottorum*.

70 *Hanc donacionem* for *Has autem donaciones et confirmaciones*, no. 141.

71 *d'norum*, *Kelso Liber*.

72 *Malcoloni*, *Kelso Liber* no. 141; presumably the copyist's error for *Malcolom*. *Malcoloni* also appears in *Kelso Liber* no. 142 (MS fol. 55^{r}).

73 This witness is omitted, *Kelso Liber* no. 141.

74 *decano*, *Kelso Liber*; *Kelso Liber* no. 141 prints *decā*, but MS fol. 55^{r} omits the abbreviation mark above *a*, which is followed by/and occurs at a line-end.

75 *d'ni*, *Kelso Liber*; *Will<elm>o de veteri ponte / iuniore fratre domini*, MS 55^{r}, entirely omitted in error, *Kelso Liber* no. 141.

76 *d'ni*, *Kelso Liber*.

77 *Domina Matilde de sancto Andrea*, added *Kelso Liber* no. 141.

There follows a full translation of *Kelso Liber* no. 140. The numbers in square brackets indicate notes, below, and for ease of reference I have divided the text into sections (1) to (5), as described in the Summary, above. Reference should also be made to the map on on the second page of this chapter, which depicts the area covered by sections (1) and (2) definitely, and section (3) probably, showing those places mentioned in both the text and the notes which can be confidently identified.

> Item, William de Vieuxpont firstborn anent the said church **[i]** and land of Langton.
>
> To all the sons and faithful of Holy Mother Church, William de Vieuxpont, **[ii]** firstborn of the sons of William de Vieuxpont, namely of those whom he had with Lady Emma of St Hilaire, **[iii]** gives eternal salutation in the Lord. May all the faithful, to whose notice this written document comes, know that by the consent of my wife, **[iv]** I have given and granted and by this my present charter have confirmed/established to God and to the church of St Mary of Kelso and to the monks serving God there,
>
> (1) the church of Langton with the land and teinds and all ecclesiastical rights pertaining to it, just as my father before me had done, with the same land which he assigned to the same church, and by the same marches, **[v]** namely: as the road lies on the east side of the church as far as Wedder Burn; **[vi]** and so by Wedder Burn as far as *Hume Pools; **[vii]** and into *Langlands as the march goes between the lord's land (i.e. demesne land) and the kirkland; and by (or through) *Langlands towards the west as far as Wedder Burn towards the north as far as Ralph's toft and Gilbert's toft; and so as far as the marches of the toft of Henry the late parson; **[viii]** and before Henry's toft beside an old ditch where the same Henry formerly set up his fold. **[ix]**
>
> (2) Furthermore I have assigned to them (Kelso Abbey) and added in the territory of that same vill (of Langton) as much land as that assigned land lacked to make it one full ploughgate, **[x]** along with that land which my father had given, **[xi]** namely Girtrig **[xii]** by these marches: as the ditches were made from Howe Burn **[xiii]** as far as in(to) *Stockfootcleugh, **[xiv]** and as the small burn (*sica*) then runs as far as into *Foulstrother; **[xv]** and then towards the east as the cultivated land and the meadow of *Foulstrother divide; **[xvi]** and then by the ditches and stones as far

as the small burn **[xvii]** which runs down between Girtrig and between *Stainkilchestre*; **[xviii]** and as the small burn runs down by three thorn-brakes [?] **[xix]** as far as the path which goes into Howe Burn on the west side of *Chimbelawe*; **[xx]** and so down by Howe Burn as far as the aforementioned ditches which stretch as far as *Stockfootcleugh. **[xxi]**

(3) Furthermore, I have given and granted to them a certain parcel of land in the territory of Langton which is called *Colemannesflat* [xxii] by its right marches, in which the church of Langton had once been seised. **[xxiii]**

(4) Also, I have granted to them and by this my present charter confirmed the church of Horndean **[xxiv]**

(5) and certain shielings in Lammermuir which are called (the) Dirringtons **[xxv]** by their right marches as fully as they possess them and the charters of my grandfather and father testify and confirm.

For this reason I want the said church and the said monks of Kelso to hold and possess the said churches and lands for ever in free and pure and perpetual alms, and with churches granted as freely, quit, honourably and fully as they hold or possess any other churches in the kingdom of Scotland equally freely, quit, honourably and fully. **[xxvi]** These donations and confirmations **[xxvii]** I have made to them for the salvation of my lords, King William and the Queen, **[xxviii]** and their son, Alexander, **[xxix]** and the rest of their children, and for my own salvation and for that of my wife and of my heirs, and for the souls of King David and Malcolm and Earl Henry, and for the souls of my father and of my mother and of all of my ancestors and successors. With these witnesses: John of Maxwell sheriff of Roxburgh, **[xxx]** John dean of Fogo, **[xxxi]** Hugh chaplain of the lord king, **[xxxii]** Ingelram clerk of the lord king, **[xxxiii]** Andrew Mansel, **[xxxiv]**, Roger de la Léqueraye, [xxxv] William de Vieuxpont junior, brother of the lord, William junior son of the lord **[xxxvi]** and many others.

NOTES

[i] 'The said church' refers back to the two preceding charters (*Kelso Liber* nos 138, 139), both of which concern the church of Langton.

[ii] This is William de Vieuxpont (III), for whom see above, The de Vieuxponts.

[iii] For Emma, a woman of importance, see above, The de Vieuxponts.

[iv] Probably Matilda of St Andrew, for whom see above, The de Vieuxponts.

[v] The details of the grant of the church along with the land within the boundaries in section (1) (as far as Henry's fold) is also found practically word for word in a charter to Kelso Abbey purporting to have been issued by William's father, William de Vieuxpont (II), along with the church of Langton (*Kelso Liber*, I, no. 142, MS fols 55^{r}–55^{v}). See also above, under subheading 'The Charters'. This charter is discussed in detail by Andrew Smith, who concludes that it is a 'cut-and-paste' production using various charters issued by his son (namely nos 140, 141).[78] From the witness list it would be dateable to 1191 × 29 July 1198.

[vi] This seems to be the older name for the burn now known as Langton Burn. It flows eastwards past Wedderburn Castle and Wedderburn Mains in the parish of Duns, and presumably gave its name to this estate. It means 'burn associated with wedders', i.e. castrated rams, and testifies to the importance of sheep-rearing along its course at the time of its coining.

[vii] For a charter settling a dispute between Kelso Abbey and William lord of Hume over 'the pools' (*le pullis*) in 1268, see *Kelso Liber*, I, no. 132.[79] These and their associated land lie on the Eden Water, and are probably represented by the detached part of Hume parish (until 1891) which lay around NT697396, just north-east of Mellerstain House. They cannot be the *Hume Pools of this Langton charter, which must lie east of Langton old kirk.

[viii] For Henry the parson of Langton, see above, under subheading 'Langton'. If *Kelso Liber*, I, no. 142 can be trusted, he may have still been alive in the 1190s, but see Note [v], above, for the likelihood that no. 142 is a confection.

[ix] *faldum* is a Latinized form of Scots *fald*, later *fauld*, 'a fold, an enclosure for cattle, sheep, or other domestic animals; an enclosed piece of ground used for cultivation; a small field'. This is the definition in *A Dictionary of the Older Scottish Tongue (up to 1700)*,[80] which gives this as the earliest attestation of the word in Scotland.

78 Smith, 'The Kelso Abbey Cartulary', pp. 267–70, including the Latin text.

79 Called *Pollys* in 1268, *Kelso Liber*, I, no. 291.

80 Hereafter *DOST*.

[x] Later a ploughgate was 104 Scots acres of arable (a Scots acre being slightly larger than an Imperial acre, viz 1.26 Imperial acres). This would make a ploughgate (also known as a carucate) roughly 130 Imperial acres. However, there was much regional variation in this unit depending on the quality of the soil, the lie of the land, etc. It got its name from the fact that it was notionally an area that could be ploughed by a team of 8 oxen in one year.

[xi] That is the land confirmed and defined in the previous section.

[xii] A small settlement shown on OS 6-inch 1st edition at NT775511, now completely ploughed over. The first element is Old English *grēot* or Older Scots *grete, greit* 'gravel; sand', so meaning 'gravel ridge'.[81] It has survived as the field-name Gartrig on the farm of Ladyflat.[82]

[xiii] Still on modern OS maps, this is a Scots name containing *howe* (also *how*, earlier *hol(l)* and *hole*), 'a hollow or low-lying piece of ground';[83] it is used here attributively to mean 'a burn running through a hollow'. The eponymous hollow may be the low land south-west of modern-day Chatterdeanford Bridge (NT773504). A tributary, the Kirk Burn, joins it at Chatterdeanford Bridge. For the ditches, see Note [xiv], below.

[xiv] The final element in *Stockfootcleugh (*stocfotcluh*) is Scots *cleugh* 'a small steep valley or gorge', and it may refer to the small valley through which the Kill and Kirk Burns flow south-west of the farm-steading of Ladyflat. Stockfoot itself survives as a name into the modern period, with three contiguous fields, *East Stockfoot, West Stockfoot*, and *South Stockfoot* marked on a plan of Ladyflat Farm from *c.* 1774 at around NT767510,[84] with the southern boundary of the field of South Stockfoot *c.* 300 m north of the feature proposed as the eponymous cleugh, above.[85] The meaning of the name is obscure. *DOST* defines *stock, stok(k)* as 'the trunk of a tree, chiefly as stripped of its branches; a tree stump; a block or length of wood; a log'. It occurs several times in the compound *stockbrig*, later *stockbridge*, 'a bridge made out of logs', the earliest example being *Stokbryg* 1295,[86] with reference to a bridge near Paisley. However, it is not clear what this might mean when compounded with Sc *fu(i)t* 'a foot', in the sense of a feature at the foot of something, unless *stok* stands

[81] *DOST, s.v. grete, greit.*

[82] I am grateful to Rhona Darling of Ladyflat for this and other information concerning the modern field-names of the farm.

[83] *DOST, s.v. howe.*

[84] RHP142397, Sheet 6; available on *ScotlandsPlaces.*

[85] Stockfoot has survived as the name of a field on Ladyflat farm immediately north of *East Stockfoot*, and bounded on its north side by the Back Burn.

[86] *Registrum Monasterii de Passelet*, p. 94.

for something made out of logs, perhaps a stockade, wooden steps, or a piece of land cleared of trees.[87] The word derives ultimately from Old English *stocc*, with the same meaning. Alternatively, but less likely, we might be dealing with the Old English *stoc* 'a place, a secondary settlement, an outlying farm',[88] very common in English place-names such as Woodstock or Basingstoke, but not (so far) identified in a Scottish place-name. If *Stockfootcleugh has been correctly identified, then the ditches to there from Howe Burn may have run along what is later shown as the parish boundary south of Kirk Burn between Langton and Polwarth.

[xv] The 'small burn' would seem to refer to the Kill Burn,[89] flowing eastwards into the Kirk Burn, which in turn flows into the Howe Burn. *Foulstrother means 'a foul or muddy marshy place' (Older Scots *strother*). It cannot be the present-day Foul Burn, the upper reaches of the Langton Burn, with Foulburn Bridge at NT719517, lying as it does over 5 kilometres to the north-west, at the very western edge of the parish. For its probable position, see next note [xvi].

[xvi] The burn is described as running as far as the division between the two components of *Foulstrother, with its arable and the grazing. From this it is clear that the name *Foulstrother had come to be applied to a settlement, or at least to much more than a marsh. It probably lay on the lands of present-day Middlefield farm, with the eponymous *strother* bounded by the bend in the Howe Burn on the west and south, and by the small burn known as the Back Burn on the east (which is also the parish boundary between Langton and Duns) and centred at around NT786507. This probably represented the meadow or grazing land, while the cultivated or arable land lay on the slightly raised ground sloping southwards down from modern Middlefield farm-steading.

[xvii] This burn is very likely the Back Burn (all OS large scale maps), which rises north-west of Polwarth and zigzags its way eastwards, for part of the way forming the northern march of the lands of Girtrig as far as the now lost farm of Crease (NT786512). It then turns southwards to flow into the Howe Burn at NT790507.

[xviii] Perhaps 'A chester (i.e. an old fort or a site perceived as such) associated with kestrels'? Older Scots *stanchall, stanchel(l), stainȝell, stainchel(l), stenchel* (later Scots *stanchel*) 'a kestrel'.[90] This belongs to a recognisable sub-group of *chester*-names combined with bird-names, such as: *Cauchesterlawe* (Greenlaw, Berwick-

[87] Compare ON *stokkland* with this meaning (Smith, *English Place-Name Elements*, II, p. 156). Note also Stockstruther ROX (*Stocksturder* 1549), which May Williamson interprets as 'marsh with tree-stumps in it': 'Non-Celtic Place-Names', p. 259 [original pagination].

[88] Smith, *English Place-Name Elements*, II, pp. 153–56.

[89] Kill Burn 'a burn associated with one or more kilns' (Sc *kil(l)* 'a kiln').

[90] *DOST*, *s.v. stanchall.*

shire) (*Kelso Liber*, I, no. 78), in which the first element is probably Older Scots *ka* 'a jackdaw'; and *Laverockchester (*Lauerocchester* 1229 × 1234), Older Scots *laverok* 'a lark', on or beside Laverock Law near Coldingham.[91] There is no obvious candidate for this chester, which has almost certainly been ploughed out. It presumably lay a short distance north of the Back Burn, and by the time of this charter may have applied to a settlement perhaps including land on which the planned village of Gavinton was built in the later eighteenth century.

[xiv] Compare *fruticetum* 'a thicket', *DMLBS*.

[xx] At the point the boundary turns southwards, leaving the Back Burn and heading for the Howe Burn. It is impossible to say exactly where this line ran, but it may have followed roughly the western march of the later farm of Ladyflat, meeting the Kill Burn at around NT761510 or NT765506. In this case the hill of *Chimbelawe* would be the extended raised area on which Boglands Plantation now stands. While the second element is clearly Older Scots *law* 'a (conspicuous) hill or hillock', the first element may be the rare Old English *cimb* 'edge, rim', attested only in compounds and in *cimbing* 'joint, conjunction'; it 'seems to appear in Chimhams (*Chimbeham* 1203) [...] where it may refer to a ridge'.[92]

[xxi] The boundary then follows the Howe Burn down to the ditches where it started. This only works if we assume that the name Howe Burn was applied loosely not only to the present-day one, but also to its western tributaries the Kill Burn and the Kirk Burn. Here the close correspondence between nos 140 and 141 ends.

[xxii] A very early attestation of Scots *flat* 'a level piece of ground; a level field'. It may refer to the eponymous *flat* of Ladyflat, though given the boundaries as described in section (1), above, it is difficult to identify exactly. In its favour is the specific element *lady*, which in Scotland frequently refers to St Mary,[93] to whom Kelso Abbey was dedicated. However, it could be any level piece of ground in the parish. The first element is clearly a personal name. The name Col(e)man is relatively frequent in the Coldingham area around this time, for which see the *People of Medieval Scotland* [PoMS] website, where it is assumed that it derives from Old Gaelic *Colmán*, a diminutive of Colm (Columba). For example, the name occurs twice in the witness list of a later twelfth-century charter (1175 × 1189), namely Uchtred son of Colman (Vchtred[o] filio Coleman) and Colman the grieve (Coleman preposito).[94] While we cannot be certain, it is likely that these two men called

[91] Raine, ed., *The History and Antiquities of North Durham*, Appendix no. CCLVI.

[92] Parsons, *The Vocabulary of English Place-Names*, *s.v. cimb*.

[93] See Hough, '"Find the Lady"'.

[94] Raine, ed., *The History and Antiquities of North Durham*, Appendix no. CCI.

Col(e)man are two separate individuals. There are also three twelfth-century occurrences of this name in the Durham *Liber Vitae*, twice as *Coleman*, once as *Colman*.[95]

[xxiii] i.e. the church of Langton had once been given full possession of *Colemannesflat*.

[xxiv] Horndean,[96] a medieval parish on the Tweed, now part of Ladykirk parish. Already in 1160 × 1161 William de Vieuxpont had granted the church of Horndean to Kelso Abbey in the presence of Bishop Arnald of St Andrews.[97] The de Vieuxponts are mentioned as holding Horndean 1165 × 1170, along with Carriden, West Lothian, and lands in England,[98] when King William re-granted all these lands to William de Vieuxpont (I), following a dispute.[99]

[xxv] Presumably the conspicuous hills of Dirrington Great Law and Dirrington Little Law, these now lie in the parish of Longformacus just beyond the west end of the parish of Langton. The 'multure of the shielings of Dirrington (*Diueringdon*')'[100] is mentioned in a charter of 1203, when a dispute about it was settled between Kelso Abbey and William de Vieuxpont (III).[101] The fact that they were due multure (mill-dues) means that these 'shielings' (Latin *scalinga*, plural *scalinge*) were more than just summer pasture for sheep and cattle. This was presumably after William (III) (*de vyerpunt*) had, with permission of his wife Matilda (*Matildis*), granted to Kelso 'certain shielings in the Lammermuirs which belonged to Horndean' (*quasdam eschalingas in* ***lambremore*** *. que pertinebant* ***ad hworuorden*** [emphasis added]).[102] The distance as the crow flies from Horndean to the Dirringtons is almost 10 kilometres and is a good indication of how complex and well-organized the system of transhumance was by this time, with communi-

[95] This name is discussed by Russell, McClure, and Rollason, 'Celtic Names', p. 37, in the following terms: 'If this is Irish [read Gaelic], it reflects Old Irish *Colmán*, a derivative of *Colum* (Columba) with the hypocoristic suffix, *-án*; see Russell (2001, pp. 246–47). It was a very common ecclesiastical name in early Ireland; see the long list in *Corpus Gen. Sanct. Hib.*, pp. 233–35. An additional source of *Coleman* in *LVD* may be Continental Germanic *Col(e) man*, although in northern England Old Irish *Colmán* was probably the more usual source even in the post-Conquest period (Feilitzen 1937, p. 218)'. See the publication itself for full details of the references.

[96] For a discussion and early forms of this Old English name, see BPNR Database, *s.n.*

[97] Shead, ed., *Scottish Episcopal Acta*, I, no. 148, first printed *Kelso Liber*, II, no. 417.

[98] Barrow, ed., *Acts of William I*, no. 84.

[99] For details of this dispute, see Barrow, ed., *Acts of William I*, p. 182.

[100] For a discussion and early forms of this Old English name, see BPNR Database, *s.n.*

[101] *Kelso Liber*, I, no. 143. For more details, see Appendix.

[102] Dated to *c.* 1200. *Kelso Liber*, II, no. 319.

ties throughout the Merse exploiting the good upland grazing of the Lammermuirs despite considerable distances.[103]

[xxvi] There would appear to be an omission in this clause, as it does not make complete sense as it stands. The equivalent clause in no. 141 is considerably longer and probably supplies roughly what is missing. This can be translated: '... perpetual alms, with tofts and crofts, with grazings and fuel and all other common easements of that same toun according to an assize of the bishopric, as much as belongs both to the church and to one carucate of land given in pure and perpetual alms to the church as freely, quit, honourably and fully as they hold or possess any other church in the kingdom of Scotland equally freely, quit, honourably and fully'.

[xxvii] Simply 'This donation I have made', no. 141.

[xxviii] Queen Ermengarde (de Beaumont), whom King William married in 1186. She died in 1234.

[xxix] The future Alexander II, born August 1198. This supplies the earliest possible date for this charter.

[xxx] Omitted from the otherwise very similar witness-list in no. 141, as well as from *Sheriffs of Scotland*.[104]

[xxxi] John was dean of the St Andrews deanery known variously as Fogo, Stichill, Merse and Nenthorn, with Merse becoming its established name from 1268 onwards. John dean of Fogo appears in several charters and can be dated 1194 × *c.* 1220.[105]

[xxxiii] Probably Hugh 'the Little' (*parvus*).[106] He witnesses no. 317 (1189 × 1195) as 'Hugh my (King William's) chaplain (and) clerk' (*Hugone capellano Clerico meo*).[107]

[xxxiii] Ingelram or Ingram, clerk of King William, appears only in the three de Vieuxpont Langton charters *Kelso Liber*, I, nos 139–41. He is in PoMS as Ingram, with the floruit 1198 × 1214.

[103] The transhumance economy in the Scottish and English borders in the 12th and 13th centuries is the subject of Winchester, 'Shielings and Common Pastures'. He discusses the considerable distances between a settlement and its assigned grazing, citing, amongst others, the Horndean-Lammermuirs example, see pp. 292–93.

[104] Reid and Barrow, eds, *Sheriffs of Scotland*.

[105] Watt and Murray, eds, *Fasti Ecclesiae Scoticanae Medii Aevi*, p. 416.

[106] For whom see Barrow, ed., *Acts of William I*, pp. 32, 61–62 nn. 33–34.

[107] Barrow, ed., *Acts of William I*, no. 317.

[**xxxiv**] Andrew Mansel or Maunsel (PoMS form of name); this is presumably the same Andrew Mansel (*Andreas Mansel*) who grants Kelso Abbey permission to build a mill-pond on his land on the east side of the town of Roxburgh 1180 × 1221, from which charter we know that he had a son and heir named Walter, and a brother named Gregory, who was a chaplain.[108]

[**xxxv**] No modern form of this name appears to have survived. It derives from the small Normandy commune St-Jean-de-la Léqueraye, about 25 kilometres north-east of Vieuxpont-en-Auge, where the de Vieuxponts probably hail from. Barrow discussed this family, stressing this geographical connection, which is reflected in the de la Léquerayes' association with de Vieuxponts in Scotland.[109]

A close reading of the above charter provides what can almost be described as a virtual walk through the landscape of the Merse over 800 years ago. We have encountered not only ditches, enclosed fields, arable and grazing side by side, prehistoric remains and a sophisticated transhumance economy. We have also encountered several of the individuals who inhabited that landscape, not only the aristocracy such as the de Vieuxponts and their circle, but also a farming parson and three peasants or small tenants called Ralph, Gilbert and Colmán. And finally, despite the Latin medium in which all this is conveyed, we have also encountered some of the earliest traces of the Scots language in the form both of lightly-clad loan-words into Latin and of place-names, as well as place-names which reach back into the Bernician (Old English) period.[110]

[108] *Kelso Liber*, II, no. 507.

[109] Barrow, *The Anglo-Norman Era*, pp. 94, 181–82 n. 6, where he spells the name 'de la Lecqueraye'.

[110] I would like to thank Dauvit Broun, Dàibhidh Grannd, Carole Hough, Peter McNiven (map), and Eila Williamson for help and advice in writing this chapter.

Appendix: A Note on Dating

As stated in the opening sentence of this chapter, the charter which forms its core, *Kelso Liber* no. 140, can be dated broadly to between the birth of the future King Alexander II, in August 1198, and the death of his father, King William I (the Lion) in December 1214. However, there is circumstantial evidence to suggest that the earliest possible date for it is late May 1203. On 21 May various disputes between William de Vieuxpont (III) and Kelso Abbey were settled in the presence of William bishop of St Andrews and an array of important men, both ecclesiastical and lay (*Kelso Liber* no. 143). The root cause of the trouble was a remarkable one: on the death of William's father, William de Vieuxpont (II), in England, Kelso Abbey had promised to bring back his bones and bury them in the abbey's cemetery. However, they had failed to do this, and this seems to have contributed to generally worsening relations between the Vieuxponts and Kelso Abbey over an unspecified period of time. At the above-mentioned hearing, a compromise was reached, which appears to have satisfied both parties. William declared them released both from the obligation to repatriate his father's remains, and from all other pleas or plaints (unspecified) which he had against them. Only one plaint is excepted from this: that is the plaint anent the multures (mill-dues) of the shielings of Dirrington, which was settled there and then by Abbot Osbert agreeing to pay William the 30 shillings which he owed him, with an additional 40 shillings. Furthermore, Osbert promised that William's father's soul would be forever specially included by name amongst the other special benefactors of Kelso in the mass celebrated in the monastery for the faithful.[111]

[111] The following extract from *Kelso Liber* no. 143 has been checked against the manuscript (fol. 55^{v}) and, apart from very small details such as the occasional capital letter in the printed version for small letter in the MS, the printed version is remarkably accurate. '[…] facta est hoc amabilis composicio inter . O<sbertum> . Abbatem Kelchoens<em> et monachos eiusdem loci et Will<elmum> de veteri ponte super controuersiis placitis et querelis quas iam dictus . W<illelmus> . aduersus eos habebat . videlicet quod prenominatus . W<illelmus> . prenominatos Abbatem et monachos Kalchoenses de ossibus patris sui de Anglia reportandis et in cimiterio Kalchoensi tumulandis quietos clamauit inperpetuum / et de omnibus aliis querelis quas usque ad illum diem aduersus eos se habere dicebat Excepta querela de multura scalingarum de Diueringdon' que communi parcium consensu iudicio ecclesie tunc erat sopienda / Et prefatus abbas consilio domini Episcopi prenominati et Conuentus et aliorum virorum prudentum [*sic*, for the grammatically more correct *prudentium*] pro bono pacis remisit prenominato . W<illelmo> xxx . solidos . quos ei debebat / et insuper dedit ei xl . solidos . Et pietatis

It is inconceivable that William de Vieuxpont (III) would have issued his generous charter (no. 140) to Kelso Abbey while it refused to honour its promise to bring back his father's bones from England and to bury them in its cemetery, with all the spiritual benefits for the deceased that this would imply. The grants and confirmations made by William (III) in no. 140 only make sense once this and other contentious issues between him and Kelso had been settled.

intuitu ad sepe dicti W<illelmi> . supplicacionem concessit ut anima patris eius specialiter et nominatim inter ceteros speciales benefactores monasterii Kalchoens<is> in missa in eodem monasterio pro fidelibus celebranda inperpetuum comprehendatur . […]'.

The Tale of Two Wandering Charters: Towards the Political and Environmental Background of the Mac-Dòmhnaill-Mhic an Tòisich Alliance in the 1440s

Philip Slavin*

History, Heritage and Politics, University of Stirling

Introduction

In a city settled by the Scots (Montreal), in a university bearing a Scottish name (McGill), in a library bearing a Scottish name (McLennan), situated on a street bearing a Scottish name (McTavish), there is a disarrayed collection of over 100 late medieval and early modern documents, containing fifteen Scottish charters, from the fifteenth through the seventeenth century. I was able to consult and tentatively catalogue, number and briefly describe eighty-nine of these documents during my postdoctoral tenure at McGill (2010–2012), but, to my best knowledge, my catalogue and notes have not been inputted into the Rare Books and Special Collections Library catalogue — or indeed, any other McGill library catalogue. According to the library's website, the collection boasts some 225 medieval European manuscripts, in addition to 'a considerable number of charters, papal letters and legal documents primarily from the

* The present humble tribute to the late Alasdair Ross explores a series of late-medieval and early modern Scottish documents at the McGill University Rare Books and Special Collections Library, focusing on their provenance and acquisition. It thus reflects not only our common interests in hunting for original archival documents, but his own interests in late-medieval palaeography and charters through which West Highland lordship expressed political and social power in the late Middle Ages. These interests have manifested themselves in several of Alasdair's publications, including a study of lordship in Lochaber under MacDhòmhnaill overlordship, and, most recently, in his important monograph on late-medieval *dabhaichean*.

'With Our Backs to the Ocean': Land, Lordship, Climate Change, and Environment in the North-West European Past. Essays in Memory of Alasdair Ross, ed. by Richard Oram (Turnhout: Brepols, 2023)
EHNAW 5, pp. 205–229 BREPOLS PUBLISHERS 10.1484/M.EHNAW-EB.5.131276

fifteenth, sixteenth and seventeenth centuries'.[1] Of all these documents, only one, a 1531 charter from Bavaria, is listed in Seymour de Ricci's 1937 *Census of Medieval and Renaissance Manuscripts in the United States and Canada.*[2] Ricci's census lists an additional fourteen charters, consulted by him during his research trip to Montreal in 1932, but none of these seem to be a part of the collection.[3]

Hence, this collection remains obscure and unexplored by scholars. Equally obscure are the origins of this collection — and of the Scottish charters, in particular. According to Dr Richard Virr, the then curator of the Rare Books collection, the collection may have made McGill its home sometime in the 1920s or the 1930s, but he was unable to provide any further details. In any event, I was unable to trace the accession number of the collection — or, indeed, any other information to shed light on its acquisition by McGill. A close look at the contents of the collection, on the one hand, and the history of acquisition of manuscripts and rare books by the McGill on the other, may provide some important clues.

There is neither geographic, nor chronological uniformity to the collection. The vast majority of the charters examined and described by me (fifty out of eighty-nine, or 56 per cent) come from England. The fifteen Scottish documents represent the second largest geographic group (17 per cent). All but one Scottish document are related to Inverness-shire and Ross-shire, either in conjunction with the town of Inverness or with landed estates (and their administration) of clan-chieftains, including two charters of Alasdair MacDhòmhnaill (Alexander MacDonald), Lord of the Isles and Earl of Ross (*c.* 1423–1449) to Máel Coluim Mhic an Tòisich (Malcolm Macintosh) (See Appendices 1 and 2), representing the core of this paper. The main reason for choosing and focusing on these two chapters is that their contents reflect Alasdair Ross' keen interest in and manifold contributions to the history of late-medieval northern Scotland in general, and highland lairds' charters in particular.[4] A further twelve charters originate in Germany, in addition to two French, five Italian,

[1] McGill Library, 'Archival Collections and Manuscripts – Medieval European Manuscripts'.

[2] De Ricci, *Census of Medieval and Renaissance Manuscripts*, II, p. 2222.

[3] De Ricci, *Census of Medieval and Renaissance Manuscripts*, II, pp. 2221–22.

[4] I revealed my discovery of the two MacDhòmhnaill charters (alongside with other Scottish charters) at the McGill University library to Alasdair, and we agreed to publish their original form in a co-authored paper. Unfortunately, his untimely death in August 2017 prevented our plan to edit and publish the charters, in their original form, from materialising. In his paper,

four Dutch and one papal document. Of the total eighty-nine examined documents, only nine charters were produced in the late medieval period (the fourteenth and fifteenth centuries); nineteen come from the sixteenth century, thirty-six originate in the seventeenth century and a further two documents were issued in the early nineteenth century. Such a geographic and chronological disarray suggests two possibilities: (1) the collection was either acquired *en masse* by the McGill library; (2) or, it is, in fact, made-up of several existing collections, donated to or purchased by the library. The sheer size of the collection and the fact that all its contents have been stored together (in six boxes) may lend more plausibility to the former hypothesis.

McGill University Library has an extensive history of manuscripts and rare book acquisition, either through private endowments, or through business dealings with European and North American traders. The preponderance of English and Scottish charters suggests that at least a large portion of the collection must have come from the United Kingdom. If we stick to the hypothesis that the entire collection was acquired *en masse*, then it must have been purchased from some British manuscript/rare book dealer. In fact, in the 1910s, 1920s, and 1930s, McGill Library had extensive dealings with at least two London-based manuscript/rare book dealers, from whom it acquired a substantial number of items, now deposited at the Rare Books and Special Collections department of the library. The traders in question were the Maggs Bros of London (still functioning) and Reginald Atkinson. Thus, in 1921, McGill Library purchased a number of manuscripts and early printed books from the Maggs, including a unique nineteenth-century Samaritan manuscript from Syria and early sixteenth-century Italian commentaries on Aristotle's *De Interpretatione*.[5] The following year, the Maggs sold McGill a fragment of Lydgate's *The Fall of Princes*.[6] At some point, the McGill library purchased an original charter of Queen Mary of England, creating John Bridges, Baron Chandos of Sudeley.[7] Likewise, between 1921 and 1927, the library acquired a series of late medieval and early modern manuscripts, individual leafs and deeds from R. Atkinson, including an early-seventeenth-century Armenian liturgical book, a fifteenth-century English missal, a dozens

'Ghille Chattan Mhor and Clann Mhic an Tòisich Lands', he did mention that the originals of the charters had 'recently been located in Canada' (pp. 117–18, nn. 59 and 60).

5 Pummer, 'A Samaritan Manuscript'; see entry on *SDBM*.

6 Edwards, 'The McGill Fragment'.

7 De Ricci, *Census of Medieval and Renaissance Manuscripts*, II, p. 2221 (no. 167).

of initials and borders from late medieval choir books, a *c.* 1440 French Book of Hours and (most importantly for our purposes) several late sixteenth-century English deeds. All this information regarding McGill's dealings with Atkinson derives from De Ricci's catalogue, rather than from the library records. It should be pointed out that the McGill library was by no means the only university library that Atkinson was supplying with late medieval and early modern manuscripts, documents and books. In particular, in 1935, the dealer sold eight fifteenth- to seventeenth-century English deeds to the University of Virginia Library.[8]

Both the Maggs and Atkinson would issue sales catalogues, and it is through these catalogues that McGill Library would have learnt what was available for sale and subsequently order the items of interest. In both instances, there were several sales of large collections of private deeds and charters in the 1920s and 1930s. In particular, in 1930 the Maggs issued a sales catalogue (Catalogue No. 542, entitled *The Art of Writing, 2800 BC to 1930 AD*) featuring almost 300 original manuscripts and other documents from antiquity to the early twentieth century, among which were several dozens of charters written on vellum, including a 1475 order of James III of Scotland to the sheriff of Edinburgh to destrain the goods of John Napier for a debt of eighty-six marks to David Kincaid.[9] Obviously, to attract potential buyers, the Maggs had to be selective in their advertising of manuscripts and, thus, chose to exhibit only illustrated manuscripts and royal charters (chiefly from England and Spanish kingdoms), apparently omitting private deeds.[10] Likewise, in the course of a systematic reading of all other catalogues of the Maggs Bros for the period 1920–1936, held at the British Library, I was unable to find any mention of the two McGill charters. It should be noted that some of these catalogues advertised late medieval royal Scottish charters, which may indicate the Maggs provenance of the McGill charters, omitted on the account of their private, rather than royal origin.[11] No Scottish charters are found in the Atkinson catalogues for that period.

[8] de Ricci, *Census of Medieval and Renaissance Manuscripts*, II, p. 2172 (nos 6–13).

[9] Maggs Bros, *Art of Writing*, pp. 331–32 (no. 187).

[10] Maggs Bros, *Art of Writing*, pp. 331–32 (no. 187).

[11] For instance, Maggs Bros, Catalogues no. 449 (1924), nos 51–52 and no. 459 (1925), no. 79 (Robert I); no. 445 (1923), no. 2599 (James I); no. 396 (Autumn 1920), nos 2279–81, no. 425 (Summer 1922), nos 1341–1341a, no. 439 (123), no. 740, no. 464 (1925), nos 1229–30, no. 471 (1925), no. 2857, no. 473 (Spring 1926), no. 303, no. 494 (Autumn 1927), nos 1601–02 (James III).

Tracing the History of the McGill MacDhòmhnaill Charters

How did the two MacDhòmhnaill charters (and additional Scottish documents) end up in an antiquarian bookshop in London? Fortunately, it is possible to reconstruct the whereabouts of the two documents since their production. Given their bilateral nature, it is likely that the charters were produced in two copies, the one to be in possession of Alasdair MacDhòmhnaill and his heirs and the other one to be stored among the family muniments of Máel Coluim Mhic an Tòisich (Macintosh) and his successors. While it was impossible to trace the Mhic an Tòisich copies of the charters, I was able to do so for the MacDhòmhnaill ones.

The charters are known to have been in possession of Sir Seamus MacDhòmhnaill (James Macdonald), a direct descendant of Alasdair MacDhòmhnaill, the 9th and the last laird of Dunyvaig, who died in his London exile in 1626.[12] Seamus MacDhòmhnaill lost all his possessions after his downfall following the Battles of Benbigrie and Traigh Ghruinneart (1598) and his subsequent imprisonment (1601–1615). His lands were temporarily seized by Aonghus MacDhòmhnaill (Angus Macdonald), Seamus' father and the previous laird of Dunyvaig, but the latter had to surrender them in 1608 to King James VI, who, in turn, awarded them to Gille Easbuig Caimbeul (Campbell), 7th Earl of Argyll (*c.* 1575–1638), a sworn enemy of the *Clann* MacDhòmhnaill.[13] Presumably, the MacDhòmhnaill muniments fell into his hands, too. Both charters were mentioned in the so-called Kinrara manuscript of 1679, a genealogical history of the *Clann* Mhic an Tòisich composed by Lachlan Mackintosh of Kinrara.[14] Lachlan Mackintosh partially based his work on earlier manuscripts, and it is possible that if he consulted the original charters, then these were the Mhic an Tòisich copies.

One of the two charters (the 1447 charter, McGill Library Charter 14) emerged on 23 August 1781 in Edinburgh, where it was copied into the second series of the so-called *Register of Deeds*, compiled by David Dalrymple, 3rd Baronet Lord Hailes (1726–1792), the Edinburgh lawyer and historian.[15] It is unknown how the charter made it into Dalrymple's office and it is unknown

[12] Fraser-Mackintosh, *Last Macdonalds*, title page and p. 1.

[13] *History of the Feuds and Conflicts Among the Clans*.

[14] Ross, '*Ghille Chattan Mhor*', pp. 112–13.

[15] National Records of Scotland [hereafter NRS], GB 234; the charter is found in NRS, RD2/230/2, no. 1104.

when it passed into new hands. In the late 1820s the *Register of Deeds* was in possession of Alexander Sinclair (1794–1877), the Edinburgh genealogist and family historian and it is from the notarial copy in the *Register*, rather than from the original, that the contents of the charter have been copied by Donald Gregory (1803–1836), a prominent historian of the Western Highlands and the Isles.[16] It does not mean, however, that Sinclair was in possession of the original charter.

Both McGill charters re-emerge in the late nineteenth century, as a part of an extensive collection of original late-medieval and early modern documents related to the history of the Highlands and the Isles of Charles Fraser-Mackintosh (1828–1901), a renowned Inverness-based lawyer, land developer, politician, antiquarian, local historian, and an avid proponent of Gaelic language and culture. It is unclear how and where from Fraser-Mackintosh acquired the two charters, but it is clear that all the Scottish documents at the McGill Library collection came from his vast collection. He referred to some of these documents in his work. Thus, in his 1895 monograph *The Last MacDonalds of Isla* he made use of several fifteenth-century MacDhòmhnaill charters. He described the original documents, upon which his study is based, as 'sometime belonging to Sir James Macdonald, the last of his race, now in the possession of Charles Fraser-Mackintosh'.[17]

Fraser-Mackintosh's historical enthusiasm meant collaboration with other historians and antiquaries. He is known to have been lending some of his original documents, including the MacDhòmhnaill charters, to a number of his colleagues. Thus, both documents were used and their contents described in *Historical Memoirs of the House and Clan of Mackintosh* (1880) by Alexander Mackintosh Shaw (1844–1932).[18] At some point in the late nineteenth century (certainly before 1896), Fraser-Mackintosh lent the two MacDhòmhnaill charters to an unknown historian, who copied them and a plethora of other documents pertaining to the history of the *Clann* MacDhòmhnaill, in a rather sloppy manner, into a single register, known as the *Macdonald Collections*, consisting of three volumes and currently held at the National Records of Scotland.[19] The *Macdonald Collections* served as a basis for the 1896–1904

[16] National Library of Scotland [hereafter NLS], MS 2131, f. 69.

[17] Fraser-Mackintosh, *Last Macdonalds*, title page and p. 1.

[18] Mackintosh, *Historical Memoirs*, p. 143.

[19] NRS, GD103/2/12–14; the copies of the relevant charters (Charters 40 and 14, according to my tentative catalogue) are found in GD103/2/13, pp. 1109–11 and 1437–38.

monograph *The Clan Donald* by A. Macdonald and A. Macdonald. Here, the authors printed the two charters in the first volume (1896), allowing the same errors committed by the compiler of the *Macdonald Collections* register, *plus* their own additional errors, resulting from their inaccurate copying of the register.[20] The 1447 charter (McGill Charter 14) has also been used and described, with its contents summarized, by Henry Paton in his *Mackintosh Muniments, 1442–1820* (1903).[21]

Fraser-Mackintosh died childless in 1901 and his book and document collection seems to have remained as it was for another twenty years, until his wife Eveline May Holland (died 1925) donated his library to Inverness Burgh Library in 1921.[22] A good portion of his document collection was given to the Scottish Record Office (now: the National Records of Scotland, henceforth: *NRS*), where it is presently housed under the shelfmark GD 128.[23] The Fraser-Mackintosh collection at the *NRS* contains a large number of original charters, including several late medieval ones.[24] It is unclear under what circumstances the 1444 and 1447 charters (McGill Charters 40 and 14), alongside with other late medieval and early modern documents from the Fraser-Mackintosh collection made their way into the hands of a private dealer. In theory, they could have been sold by Fraser-Mackintosh's widow; or purchased directly from the Inverness Burgh Library, shortly after their deposit there in 1921; or, procured via some illegal means. It is clear that they were purchased by the McGill University Library, at some point in the course of the expansion of its manuscripts collection in the 1920s and 1930s.

Intriguingly, the 1444 and 1447 charters are not the only late medieval documents from the Fraser-Mackintosh collection that made their way from the UK to the dominions: in the Special Collections Department of the University of Western Australia, there is a 1446 precept of Alasdair MacDhòmhnaill to Seòras Mac an Rothaich (George Munro) to give sasine to Alasdair Marshall of the lands of Dochcarty in the earldom of Ross and sheriffdom of Inverness'

[20] Macdonald and Macdonald, *Clan Donald*, I, pp. 533–34 and 535.

[21] Paton, *Mackintosh Muniments*, no. 2.

[22] High Life Highland, 'Reference and Local History'.

[23] *NRS*, 'Fraser-Mackintosh Collection'.

[24] For instance, NRS, GD 128/64/4/2 (a 1376 charter of Raibeart Chisholm selling land in Auld Castlehill to Seamus son of Steaphan, citizen of Inverness and a 1447 charter of Dòmhnall of Auld Castlehill granting land in Auld Castlehill to Uilleam de Buythe, citizen of Inverness).

(*ALI*, no. 46).[25] Just as with the McGill charters, I was, alas, unable to retrieve any information about the ways this document made its way to the University of Western Australia.

The acquisition of the two original charters by private book-dealers, their crossing the ocean and eventual disappearance at the McGill Library circulation, where they were, to my best knowledge, left uncatalogued and largely unknown to Scottish historians, meant that when Jean and Billy Munro came to work on their critical edition of the acts of the lords of Isles (*ALI*), they had to rely on later copies, rather than on the original charters. The 1444 charter (McGill Library Charter 40; no. 42 in *ALI*) was edited from the late-nineteenth-century *Macdonald Collections*, while the 1446 document (McGill Library Charter 14; no. 47 in *ALI*) was edited from the 1781 notarial copy by David Dalrymple. All the existing discrepancies between the originals and the late copies and editions are supplied in the *apparatus criticus* of Appendix 1. It should be noted that of thirty surviving documents issued by Alasdair MacDhòmhnaill 1427 and 1449 and printed in *ALI*, only six survive in their original form (*ALI*, nos 25, 31, 34–35 [in effect, the same document], 44, 46 and 50), while the remainder was edited from later copies (some of which are found in the Great Seal Register). With the addition of the two McGill charters, the total number now rises to eight.

The 1444 and 1447 Grants of Alasdair MacDhòmhnaill to Máel Coluim Mhic an Tòisich: Contents

The charters are written in a usual formulaic manner and their contents are fairly straightforward. According to the 1444 charter (McGill University Charter 40 = *ALI*, no. 42), Alasdair MacDhòmhnaill bestowed upon Máel Coluim Mhic an Tòisich forty merks (*quadraginta marcarum*) of land in the Braes of Lochaber, on the north bank of the River Spean extending from Loch Laggan in the east to Inverroy in the west. The lands came with all their appurtenances, which included 'moors, marshes, tofts, open fields, paths, arable land, waterways, ponds, meadows, pastures, mills', 'the rights of multure, fowling, hunting, fishing, access to peet-bog, turbary, charcoal collection, iron mills and breweries' and 'payments of bludewite (fine for bloodshed), heriots (entry dues), merchets (marriage dues), and carriage services'. Obviously, these legal

[25] University of Western Australia, Special Collections Department, MS Med. 2; printed in Munro and Munro, eds, *Acts of the Lords of the Isles*, pp. 69–70 (no. 46) (henceforth, *ALI*).

formulae do not necessarily reflect reality. For instance, there is no evidence, historical or topographic, that the area in question ever had any 'arable land', let alone 'open fields'. Also, there is no evidence about the existence of a mill there, which is implied by the 'right of multure'. In effect, the lands in question constitute a highland corridor, consisting primarily of pasturage for cattle, sheep, and goats. The land grant in Lochaber in 1444 was followed by the 1447 appointment of Máel Coluim Mhic an Tòisich to the bailliary of Lochaber (McGill University Charter 14 = *ALI*, no. 47). This a highly important and prestigious administrative, judicial, and military position, which implied mutual trust and association between Alasdair and Máel Coluim.

Under what circumstances did Alasdair MacDhòmhnaill grant Máel Coluim Mhic an Tòisich the lands in Lochaber in 1444, and why did he appoint him to the bailliary of Lochaber three years later? In order to appreciate the significance of the two charters, it is necessary to place them into their larger socio-political and environmental contexts.

The 1444 and 1447 Grants: Political Context

As it was a norm in the Gaelic society of the Western Highlands, the relations between two powerful clans were always complex and neither the MacDhòmhnaills nor Mac an Tòisiches, the two most powerful clans of that region, were exempt from this rule. The uneasy relationships between the two clans were studied, in meticulous detail, by Alasdair Ross.[26] For the purpose of the present paper, it will suffice to survey the most important key points in the personal relationships between Alasdair MacDhòmhnaill and Máel Coluim Mac an Tòisich. In the Battle of Lochaber in 1429, Máel Coluim Mhic an Tòisich switched sides by defecting from Alasdair MacDhòmhnaill to the royalist forces of James I.[27] The king thanked Máel Coluim by granting him lands of Alasdair of Lochalsh, Alasdair MacDhòmhnaill's uncle.[28] It was not until the late 1430s that the relationships between the MacDhòmhnaills and the Chatain confederation (and the *Clann* Mac an Tòisiches in particular) got better.

This personal association between Alasdair MacDhòmhnaill and Máel Coluim Mhic an Tòisich was strengthened by the marriage of Máel Coluim

[26] Ross, '*Ghille Chattan Mhor*'.

[27] Bower, *Scotichronicon*, VIII, pp. 262–63.

[28] Adam, *The Clans*, p. 68.

to a daughter of MacDonald of Clanranald, Alasdair's kinsman, at some point before 1444.[29] The kinship ties between the two were cemented further with Máel Coluim's son Donnchadh (Duncan) marrying Alasdair's daughter Florence — possibly between 1444 and 1447, given that only in the 1447 charter Máel Coluim is referred to as 'noster consanguineus' (our consanguineous). The confidence of Alasdair in Máel Coluim is further indicated in the latter's designation as 'confidentissimus' (most confident). Conversely, in the 1444, Máel Coluim was designated merely as 'dilectus noster' (our beloved).

The grant of the Lochaber lands between Loch Laggan and Inverroy and the installation of Máel Coluim Mhic an Tòisich in the bailliary reflects a long-term strategy of Alasdair MacDhòmhnaill. Indeed, Ross referred to the 1444 grant as 'strategic [...] because it effectively meant that his [Alasdair's] new tenant's lands sat astride and controlled the two main corridors of communication between the lordships of Lochaber and Badenoch, Glen Spean and Glen Roy'.[30] Máel Coluim Mhic an Tòisich was not only the chieftain of the *Clann* Mac an Tòisich, but also, since 1436, captain of the Chatain confederation, consisting of twelve Highland clans. The grant of vast lands and the stewardship over all of Lochaber meant to ensure that Alasdair had not only a reliable tenant, but also a mighty ally, who can be called up in hour of need.

Why did Alasdair want to join hands with Máel Coluim? It appears that Alasdair's political and socio-economic ambitions were at odds with both the centralizing power of Stewart monarchy and the expansion of local clans. The relationships between the *Clann* MacDhòmhnaill and the Stewart dynasty have never been easy, but the tensions appear to have reached new heights in the period of *c.* 1425–1452. The Battles of Lochaber (23 June 1429), and Inverlochy (1431), were by no means the only examples of the conflict between Alasdair and the royal authority.[31] In 1439 Alasdair's people were also involved in a raid at Inchmurrin, in the course of which Sir John Colquhoun, governor of the royal Dumbarton Castle, was killed.[32] In 1445 or 1446, he entered into a political bond with Alasdair (Alexander) Lindsay, 4th Earl of Crawford and Uilleam Dubh Glas (William Douglas), 8th Earl of Douglas.[33] The latter was

[29] Macfarlane, *Genealogical Collections*, I, p. 183.

[30] Ross, '*Ghille Chattan Mhor*', p. 118.

[31] On the relationships between Alasdair and the first two Jameses, see Nicholson, 'From the River Farrar to the Loire Valley' and Cameron, '"Contumaciously Absent"?'.

[32] *ALI*, p. lxvii.

[33] *ALI*, no. 45; McGladdery, *James II*, p. 64.

at that point the most powerful noble in the kingdom, involved in a bitter conflict with James II, which, ultimately, led to his murder at the hands of the king and his man in 1452. The political alliance of the MacDhòmhnaill–Mac an Tòisich/Chatain should be seen, therefore, in a wider context of the political contradictions between local lairds on the one hand, and the growing authority of kingship on the other. It was in this very context of political tensions that we witness the process of formation of inter-clanial alliances. Just as Alasdair MacDhòmhnaill entered into alliance with Mac an Tòisich/Chatain in 1444, so he did with the earls of Douglas and Crawford in 1445 or 1446.[34] This tripartite alliance would be renewed after the death of Alasdair in 1449 by his son Iain (John) and it would, ultimately, cost Douglas his life.

The royal authority, however, was only one political challenge to Alasdair. Here, we also have to account for the rise and expansion of several great Highland clans. Firstly, there was *Clann* Gòrdan (Gordon) in the North-East, which led eventually to inevitable tensions and conflicts with the MacDhòmhnaills. In their expansion, the Gòrdans were supported by the royal authority of James I and James II.[35] Although by *c.* 1440, the Gòrdans still held very limited Highland landholdings, there is indirect evidence for their attempts to expand their authority westwards, through building up a network of Highland allies. Perhaps the single most important strategic ally was the *Clann* Chatain confederacy, as it is reflected in a 1442 charter issued by Alasdair Seton, 1st Earl of Huntly and chieftain of the Gòrdans (1440–1470).[36] According to the charter, Alasdair Seton granted Máel Coluim Mac an Tòisich, the captain of the *Clann* Chatain, lands in Raits and Meikle Geddes, in the sheriffdom of Nairn, in recognition for Máel Coluim's faithful services there. Alasdair Seton must have undoubtedly appreciated the potential of an alliance with such a powerful leader as Máel Coluim Mac an Tòisich, in his attempts to expand his authority in the Highlands. Could the 1444 grant by Alasdair MacDhòmhnaill be a natural reaction to the threat posed to his authority by the potential Gòrdan-Mhic an Tòisich (and, by extension, Chatain) alliance? Could the grant of lands in Lochaber, all the way from Loch Laggan and Inverroy, be seen as an attempt to hinder the territorial expansion of the *Clann* Gòrdan?

It should be noted that using Máel Coluim Mhic an Tòisich as a possible buffer against Alasdair Seton was only one approach undertaken by Alasdair

[34] *ALI*, no. 45; McGladdery, *James II*, p. 64.

[35] Oram, 'Introduction. A Celtic Dirk at Scotland's Back?', p. 33.

[36] NRS, GD176/7 (5 October 1442).

MacDhòmhnaill in his relationships with the *Clann* Gòrdan. In a 1442, Alasdair granted Seton a life-rent of the barony of Kingedward in the earldom of Buchan, with all its appurtenances, services, and income.[37] This grant may indicate that Alasdair not only attempted to prevent the rise of the *Clann* Gòrdan in peaceful terms, but that he also sought to keep him as far as possible from the Highlands. After all, it was Máel Coluim and not Seton that got the lands and stewardship in Lochaber in 1444 and 1447, respectively.

The *Clann* Gòrdan was not the only threat to Alasdair. One also has to account for a conflict between the *Clann* MacDhòmhnaill and the *Clann* Camshron (Cameron), another powerful Highland clan. Just as the *Clann* Mac an Tòisich, the *Clann* Camshron, too, switched side in Battle of Lochaber, defecting to James I's side. In 1431, the two clans clashed in the Battle of Inverlochy, which ended with the defeat of the pro-royalist forces at the hands of the MacDhòmhnaill.[38] Unlike with the Mac an Tòisiches, there was no immediate reconciliation with Camshrons. What may have played a role in the reconciliation between Alasdair and Máel Coluim was the ongoing conflict between the Camshrons and Mac an Tòisiches. The two clans clashed violently on 20 March 1429 (the Palm Sunday Battle) at an unknown location, resulting in heavy causalities on both sides. In 1441, the two clans were engaged in a bitter conflict, culminating with the Battle of Creag Cailleach, followed by a series of Mac an Tòisiches raids into the Camshron territories and the forced exile of Dòmhnall Dubh, the clan's leader.[39] The anti-Camshron stance of Máel Coluim Mhic an Tòisich may have encouraged Alasdair MacDhòmhnaill to seek an alliance with him.

To that we should also add the expanding powers of the *Clann* Caimbeul (Campbell). Thus, Donnchadh (Duncan) Caimbeul of Loch Awe, Lord of Argyll (d. 1453/4), one of the most powerful figures in Gàidhealtachd west, rose to a considerable prominence during the minority of James II. In addition to his powerful position as Justiciar of Argyll, he was knighted at some point before March 1440 and created a Lord of Parliament as Lord Campbell of Lochawe in 1445. Donnchadh's nephew, Cailean (Colin) of Glenorchy was be endowed with similar honours, including a crown annuity of forty merks from the lands around Loch Tay.[40]

[37] *ALI*, pp. xxxv–vi.

[38] Brown, *James I*, pp. 138–40.

[39] Mackintosh, *Historical Memoirs*, p. 145.

[40] McGladdery, *James II*, p. 133.

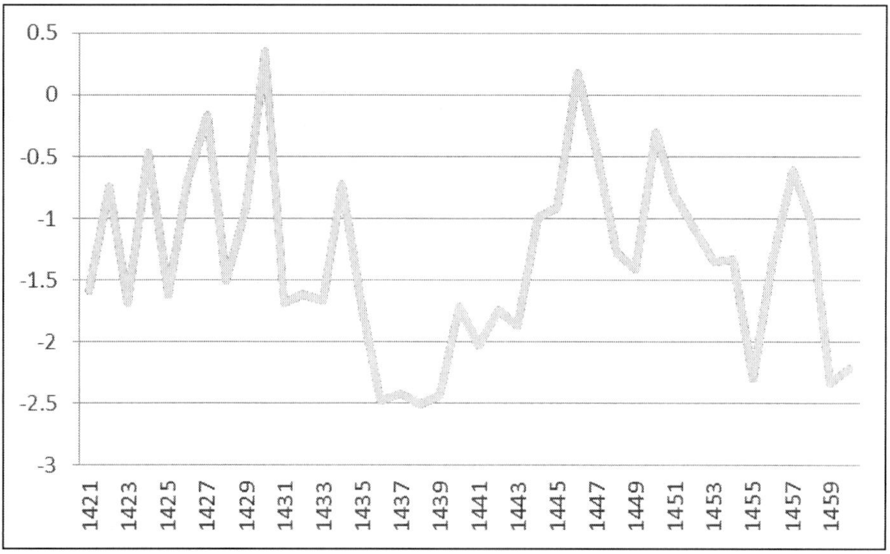

Figure 9.1. Summer temperatures in Highland Scotland, expressed in a deviation from the 1961–1990 average, in degrees centigrade (1961–1990 = 12.6). *Source*: Rydval and others, 'Reconstructing 800 Years'.

The 1444 and 1447 Grants: Environmental Context

The political circumstances behind the 1444 and 1447 grants were only one side of the story. To appreciate the charters even better, it is necessary to consider their environmental context. The period of *c*. 1430–1460 was certainly not the best times for Highland chieftains. It was around that time that northern Europe (and indeed, the rest of the world) was entering into a new long-term climatic phase, commonly known as the 'Little Ice Age' (LIA). The LIA opened with a particularly cold sub-phase, known as the 'Spörer Minimum', reaching its lowest point between *c*. 1437 and 1460, when the levels of solar irradiance and air temperatures plunged to their lowest level in the past 1500 years.[41] This fact is reflected in the dendrochronological record from the Highlands, most recently reconstructed by Rydval and Wilson (Fig. 9.1). In particular, the years 1431–1444 were excessively cold, with summer temperatures standing at between 1.7 and 2.5 degrees Centigrade below their 1961–1990 level. The impact of late-medieval climate change (for the period of *c*. 1300–1400) on

[41] Campbell, *Great Transition*, pp. 337–39; Ogurtsov, 'The Spörer Minimum Was Deep'.

Highland economies has been already discussed by Richard Oram and Paul Adderley.[42]

The palaeo-climatic evidence is fully corroborated by textual evidence. According to a number of Irish and Scottish chronicles, 1438–1440 were particularly bad years, marked by subsistence crisis and mortality.[43] Evidence from English manorial accounts also indicates high mortality and low fertility levels of sheep. Also, as data from south England indicate, average annual wool fleece yields reached their lowest point in the period *c.* 1210–1450, standing at between 1.05 and 1.10 lbs per animal, in contrast with the average 1.4 for the entire period.[44] The situation in the British Isles reflects the situation elsewhere all over Europe, which experienced a short-term weather anomaly and a subsistence crisis (which, in some parts unfolded into a full-fledged famine).[45]

Given that the Highland economies were based predominantly on livestock rearing (cattle, sheep, and goats in particular), with occasional pockets of land devoted to oat cultivation, grassland was the single most important type of land-use in that area. Gradual cooling and the decreased solar irradiance reduced the grazing season by about one month, with early May and late September being largely eliminated.[46] The shortening of the grazing season meant that the annual biomass for grazing animals was now reduced by some 10–15 per cent. To make things even worse, however, there is some indirect evidence about the gradual decrease in the levels of precipitation, as it is reflected in the annual chronology of speleothem band-widths from Uamh an Tartair cave (Sutherland) (Fig. 9.2). It should be borne in mind that speleothems tends to contain a mixture of temperature and precipitation signals, thus rendering 'noisy' results, based on unclear relationship between the two types of signals. Thus, the quality of this data can be questioned. For what it is worth, it does indicate a gradual decline in the levels of precipitation from *c.* 1433 onwards, reaching its peak in 1452. If the speleothem record indeed provides at least an indirect indication of precipitation levels, then its chronology may indicate relatively dry 1440s. Grass growth strongly depends on precipitation, and several dry seasons can depress pasture availability. Insufficient pasture can be disastrous not only for wool production, but also for the physi-

42 Oram and Adderley, 'Lordship and Environmental Change'.

43 Oram, '"The Worst Disaster"', p. 244.

44 Stephenson, 'Wool Yields in the Medieval Economy'.

45 Camenisch and others, 'The 1430s'.

46 Oram and Adderley, 'Lordship and Environmental Change', p. 79.

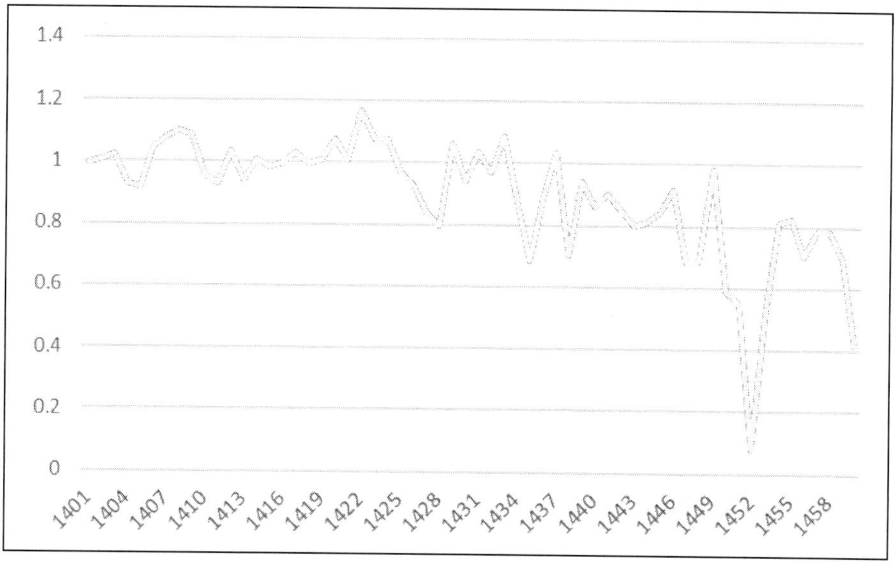

Figure 9.2. Approximate precipitation levels in northern Scotland deriving from annual speleothem bands of Uamh an Tartair cave (Sutherland), 1401–1460 (Indexed on 1401–1430). *Source*: Proctor and others, 'A Thousand Year Speleothem Proxy Record'; Proctor, Baker, and Barnes, 'A Three Thousand Year Record'. The dataset is available online at Proctor, Baker, and Barnes, 'Northwest Scotland Stalagmite Data to 3600 BP'.

cal well-being of horses, cattle, and sheep. This fact may have made livestock husbandry more challenging and costly than before, and this was especially true in Gàidhealtachd areas, where this type of agricultural enterprise formed the backbone of local economies.

The crisis of livestock husbandry undoubtedly had some far-reaching consequences for the social order in the Highlands. As Oram and Adderley have shown, in the post-Black Death era, it gave a rise to the widespread activities of *ceatharn* (caterans, in their Anglicized form). *Ceatharn* were, in effect, bands of men, engaged in occupying lands and raiding possessions of local Highland landowners.[47] *Ceatharn* were employed on numerous occasions by clan chieftains, as it happened in a 1396 conflict between the *Clann* Chatain and the *Clann* Qwhele.[48] Although there is no direct evidence about the activities of *ceatharn* during the 1430s and 1440s, the omnipresent pressure and dearth created by the malevolent environmental situation may have had a profound

[47] Oram and Adderley, 'Lordship and Environmental Change', pp. 76–77.

[48] Bower, *Scotichronicon*, VIII, pp. 7–9.

impact on social climate, with more deprived members of local communities prompted into lawless behaviour, consisting, first and foremost, of cattle raids. Some Irish evidence, reflecting the larger Gaelic environmental context, may provide a partial corroboration. According to the *Annals of Four Masters*, in 1434, in the course of the conflict between the brothers Niall and Naghtan Ó Domhnaill (O Donnell), both sides committed a number of raids in the lands of Moy and Tirhugh (north-western Ireland), burning houses, seizing loot and raiding livestock.[49] The following year saw the invasion of the Uí Néills into Fermanagh. Local inhabitants were ready for that and, to avoid any depredation, shipped all their cattle and movables westward across Lough Erne. The annalist adds that 'it was not in boats that they conveyed them, but over the ice, which was then so great that steeds and horses carrying burdens were wont to cross the lake upon it'.[50]

Could it be that it was in these challenging circumstances of cold and dry years that Alasdair MacDhòmhnaill sought an alliance with another great chieftain, namely Máel Coluim Mac an Tòisich? After all, that would have boosted the chances that his possessions, which undoubtedly included vast amounts of pasturage, and a multitude of livestock, would not have been raided either by independent *ceatharn*, or at the instigation of a hostile clan leader. Putting Máel Coluim in charge of Lochaber lands would be the optimal strategy to achieve this goal. Moreover, this would enable Alasdair to use Máel Coluim's forces, in the event of a punitive excursion into enemies' territories, had such need arisen. As we have seen, Máel Coluim was undoubtedly experienced in conducting such punitive raids: in 1441 Máel Coluim plundered the territories of Camshrons, after the latter's defeat at the Battle of Creag Cailleach.

Conclusions

The 1444 and 1447 charters of Alasdair MacDhòmhnaill to Máel Coluim Mac an Tòisich, whose originals have re-emerged in the McGill Library collections, reflect the ways in which local Highland lairds were exercising their authority in the context of their political and socio-economic ambitions and struggles with other political forces. In the Gaelic-speaking world of the West Highlands, characterized by decentralized authority, loose administration and lack of central record-keeping, written documents were not omnipresent real-

[49] 'Annals of the Four Masters', p. 898.

[50] 'Annals of the Four Masters', p. 903.

ity — in contrast with the more 'Anglicized' East, marked by the stronger presence of central authority, developed bureaucracy, and a culture of literacy.[51] One can suspect that most daily transactions and agreements would be conducted among local Gaelic inhabitants orally, and hence were based on word, memory, and honour, rather than text, record, and law. Production of written documents, written primarily in Latin,[52] can be interpreted, therefore, as the attempts by local lairds not only to deploy a system of legally binding agreements, rules, and networks to underpin their authority, but also to ensure that they would be respected and enforced. In doing so, they were mimicking not only the royal authority, but also English and continental nobility that represented, at least to a certain degree, a cultural and political model for their lordship.[53] In doing so, one can argue, they were distinguishing (if not segregating) themselves from lesser landlords and commoners and stressing their authority over the latter.

Legally binding documents underpinning local authority were especially crucial during the political, socio-economic, and environmental instability of the period *c.* 1430–1450. In a time of unpredictable vagaries of nature, pressure on resources, cattle raids, territorial encroachment, frequent armed conflicts and skirmishes, and changes of political allegiance, it was becoming increasingly difficult for local lairds to maintain and exercise their authority. One obvious way to do so was to form political and military alliances with other lairds, on the basis of commonly shared interests and enmities. To ensure that mutually binding obligations imposed by those alliances were respected in the context of crisis and instability, local clan leaders had to go beyond the 'traditional' system of oral contracts and produce more robust forms of contracts, namely written and sealed charters. Throughout his lordship (*c.* 1423–1449), Alasdair MacDhòmhnaill seems to have produced a fair number of charters, through which he was exercising his power. So far, only thirty charters issued by him have been found, and this may well be just a fraction of their total number. Still, this number is higher, when compared to his predecessor, Dòmhnall MacDhòmhnaill (1397–*c.* 1423) (twelve surviving charters in total). The thin

[51] On clerical administration of late-medieval lords of the Isles, see Thomas, 'Bishops, Priests, Monks and Their Patrons', pp. 142–43.

[52] Some surviving documents were issued in Scots (for instance, *ALI*, nos 20, 22, 29, 32) and one in Gaelic (*ALI*, no. 16).

[53] On the material culture of late-medieval lords of the Isles (including the adoption of 'mainstream' European fashions and practices), see Caldwell, 'The Lordship of the Isles'.

and patchy survival of the original documents may, of course, conceal the truth and we will never know the total number of the charters issued by both MacDhòmhnaill lords. But if this gap in numbers of surviving charters can indeed be taken as an indicative, then it may reveal that the practice of charter production became a legal and political commonplace precisely in the context of the fifteenth-century crisis (*c.* 1430–1450). Obviously, this hypothesis can only be strengthened on the basis of a comparative analysis of similar charter collections from other local lordships from that period — something that cannot be undertaken in the scope of the present paper.

Appendix 1: A Critical Edition of Two Alasdair MacDhòmhnaill Charters at the McGill University Library

McGill University Rare Books Library, Charters Collection, Charter 40

Grant by Alasdair MacDhòmhnaill, earl of Ross and lord of the Isles and of Lochaber to Máel Coluim Mhic an Tòisich of lands in Lochaber in the sheriffdom of Inverness.

Inverness, 11 February 1444

1 Omnibus hanc cartam uisuris uel audituris, Alexander de Ile[54] comes Rossie, d[omi]n[u]s Insular[um] et Lochabrie[55] — et[er]nam in Do[mi]no salute[m]. Nou[er]itis

2 nos ex matura deliberac[i]o[n]e consilii nostri dedisse concessisse et hac p[rese]nti carta nostra confirmasse dilecto nostro Malcolmo McKy[n]tosch[56]

3 totas et integras t[er]ras nostras quadraginta marcaru[m] infra-sc[ri]ptas,[57] ui[delicet] de Dalidundelg,[58] Braenathan,[59] Tullothaird,[60] Inuercany,[61]

[54] *Yle* McDC.

[55] *Lochaber* McDC.

[56] *McKintoch* McDC.

[57] *infra scriptas* McDC.

[58] *Daliemeleg* McDC.

[59] *Braenuchan* McDC.

[60] *Tullochairder* McDC.

[61] *Invercam* McDC.

4 Murbalgane, Glenglastour, Kilkaraill,[62] Bothsyny, Bothchasky,[63] Achaddoire,[64] Collicharam, Bothynton, Blayrnofyngone,[65]

5 Bothynton Moyr,[66] Cranothan,[67] Keppath,[68] Achadmaddy, Achadnocroisse,[69] Breagath, Inuerrowaybeg, Bothlam[70] et Inuerroy

6 Moyr,[71] pro dimidia p[ar]te cu[m] p[er]tine[n]ciis iacen[te][72] in dominio nostro de Lochabr'[73] infra uicecomitatum de Inu[er]nys, pro suo fideli s[er]uic[i]o nobis

7 gratant[er] impenso et impende[n]do. Tenendas et habendas totas et integras p[re]no[m]i[n]atas terras cu[m] p[er]tinenciis[74] p[re]fato Malcolmo et

8 heredibus suis masculis de nobis et heredibus nostris et successoribus in feodo et hereditate imp[er]petuum[75] p[er] om[n]es rectas metas

9 suas antiquas et diuisas in moris maresiis toftis planis uiis semitis aquis stagnis pratis pascuis et pasturis, molendinis

10 multuris et eor[um] sequelis aucupationibus uenationibus piscariis petariis turbariis carbonariis fabrinis[76] et brasinis cu[m] curiis[77]

[62] *Kilkaraith* McDC.

[63] *Bothasky* McDC.

[64] *Achadoire* McDC.

[65] *Blairnafyngon* McDC.

[66] *Bothynton Moir* McDC.

[67] *Cranothan* McDC.

[68] *Keppach* McDC.

[69] *Achadnacroise* McDC.

[70] *Bothlan* McDC.

[71] *Inverroymoir* McDC.

[72] *jacentes* McDC.

[73] *Lochaber* McDC.

[74] *ac pertinenciis* McDC.

[75] *in perpetuum* McDC.

[76] *farbrilibus* McDC.

[77] *ac curiis* McDC.

11 et curiar[um] eschactis[78] cu[m] bludeuetis[79] hereȝeldis[80] mulier[um] marchetis[81] arragiis[82] et cariagiis[83] omnibusq[ue] aliis [com]moditatibus lib[er]tatibus

12 et aisame[n]tis ac iustis suis p[er]tine[n]tiis quibuscu[m]que ta[m] subtus t[er]ra[m] q[uam] supra terra[m], ta[m] non no[m]inatis q[uam] no[m]inatis ad d[i]c[t]as terras

13 cum p[er]tine[n]ciis spectantib[us] seu q[uo]modolibet spectare ualentib[us] in futur[um]; adeo lib[er]e quiete integre honorifice bene et in pace

14 sicut aliqua terra i[n] dominio[84] nostro insular[um] datur et concedit[ur]. Faciendo inter nobis heredibus nost[ri]s et successoribus d[i]c[t]us

15 Malcolm[us] et heredes sui masculi s[er]uitiu[m] warde et releuii. Et nos Alexander de Ile[85] comes ac d[omi]n[u]s prefatus heredes nostri et

16 successores predictas terras cu[m] p[er]tine[n]ciis p[re]fato Malcolmo et heredibus suis masculis ut p[re]fertur in om[n]ibus et p[er] om[n]ia ut p[re]dictum[m]

17 est cont[ra] om[n]es mortales homi[n]es et feminas warantizabim[us] acquietabim[us] et imp[er]petuu[m][86] defendem[us]. In cui[us] rei testi[m]o[niu]m

18 sigillu[m] nostru[m] appendi fecimus, apud Inuernys[87] undecimo die mens[is] Februarii anno Domini millesimo quadringentesimo qua-

[78] *et curiarum [exitibus] eschetis* McDC.

[79] *et bludwitis* McDC.

[80] *herezeldis* McDC.

[81] *merchetis* McDC.

[82] *arriagiis* McDC.

[83] *carriagiis* McDC.

[84] *de dominio* McDC.

[85] *Yle* McDC.

[86] *in perpetuo* McDC.

[87] *Innernys* McDC.

19 dragesimo tertio. P[re]sentibus ibid[em] Lachlano McGilleon d[omi]no de Dowart, Iohanne[88] Murchardi McGilleon d[omi]no de Cannlochboyg,[89]

20 Iohanne[90] Lachlam[91] McGilleon d[omi]no de Cola,[92] Uylando de Cheshelm, Georgio M[un]ro d[omi]no de Foulis, <Hectore Torquelli>[93] et

21 Nigello McLoyd consiliariis nost[ri]s et plurib[us] aliis.

McGill University Rare Books Library, Charters Collection, Charter 14

Appointment by Alasdair MacDhòmhnaill, earl of Ross and lord of the Isles and of Lochaber, of Máel Coluim Mac an Tòisich, to the bailliary and stewardship of Lochaber, with all its appurtenances

Dingwall Castle, 13 Nouember 1447

1 Om[n]ibus hanc cartam uisuris uel audituris, Alexa[nder] de Yla[94] comes Rossie et dominus Insularum — et[er]nam in Domino

2 salutem. Nouerit[is] nos dedisse, concessisse et hoc p[rese]nti sc[ri]pto nostro confirmasse confidentissimo n[ost]ro consanguineo

3 Malcolmo McIntosche[95] p[resent]iu[m] conseruitori[96] totum et integru[m] officiu[m] balliatus seu senescallie om[n]i[um] et singularu[m]

4 terraru[m] dominii nostri de Llochabbor.[97] Tenendum et habendum dictu[m] officiu[m] cum om[n]ibus et singulis p[er]t[inenciis]

[88] *Joanne* McDC.

[89] *Canlochbouye* McDC.

[90] *Joanne* McDC.

[91] *Lachlani* McDC.

[92] *Colla* McDC.

[93] *Hectore Torquelli* deest McDC.

[94] *Yle* DG.

[95] *MacIntosche* McDC, DG.

[96] *pro conseruatori*; *conservatori* McDC, DG.

[97] *Lochabber* McDC, DG.

5 ad dictu[m] officiu[m] sp[ec]tan[tibus], seu iuste sp[ec]tar[e] ualentib[us] q[uo]m[odo]libet in futuru[m], dicto Malcolmo McIntosche[98] ac om[n]ib[us]

6 suis heredibus masculis genetis seu generand[is] de nobis et om[n]ibus h[er]edib[us] nostris in feodo et hereditate

7 imp[er]petuum,[99] adeo libere pacifice bene et in pace, sicut aliquod officiu[m] balliatus uel senescallie in toto reg-

8 no Scotie alicui balliuo conceditur seu pro perpetuo in carta confirmatur. Quod quid[em] officiu[m] ut p[re]fert[ur],

9 nos Alexa[nder] comes et do[mi]n[u]s an[te]dict[us] et heredes nostri an[te]dicto Malcolmo et heredibus suis ut p[re]dicit[ur]

10 contra quoscu[m]q[ue] mortales warantizabim[us],[100] acq[ui]etabim[us] et imp[er]petuu[m][101] defendem[us]. In om[n]i[um] p[re]missoru[m] testi-

11 moniu[m] sigillu[m] nostru[m] p[rese]ntib[us] appendi fecim[us], apud castru[m] nostru[m] de Dinguale decimo tertio die me[n]sis

12 Nouembr[is] anno D[omi]ni millesimo quadringe[n]tesimo quadragesimo septimo. His testibus: Torquello

13 McLeoid domino de Leoghos, Iohanne[102] McLeoid[103] domino de Glenelg, Celestino de Insulis filio n[ost]ro nat[ur]ali, Nigello

14 Flemyng secretario n[ost]ro et Donaldo iudice cu[m] diu[er]sis aliis p[rese]ntibus.

[98] *MacIntosche* McDC.

[99] *in perpetuum* McDC, DG.

[100] *warrantizabimus* McDC, DG.

[101] *in perpetuum* McDC, DG.

[102] *Johanni*, DG.

[103] *MacLeoid* McDC, DG.

Appendix 2: A Short Description of Scottish Documents in the McGill Library Collection

McGill Library Charter 12. *Land grant by Uilleam Paterson (William Paterson) to Aindrea MacDhòmhnaill (Andrew MacDonald) in Inverness (1.11.1595).* Written in Latin, in secretary script. Two seals attached. Measurements: 301 × 491 mm.

McGill Library Charter 13. *Petition by the burghesses of Inverness (1590).* Written in Scots, in secretary script. Seals missing. Measurements: 480 × 328 mm.

McGill Library Charter 14. *Appointment by Alasdair MacDhòmhnaill, earl of Ross and lord of the Isles and of Lochaber, of Máel Coluim Mac an Tòisich, to the bailliary and stewardship of Lochaber, with all its appurtenances (13.11.1447).* Written in Latin, in secretary script. Seal attached. Measurements: 157 × 262 mm.

McGill Library Charter 17. *Land grant by Jasper Cuimeanach (Cumming) to diverse people in Inverness (1599).* Written in Latin, in secretary script. Seal(s) missing. Measurements: 271 × 393 mm.

McGill Library Charter 21. *Land grant by Ruairidh MacConaill (Roderick MacConnell) to Cailean MacChoinnich (Colin Mackenzie), earl of Seaforth in Seaforth (6.5.1627).* Written in Latin, in secretary script. Seal(s) missing. Measurements: 240 × 583 mm.

McGill Library Charter 40. *Grant by Alasdair MacDhòmhnaill (Alexander MacDonald), earl of Ross and lord of the Isles and of Lochaber to Máel Coluim Mhic an Tòisich (Malcolm McIntosh) of lands in Lochaber in the sheriffdom of Inverness (11.2.1444).* Written in Latin, in secretary script. Seal attached. Measurements: 214 × 286 mm.

McGill Library Charter 42. *Financial account of the baillies of Inverness (24.7.1592), containing revenue and expenditure parts.* Written in Latin, in secretary script. Measurements: 475 × 275 mm.

McGill Library Charter 43. *Debt acknowledgement of the burghesses of Inverkeithing (6.4.1697).* Written in Latin, in secretary script. Measurements: 147 × 257 mm.

McGill Library Charter 44. *Land grant by Iain Fòlais (John Fowlis) to Iain (John) Miller in Inverness (5.11.1533).* Written in Latin, in secretary script. Two seals attached. Measurements: 300 × 232 mm.

McGill Library Charter 46. *Petition by the burghesses of Inverness (9.8.1574).* Written in Scots, in secretary script. Seal(s) missing. Measurements: 222 × 315 mm.

McGill Library Charter 48. *Land grant by Iain (John) Leslie, Bishop of Ross to Archibald Broun in Ross (9.12.1567).* Written in Latin, in secretary script. Seal attached. Measurements: 325 × 637 mm.

McGill Library Charter 60. *Land grant by Iain Robasan (John Robertson) to Janet Simpson in Inverness (28.10.1551).* Written in Latin, in secretary script. Seal attached. Measurements: 194 × 249 mm.

McGill Library Charter 65. *Instrument of sasine concerning in Croy (in Inverness-shire) (22.12.1679). Parties involved: Séamus Ròs (James Rose) and Alasdair Ròs (Alexander Rose).* Written in Latin, in secretary script. Seal(s) missing. Measurements: 320 × 472 mm.

McGill Library Charter 67. *Land grant by Cailean MacChoinnich (Colin Mackenzie), earl of Seaforth to Sìomon Cèamp (Simon Kemp) in Chanonry (28.10.1551).* Written in Latin, in secretary script. Seal attached. Measurements: 250 × 259 mm.

McGill Library Charter 68. *Land grant by Seòras Mac an Rothaich (George Munro), master, to Iain Robasan (John Robertson) in Kinnettes (Ross-shire) (5.8.1595).* Written in Latin, in secretary script. Seal attached. Measurements: 423 × 330 mm.

Salmon Variability Related to Phases of the Little Ice Age: Consilience* from Arctic Russia to Scotland?

Richard C. Hoffmann
FRSC, Professor emeritus and Senior Scholar, York University, Toronto

Introduction

What factors drove the fluctuating abundance of a culturally valued fish, the Atlantic salmon (*Salmo salar*)? How did people respond to perceptions of loss? This paper presents a portion of a work in progress on relations between climate conditions and fisheries in late medieval and early modern Europe. Its significance in the larger investigation lies in demonstration that long-term records of catches and of local price series for a particular species both parallel one another and correspond to regional chronologies of climate change. Scotland, though less richly endowed with serial quantitative data than some other regions, seems to fit the general pattern and, more significantly, provides

* Consilience: 'the principle that evidence from independent, unrelated sources can "converge" on strong conclusions' (*Wikipedia*, consulted 4 January 2019). Working with Alasdair Ross during my Leverhulme Fellowship in the Centre for Environmental History and Policy at Stirling in 2011 piqued my interest in medieval and early modern Scottish salmon and then how the records for those rivers and runs might compare with those from salmon fisheries elsewhere. With Alasdair's advice I assembled the important Scottish story for a scientific audience in Hoffmann, '*Salmo salar* in Late Medieval Scotland', and then joined with Alasdair in a close look at long-term conflicts over salmon which became Hoffmann and Ross, 'This Belongs to Us!'. Now some aspects of human relations with *Salmo salar* in Scotland turn out to be consilient with trends and responses elsewhere. Under the title 'Climatic Variability and the Fisheries of Medieval and Early Modern Europe' I presented a preliminary summary of the larger tale to the Climate Change and History Research Initiative, Princeton University, in April 2017. Due to other commitments this remains a work in progress.

'With Our Backs to the Ocean': Land, Lordship, Climate Change, and Environment in the North-West European Past. Essays in Memory of Alasdair Ross, ed. by Richard Oram (Turnhout: Brepols, 2023)
EHNAW 5, pp. 231–244 BREPOLS PUBLISHERS 10.1484/M.EHNAW-EB.5.131277

a clear example of a cultural reaction to a generational shift from resource abundance to scarcity. Taken as a whole, findings reported in this paper and evolving in the larger work in progress speak both to historic impacts of climate change on aquatic resources used by humans and to preindustrial Europeans' cultural responses to experiences of change in the natural world.[1]

Two millennia of history for the Atlantic salmon[2] might correctly be thought a narrative of anthropogenic destruction and depletion but the arc is not wholly downhill. Close observations reveal unevenesses and also cultural myths of hyperabundance, both arguably related to fluctuating climates in different parts of an extensive migratory range. Records are in some ways quite rich because *S. salar* has always suffered from high social prestige wherever its feeding territory in the North Atlantic allowed access to well-oxygenated spawning rivers where temperatures rise above 10°C for about 3 months a year and do not exceed 20°C for more than a few weeks in summer. I'll leave North American stocks aside to examine northern margins and a central zone of those European salmon waters which once extended from the Barents Sea around to southern Portugal.[3]

Genuine and meaningful catch records are vanishingly rare for all medieval and early modern fisheries. Capture sites of unusual administrative importance, however, occasionally yield defensibly representative indicators which can be connected to more impressionistic sources. For salmon this occurred on lower reaches of rivers where long-stable weir fisheries could take fairly consistent 'samples' from runs that meant considerable value to those who owned or taxed them.

In thinly-settled **early modern far north-western Russia**, specific rivers feeding the White and Barents seas belonged exclusively to certain monasteries

[1] The paper thus aims to work within two related but distinct contexts. On the historical ecology of fishes see, for example: Checkley and others, eds, *Climate Change and Small Pelagic Fish* or Brander, 'Impacts of Climate Change'. Effects of early modern climate changes on European societies are treated in, for instance: Behringer, *A Cultural History of Climate*, especially chapters 3 and 4, and Pfister, Brázdil, and Glaser, eds, *Climatic Variability*.

[2] Coates, *Salmon* offers a good popular introduction to the animal, despite needlessly conflating the meaningfully different ecologies and historic circumstances of the Atlantic salmon and the Pacific salmons (certain species in genus *Oncorhyncus*).

[3] A debate over the abundance of salmon in pre-European and colonial New England rivers seems to pay no attention to the location of these waters at the southern limit of the species' range in North America. See initially Carlson, 'Where's the Salmon?' and 'The (in)Significance of Atlantic Salmon', and most recently Jane, Nislow, and Whiteley, 'Use (and Misuse)'.

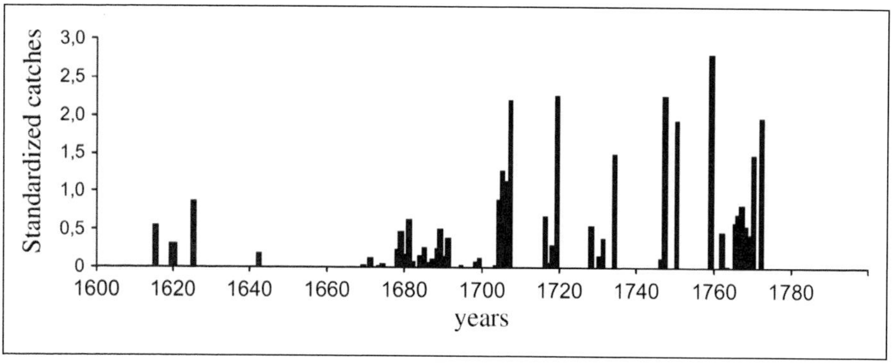

Figure 10.1. Salmon fisheries in the Russian north, seventeenth/eighteenth centuries. Pooled data from weir fisheries on Onega and Vyg Rivers, Varzuga district, and Western Murman. Data and graphics from Lajus and others, 'Status and Potential'; Lajus and others, 'Use of Historical Catch Data'; and Lajus and others, 'Atlantic Salmon Fisheries'. Reprinted with permission of Dmitry Lajus.

while other whole districts paid to the tsars' officials a share of the catch or a fixed tax per fish.[4] An interdisciplinary team of Russian historians and fisheries scientists critically assessed the surviving records and compared the results with the nineteenth and early twentieth century take of salmon from the same waters by the same methods (see Fig. 10.1). All periods show considerable year to year fluctuations, but perhaps surprising to declensionists, early catches were not only much lower than modern ones, they showed medium term variations as well. Researchers noted in particular catches during 1614–1625 that were markedly higher than those in 1685–1691 but subsequently more salmon during the 1750s followed by fewer in the following two decades.

Researchers proposed the climate of sub-arctic Russia as an agent varying returns of salmon. Ice core data suggests that for much of the seventeenth century summer temperatures in far northern Europe averaged 1.5°C lower than twentieth century means, but a warm spell in 1645–1665 preceded more severe conditions up to the 1680s. Local written sources from European Russia indicate relatively moderate winters in the second half of the sixteenth century followed by progressively cooler ones in 1601/50 and 1651/1700. Chroniclers

[4] Lajus and others, 'Status and Potential'; Lajus and others, 'The Use of Historical Catch Data'; Lajus and others, 'Atlantic Salmon Fisheries'; Haidvogl and others, 'Typology of Historical Sources'.

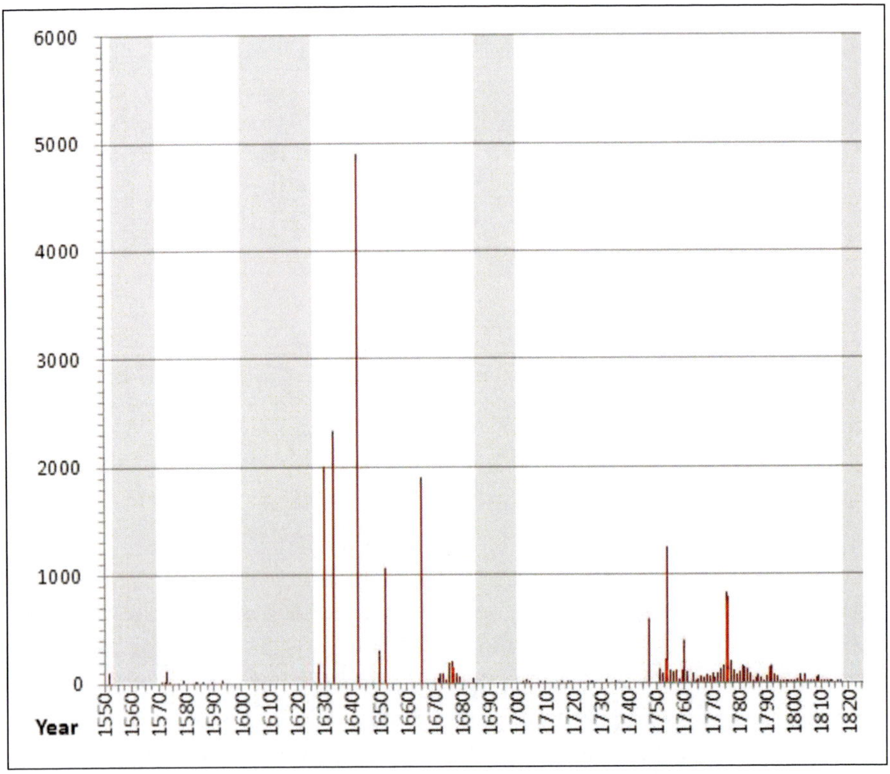

Figure 10.2. Salmon from the early modern Elbe. Atlantic salmon reported caught. Grey bars are years without quantifiable information. From Wolter, 'Historic Catches, Abundance, and Decline', figure 2. Reprinted with permission of Christian Wolter.

in the north itself mention bitter cold during 1656–1663 and in ten more years up to 1697. The authors observe that low catches correlate with cold climate phases at the northern extremity of salmon's range and that low seasonal temperatures affect both the river and oceanic phases of the salmon life cycle. Present-day research has shown that longer winters stress salmon parr (juveniles) and slow their growth rate, while colder seas reduce biological productivity in the east Greenland feeding zone.

Cold temperatures constrained salmon stocks and fisheries near the northern limit of their range. Following this lead from Russian colleagues, what of the larger salmon populations further south in western Europe?

Rivers flowing into the North Sea once hosted large runs of salmon, but as they drained large and diverse medieval landscapes, good, fine-grained indicators over the long term are few. When recoverable, however, early modern records offer interesting parallels and differences to their Russian counterparts. There is both direct and indirect evidence of salmon abundance and scarcity. Christian Wolter recently compiled all available data and information on salmon catches in the River Elbe system, to assess, inter alia, the historic size of the stock and chronologies of decline in a catchment extending from Hamburg to the southernmost boundaries of Bohemia (see Fig. 10.2).[5] His methods combined quantitative records from weirs and designated seine fisheries with wider coverage from more impressionistic comparisons of good and poor harvest years. He graphs reported catches more as indicating relative changes than the actual size of the runs. All things considered, Wolter's evidence shows low returns in the 1550s and again *c.* 1590–1600 followed by large catches in the middle third of the seventeenth century and much lower ones in the last quarter. Verbal sources not reflected in Wolter's graph further indicate mid-seventeenth century fall harvests of salmon across the upper Saale basin (perhaps the Elbe's most important nursery habitat) were ten times those of the late sixteenth century or of the early eighteenth.[6] A decade of high harvests after 1750 was again followed by lower returns in that century's last quarter.

Upper reaches of the Rhine basin contribute two useful data sets. The River Kinzig, an important spawning stream flowing from the Black Forest, had no formal weir, so adult salmon had merely to run a traditional gauntlet of dip nets, lift nets, and small seines as they passed by Offenbach, some villages, and then the town of Wolfach, principal settlement on the river's upper 20 km.[7] Wolfach residents had the right to catch salmon by those traditional means so long as they offered the fish for sale at the river bank and turned over half the receipts to the agent for their lord, Prince Fürstenberg. The annual account of receipts archived in the prince's eighteenth century Rentamt (see Fig. 10.3) has been reasonably argued to represent about twenty per cent of the catch in the Kinzig in the period before the start of river regulation in the Rhine. Behind the usual large year-to-year fluctuations, this data suggests relatively high catches roughly 1720–1735, mostly lower in the 1750s, then distinctly large harvests 1759–1771 and lower ones thereafter.

What geographers call the 'High Rhine' (*Hochrhein*) is the reach from Basel to the falls at Schaffhausen, largest in continental Europe and the upper limit

5 Wolter, 'Historic Catches, Abundance, and Decline'.

6 Schwarz, 'Nochmals'.

7 Nauwerck, 'Lachsfang in der Kinzig', notably pp. 521–52 and fig. 6.

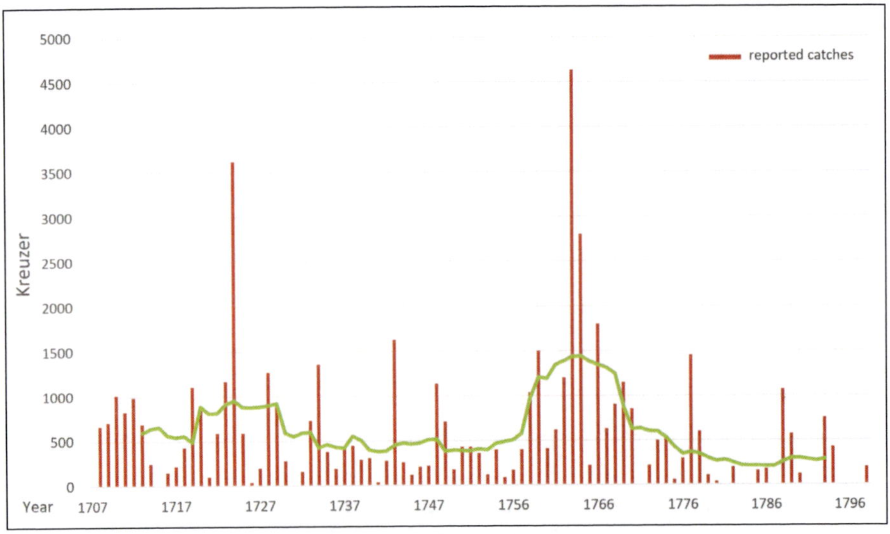

Figure 10.3. An index of salmon returns on a Rhine tributary, the River Kinzig. Annual income from fees paid per salmon caught at Wolfach, 1707–1799. Data from Nauwerck, 'Lachsfang in der Kinzig', figure 6. Used with permission of Arnold Nauwerck. Extracted and graphed by R. and K. Hoffmann.

for salmon in the Rhine. Nineteenth-century commentator Johann Vetter compiled diverse sources to assess changes in High Rhine salmon catches over the previous two centuries, finding poor returns between 1610 and 1630, especially good catches around 1640–1650, and another trough in the late seventeenth century.[8] While much of the eighteenth century may have been unremarkable, its last few decades seemed quite productive before a great decline during 1810–1830, which coinciding with major engineering works to straighten and 'tame' the river.[9] Below the High Rhine and opposite the mouth of the Kinzig, narrative and thus more impressionistic sources from Strasbourg also refer to abundant salmon around 1647.[10]

The chronologies of low and high catches in Rhine and Elbe are reasonably consistent. Catch records or proxies have the advantage of local precision and converse risk of local fluctuations unrelated to the abundance or scarcity

[8] Vetter, *Die Schiffart, Flötzerei und Fischerei*.

[9] On the nineteenth century 'regulation' of the Rhine see Cioc, *The Rhine*.

[10] Nauwerck, 'Lachsfang in der Kinzig', p. 499, citing a 1666 manuscript discussion of Rhine salmon by Strasbourg Fischermeister Leonhardt Baldner.

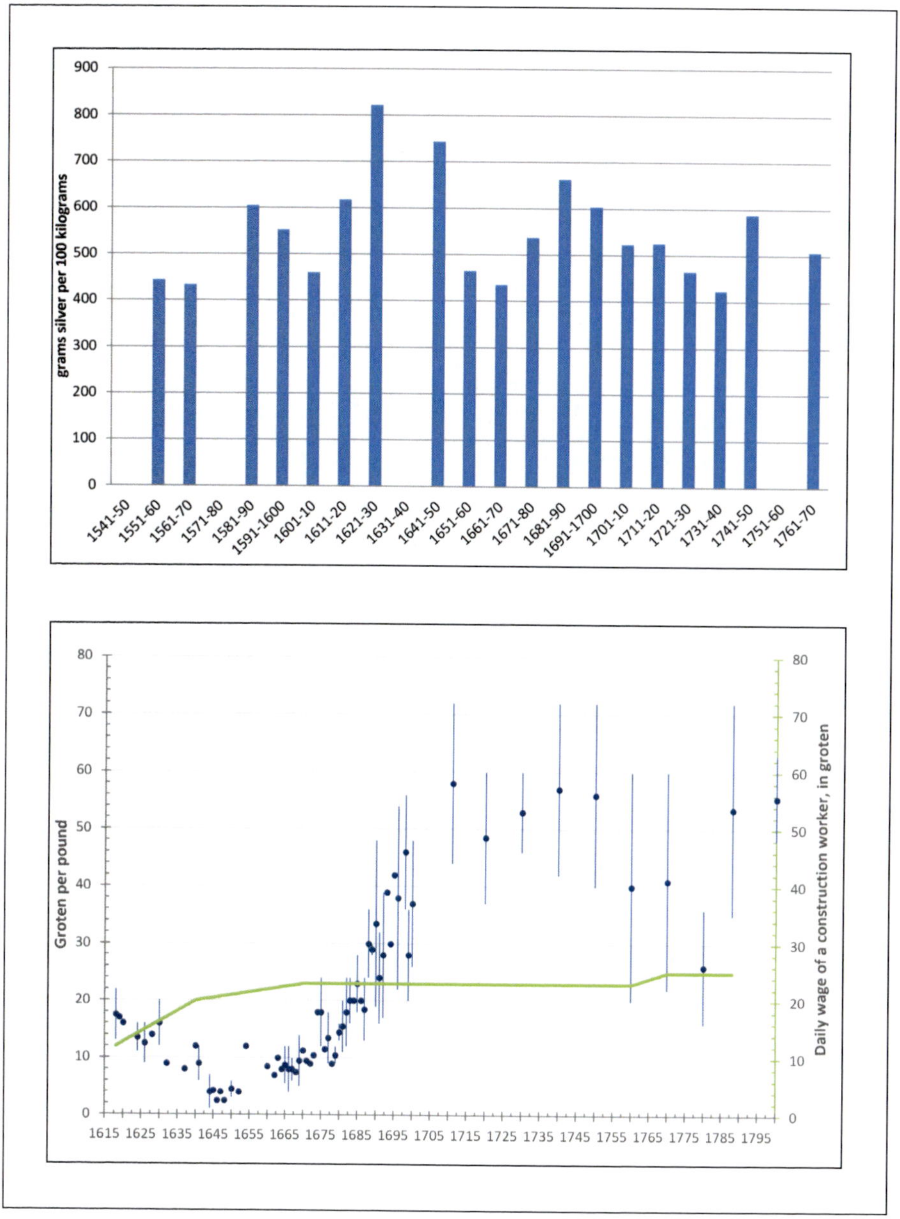

Figure 10.4. Price trends of salmon at two sites in early modern central Europe, Würzburg (top) and Bremen (bottom). Data as published by Hitzbleck, 'Bedeutung des Fisches', table 7 and Schwarz, 'Weserlachs und die bremischen Dienstboten', pp. 149–54. Tabulated and graphed by R. and K. Hoffmann.

of an entire fish population. Wolter, for instance, makes plain that catches at any one of the weirs on the Elbe depended on water levels allowing passage of lower weirs and delaying fish at those higher up; major floods reduced catches in the lower river and increased those higher up. Hence as in the Russian case, parallels among different local records or different watersheds are especially informative.

Good **price series** provide still broader indications of fluctuating regional salmon supplies, always assuming grounds for fairly stable demand and recognizing that prices move inversely to abundance. For instance in another major spawning zone for Rhine salmon on the River Main, decadal average prices for salmon at Würzburg[11] varied in what should now be a familiar rhythm; high (low supply) in the 1620s falling by fifty per cent into the 1660s; a large increase by the 1680s and another low price period in the 1720s–30s (see Fig. 10.4). The later phases here mirror the record of catches on the Kinzig. From the River Weser, principal basin between Rhine and Elbe, a price series for salmon at Bremen further corroborates the mid-seventeenth century peak of abundance and much greater scarcity in decades around 1700; subsequently less thorough information still likely indicates more salmon on the market in the mid-eighteenth century than at its end. The daily wage of construction workers in Bremen provides a useful benchmark as well as a reminder that salmon were always an elite food.[12]

In late medieval eastern Scotland, a different geology, agrarian regime, and probably superior protective regulations, kept salmon stocks much healthier than those on the opposite shores of the North Sea, although direct quantitative measures are not yet possible. While the published medieval price series for Scottish salmon is scanty and breaks off in 1541, abrupt declines during the 1450s and again at the start of the sixteenth century might suggest periods of unusually good supplies.[13] The abundance of this valued resource in especially the River Forth was subsequently remarked by successive eye witnesses,

[11] Hitzbleck, *Bedeutung des Fisches*, table 7. Hitzbleck's use of silver rather than nominal prices limits interpretation of his sixteenth-century entries in the midst of well-known silver price inflation.

[12] Schwarz, 'Der Weserlachs und die bremischen Dienstboten'.

[13] Hoffmann, '*Salmo salar* in Late Medieval Scotland'. This refers to the medieval data set in Gemmill and Mayhew, *Changing Values*, pp. 303–17. Intensive enquiries have turned up no annual price series for salmon in early modern Scotland. Some historians seem to think collecting and publishing such data not only laborious but too old-fashioned.

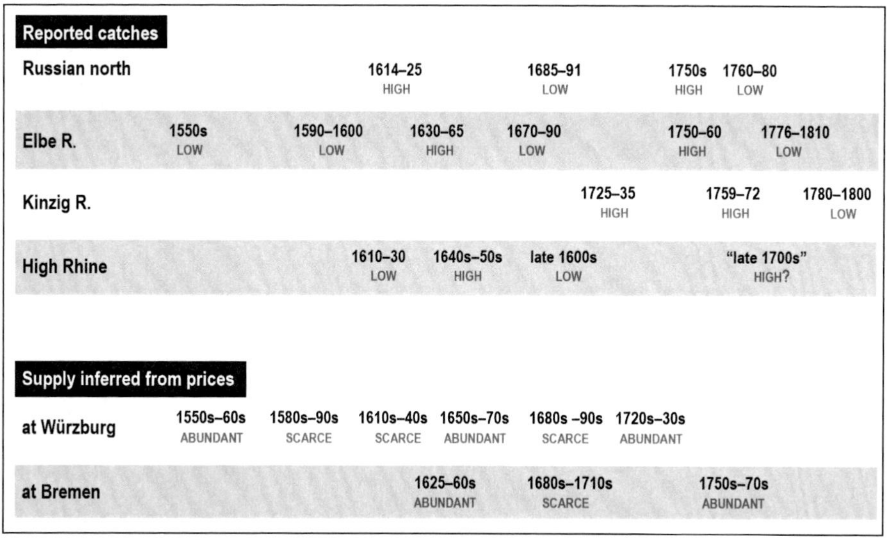

Figure 10.5. Summary: Early modern variations in abundance of salmon.

the local historian John Major in 1521; English Puritan soldier and sport fisher, Richard Franck, who visited in 1656; and French travellers five years later.[14]

Scottish reports, such as they are, tidily corroborate a prominent feature of all evidence from the central area of the salmon's European range, namely a great abundance of these fish at and immediately after the mid-seventeenth century (see Fig. 10.5). The further historical record so far recovered establishes generally low returns at the turn of the sixteenth/seventeenth centuries and again in the late seventeenth. Mid-eighteenth-century decades were again markedly productive but yields fell off well before that century's end. Western trends seem often to lag behind those in northern Russia by a decade or more and may show less extreme variation. Returns to the Elbe and Weser more closely parallel one another than either one parallels the Rhine. Recalling the findings from Russia, what about climate trends?

Historians of climate have now refined **a chronology of phases in the global cooling called the Little Ice Age** (LIA).[15] Though initially named for

[14] Smout and Stewart. *Firth of Forth*, pp. 137–41. Franck, *Northern Memoirs*, pp. 112–13, saw the rich runs in the Forth as typical of Scotland as a whole. Hodgson, '"To the abbittis profeit"' in the present volume provides evidence for the importance of salmon fisheries on the estate of a major late-fifteenth-century institutional landholder.

[15] My grasp of early modern climate history and its complexities is thanks to working with

the world-wide growth of glaciers, the LIA is now recognized by climatologists as a generally cool or cold global climate regime which set in toward the end of the thirteenth century and lasted into the nineteenth. Informed scholars and scientists understand this long-term planetary phenomenon to be manifest in regionally distinct patterns of warmer and colder conditions with medium-term fluctuations. Those were the variations in climate and weather which affected regional habitats and populations of plants and animals and people's experiences of them.

In the northwestern European centre of the salmon's eastern Atlantic range, as elsewhere during the LIA, climate and weather patterns fluctuated, with some decades warmer and others colder than the long-term average and the seasons not always responding uniformly. Variations around the North Sea drainage were not entirely synchronous with those of Arctic Russia. While the fourteenth century was most characterized by extremely volatility, the following century experienced the first of four severely cold phases within the LIA. Reduced solar radiation was a key driver of this change and the cold nadir reached during the 1450s–80s is named for the Spörer solar minimum of 1450–1530. Then, following some milder decades, a second major cold phase between the 1560s and 1620, named the Grindelwald fluctuation after an expanding Swiss glacier, hit its lowest point during the 1580s and 1590s. Volcanic eruptions and changes to the earth's atmospheric circulation likely played a greater role than did solar radiation. For some late-sixteenth-century decades, the growing season in northwestern Europe was reduced by up to six weeks. Again, warmer conditions prevailed during the 1620s–40s before the onset of a sharply colder Maunder minimum *c.* 1645–1720, with the coldest times during roughly the years 1690–1710. Temperatures rebounded by 1720, initiating a warmer mid-eighteenth century. Then colder times returned as a fourth solar minimum, the Dalton (1760–1850) emerged and volcanic eruptions reinforced the downward trend to a trough dated *c.* 1790–1815/18. As the salmon rivers providing data of interest to this study are roughly centred on the Low Countries, a seasonal reconstruction of temperatures there (Fig. 10.6) can stand for the relevant regional chronology.[16] Seasonal conditions in the

my former student Dagomar Degroot. His *The Frigid Golden Age*, pp. 31–49, provides a thorough and richly-documented overview of scholarship on early modern climate in the region of special interest here.

[16] Data from Van Engelen, Buisman, and Ijnsen, 'A Millennium of Weather', p. 112. A more elaborate climatological data set published by Moreno Chamarro and others, 'Winter Amplification', Figure 1, confirms the importance of winter cooling, in particular during the 1460s/70s,

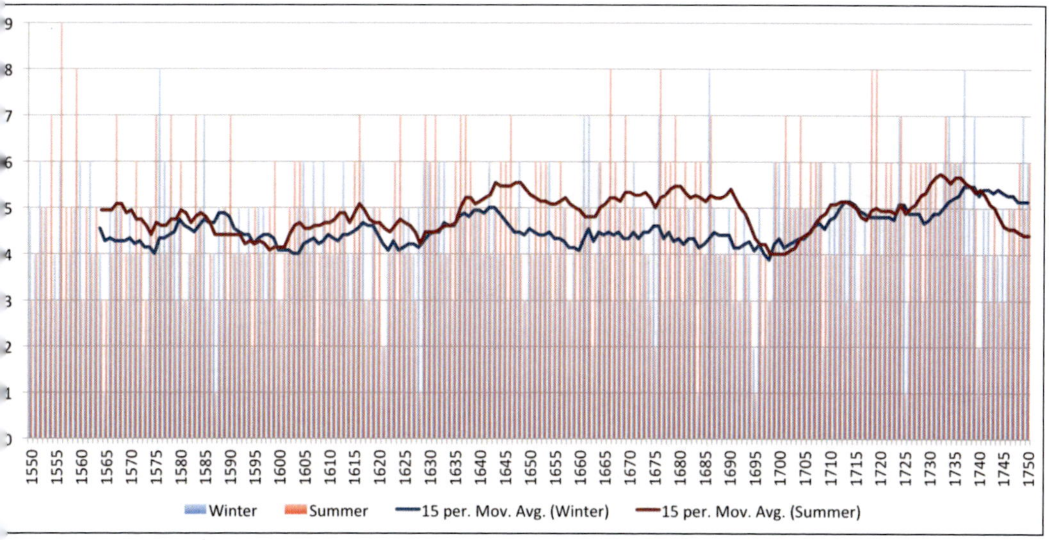

Figure 10.6. Summer and winter temperatures in the Low Countries, 1550–1750. A graph compiled from documentary evidence, showing winter and summer temperature changes in the Low Countries from 1530 to 1750. Higher readings for summer (May, June, July, August and September) and winter (November, December, January, February and March) correspond to higher temperatures. A 15-year moving average shows the trends. Data from Van Engelen, Buisman, and Ijnsen, 'A Millennium of Weather', p. 112. Graph reproduced from Degroot, *The Frigid Golden Age*, figure 1.2, with permission of Dagomar Degroot.

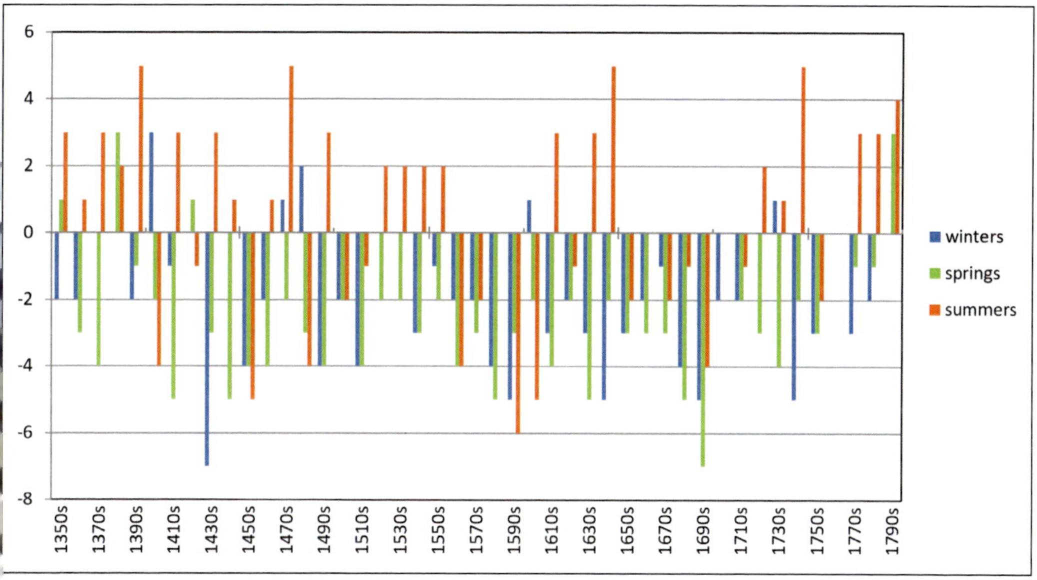

Figure 10.7. Seasonal temperature indices for Germany, annual balance by decade of cold (-) and warm (+) seasons in documentary record, 1350–1800. Data showing total monthly ratings of temperatures in winter, spring, and summer per decade from Glaser, *Klimageschichte Mitteleuropas*, Abb. 23, p. 57, extracted and redrawn by R. and K. Hoffmann.

interior zone where salmon spawned and the next generation went from egg to juvenile parr to smolts ready to go to sea are presented from German documentary sources in Fig. 10.7.

Atmospheric and terrestrial temperatures are not, however, the same as marine and aquatic ones. Large bodies of water take time to warm and to cool. North Atlantic sea surface temperatures began to decline about 1580 and the cooling lasted up to around 1640. Warmer conditions prevailed between *c*. 1640–1710 and renewed cooling from the 1760s. On a seasonal basis, temperatures in freshwater and near-shore aquatic systems lag behind the changes in air temperature by some months. Furthermore, returns of adult salmon are affected by conditions in juvenile habitats some 4–5 years earlier and by marine conditions during the ensuing 1–3 years. Insofar as decade-scale fluctuations in salmon stocks may be associated with climate variability, these lags will be significant. Well-known, too, are sometimes violent year-to-year changes in salmon returns and catches which may have causes wholly independent of temperature. In the absence of concerted human interventions, however, consilience of decade-scale trends among several different rivers should indicate a more general natural influence such as climate.

Seen against the long run of decade-scale climate reconstructions, early modern salmon stocks, in western Europe, like those in Russia, related positively to temperature. Salmon were relatively scarce during the coldest phases of the LIA: the last decades of the sixteenth and start of the seventeenth centuries, the end of the seventeenth century, and probably the end of the eighteenth century. Contemporaries even consistently commented on the cold summers at those times. In contrast, intervening warmer periods — especially the middle third of the seventeenth century and the second and third quarters of the eighteenth — saw salmon in relative, noteworthy, and noted abundance. In central

1510s, 1580–1610, and the 1690s, the latter being the coldest winter decade of the period. Summers during the 1360s, 1470s/80s, decades around 1600, the 1690s, and 1800–1820 were also notably cooler than the long-term average. Data tabulated from documentary sources provide generally similar chronologies. Conditions in salmon nursery habitats across the headwaters of the Elbe, Weser, and Rhine tributaries are reported in Glaser, *Klimageschichte Mitteleuropas*, pp. 56–194, and then updated and recalibrated in Glaser and Riemann, 'A ThousandYear Record'. This work draws attention to notably cold periods in the 1360s, 1450s, 1580s–1600s, 1690s, and 1800–1820. Further to the west at Metz, in the heart of spawning and nursery areas for Rhine salmon along the upper Moselle, Litzenburger, *Une ville face au climat*, pp. 95–116, also relates fierce winters accompanied by chilly summers during the 1450s through 1480 and again in the 1530s. Scotland in particular experienced fierce cold and concomitant loss of agricultural productivity during the 1690s (D'Arrigo and others, 'Complexity in Crisis').

Europe (and less distinctively in the Low Countries) those decades experienced a high proportion of distinctly warm summers. Other seasons between 1550 and 1800 departed less from consistently cool conditions. This pattern suggests that reproduction and recruitment during the riverine phase of the salmon life cycle — i.e. hatching and growth — were the stages more vulnerable to LIA conditions. Some years later returns to fishers, to rights and property owners, and to traders in fresh or preserved salmon varied accordingly. So, presumably, did consumption of salmon in some socio-economic strata.

What were, therefore, likely significantly climate-driven early modern variations in the abundance of salmon, also triggered an intriguing **cultural reaction**, construction of **a long-lasting myth of past hyperabundance**. By the nineteenth and twentieth centuries people from Norway to France, across the British Isles, in New England and the Canadian maritime provinces, and even on the shores of Lake Ontario, believed salmon were 'once' so numerous that servants, apprentices, and other subordinated groups had contractual or statutory protection against being served salmon more than twice (or was it three times?) a week.[17] Despite the offer in Victorian Britain of a substantial reward for contemporary evidence of any such contract or statute and concerted official enquiries in France in 1972 and 1997, no such document has been found. What appear to be the earliest known assertions of such cultural measures have, however, now been tracked back to the 1690s in Britain and in northwestern Germany. Englishman Richard Franck was in Scotland in 1656/57 and likely drafted his *Northern Memoirs* in 1658. He revised this work in 1685 and eventually published it only in 1694, there asserting that the town council of Stirling, Scotland, had in the past to 'reinforce' an old — and seemingly unrecorded — municipal statute to that effect.[18] In the ensuing two years two Hamburg school-

[17] Schwarz, 'Der Weserlachs und die bremischen Dienstboten'; Schwarz, 'Nochmals'; Danker-Carstensen, 'Wiefiel Störfleisch verträgt die Dienstmagd', and somewhat more extensively, Danker-Carstensen, 'Stör oder Lachs'. Thibault and Garçon. 'Un problème d'écohistoire', refer to more than eighty independent mentions of this 'fact' from North America, the British Isles, and continental Europe from Norway to France, mostly dated to the nineteenth/twentieth centuries but going back to the seventeenth. Only one thoroughly peculiar actual 'contract' from the Dordogne in 1842 has been found, which is itself likely to be understood as referring to 'black salmon', the exhausted, even self-consumed, post-spawn organisms with marginally palatable flesh. The French and German studies are mutually unaware and neither mentions Scotland.

[18] Franck, *Northern Memoirs*, pp. 112–13. Among others, Walter Scott, who himself published a second edition of Franck's memoir in 1821, was repeating this anecdote in his *Tale of Old Mortality* as early as 1816, perhaps tellingly also a time of generally diminished salmon stocks.

masters, one compiling a chronicle still available only in manuscript form, the other publishing a 'Kurtze Historische Beschreibung der [...] Stadt Hamburg', each situated the servants' demands in the year 1454. A version of the tale set in '1320' appeared shortly thereafter in Lübeck, and others had by the early 1800s spread up the Elbe and along the Baltic coast. French parallels — all placing the event in Holland — had appeared by then as well. Salmon were never in medieval or early modern Europe an inexpensive foodstuff given out as rations to subordinates. But precisely in decades when a human generation of relative abundance had suddenly turned to relative scarcity, this fish became a symbol of a better past that never was.[19]

Tales of early modern salmon and climates remind us that environmental change has greater impact at the margins of an organism's range and of the need to identify sensitive parameters and stages of life. In terms of method, these findings illustrate the potential corroborative value of indirect evidence (e.g. prices in lieu of catch records) and, for the future, the desideratum of looking at the southern and not just northern edge of a distribution or ecosystem. Were Iberian stocks less affected by the LIA? Effects of climate fluctuations on fish will not be uniform but constrained by habitat requirements of individual species. Readers with scant interest in fish as such can nevertheless recognize that climate change, the LIA cooling in particular, affected early modern societies through disturbance of natural productivity as well as cultivated agricultural yields. At least in retrospect, contemporaries were aware of the losses, if not in this case, their deeper causes.

[19] While piecing together this largely early modern discussion I chanced to look again at what Alasdair and I had learned about conflicts over salmon along also the late medieval River Forth, where we have no records or proxies for actual catches (Hoffmann and Ross, 'This Belongs to Us!'). Surviving documentation places violent and judicial confrontations in 1365 (coldest decade of the fourteenth century), in 1494–1504 but said to reach back to 1470 (a cold phase of the Spörer minimum and a high point in Scottish salmon prices [Gemmill and Mayhew, *Changing Values*, p. 310]), and in 1682 (as temperatures slid to their Maunder minimum nadir). This is but a small sample, but resource conflicts are predicted to rise with climate change. Judicial records in Scotland and elsewhere may be worth systematic exploitation to see how they developed during the fluctuating LIA.

Pathos and Poverty: Fuel Economies of the Poor in the British Isles in the Late Medieval and Early Industrial Periods

Ian D. Rotherham
Sheffield Hallam University

Introduction

This research is based on long-term landscape history studies into resource utilization. Approaches applied include detailed field surveys, ethnology surveys and interviews, archival and literature review, and related action research. The work is supported by case-studies which consider the specific issues of fuel allotments and associated common rights.

In traditional agrarian, early industrial, and subsistence societies both urban and rural, local countryside provided most of a community's needs. Generally, only the rich in more sophisticated economies accessed imported goods and luxuries. For centuries, most people depended for arable, pasture, building materials, and fuel, on their own limited resources of land or that held in common. Uses evolved until by medieval times a web of sophisticated cultural and legal rights and restrictions controlled them. Across much of England but also in the wider British Isles, fuel for many people was gathered as of right from commonland held as such at a village or manorial level. However, much traditional utilization of rural resources ended over a period of less than two hundred years through processes described as 'cultural severance'.[1] Furthermore, when commons were subject to widespread enclosure (as was the case in England throughout the 1700s and 1800s), it was often the case that 'fuel allotments' were set aside for the support of the poor. These were either areas of land where the poor might gather fuel or other necessities of life, or else which could be let

[1] Rotherham, 'Importance of Cultural Severance'.

'With Our Backs to the Ocean': Land, Lordship, Climate Change, and Environment in the North-West European Past. Essays in Memory of Alasdair Ross, ed. by Richard Oram (Turnhout: Brepols, 2023)
EHNAW 5, pp. 245–275 BREPOLS PUBLISHERS 10.1484/M.EHNAW-EB.5.131278

commercially. The money generated was used to pay for fuel such as coal to be given to the poor of the village or manor. Such systems and approaches are well documented across the whole of England and much of Scotland, Wales, and Ireland too. The main driver of the changes was the idea of 'improvement' and hence the generation of wealth for the larger landowners; and the mechanism was 'enclosure'.

'Improvement' of commonland in England, especially the parliamentary enclosures, achieved major agricultural improvements but at a cost. The effects on the poor and needy were often dramatic and sometimes catastrophic. The particular interest of this chapter relates to fuel acquisition and use especially by commoners and the poor. Many texts on commonlands such as Stamp and Hoskins, Lord Eversley, or Lefevre make almost no reference to fuel gathering, fuel rights, or compensation for their loss.[2] This omission is remarkable both in terms of historical interest and significance, but also for understanding ecological and social dynamics. When commons were enclosed and traditional rights lost, there were attempts to compensate cottagers and the landless poor of the parish for the loss. The main motivation was not necessarily benevolence but more that the poor were the responsibility of the parish in which they resided and it was cheaper to allow access to at least some minimal resources than to fund their upkeep. This access to necessary subsistence was often as small plots of land held in trust for the benefit of the poor as charitable 'poor allotments'. Sometimes this operated through existing charitable mechanisms within the parish or township, but in other cases new charities were established. Some institutions were specifically to manage allotted land in compensation for lost fuel-gathering rights on now enclosed commons. Parliamentary acts sometimes specifically excluded anyone who occupied land about to be enclosed. Presumably occupiers were compensated in other ways, so this was to benefit those with rights over, but not residence on, commonland. Pinches gives the example of Tysoe Fuel Land in Warwickshire where

> for such poor people residing in the said parish as should not occupy any part of the lands intended to be enclosed, in lieu of a right to cut furze or gorse upon the waste land of the said parish.[3]

[2] Stamp and Hoskins, *Common Lands*; Eversley, *Commons, Forest and Footpaths*; Lefevre, *English Commons and Forests*.

[3] Pinches, 'From Common Rights to Cold Charity'.

In smaller rural parishes the poor often had direct access to such fuel allotments, but in other cases, such as Frimley, the sale of commercial rights from allocated lands was used to provide money to pay for fuel, often mineral coal, for the poor. There are numerous examples of such fuel allotments and associated charities across for example, East Anglia, and Birtles gives a detailed account of the one at East Ruston in Norfolk.[4]

In some cases, enclosure awards did not establish specific allotments but set down rights to gather fuel in identified restricted areas. For example, the enclosure award for Kenilworth in Warwickshire preserved as common, 35 acres 36 perches of waste on Tainter Hill Common, and 4 acres, 3 roods, 4 perches near Burton Green. The act decreed the land

> shall remain common and unenclosed to the intent and purpose that the poor belonging to the said parish of Kenilworth shall from time to time for ever hereafter use exercise and enjoy a free and constant right to get furze goss or fern.[5]

For medieval and early industrial communities it was necessary to provide for varying populations, the vagaries of weather, and natural catastrophe. They developed systems to protect and divide resources within and between communities not of equals. However, until relatively recently, unless systems were generally reliable and sustainable, the community itself was at risk. It was therefore imperative to maintain access to local resources. Such demands for resources were critical in forming many landscapes and their ecologies, which today are valued for conservation. In this context, understanding the implications of land-use and impacts, both drastic and subtle, on soils, water, and vegetation can inform contemporary conservation management. However, many conservation sites are managed in ignorance of former cultural uses, and of the implications of severance from former subsistence usage. Abandonment of earlier usage may trigger ecological successional changes, and for many species of conservation significance and for landscape heritage, the approaches are not sustainable.[6]

Pinches noted the allocation caused confusion with former commoners, and later other local residents, feeling these allotments were theirs by right.[7] This also applied to supposed common rights such as cutting furze on a strip of

[4] Birtles, 'Common Land, Poor Relief and Enclosure'.

[5] WCRO Y1/50, transcript of Kenilworth Enclosure Award, 27 January 1757.

[6] Rotherham, 'History and Heritage in the Bog'.

[7] Pinches, 'From Common Rights to Cold Charity'.

land parallel to the public road at Lawford Heath; a right disputed by the Lord of the Manor upheld by the Commissioners. Similar situations occurred with the rights at Frimley Common near London. It was important that allotments were seen not as charity but commuted rights in compensation for loss of common. Around settlements such as smaller towns and villages, allotments were often called Town Lands, Poor Lands, Poor Man's Piece, or Fuel Allotments.

The situation of fuel demand, access and competition in Scotland was discussed in detail by Oram and this work demonstrated from written sources the reasons for peat-fuel exploitation and its relationships to alternatives like wood and coal.[8] Furthermore, Oram explored the socio-economic drivers behind fuel uses and their transitions and the consequences of usage, demand and resources competition between communities. Oram also notes that this is a subject that is widely neglected by historians.[9]

Common Rights

In most rural areas particularly, which until the advent of industry, constituted the bulk of the population, rights of common were hugely important to rural communities until the widespread enclosures from the 1700s onwards. Larger urban settlements were the first to move to the commercial acquisition of fuels and here both lack of technology for effective combustion and the often crowded conditions of the poor and hence associated gross air pollution made coal problematic. Context for these widespread uses is provided by the nature of common rights under medieval manorial systems in two broad categories, **1) appendant** (with land-ownership) and **2) appurtenant** (related to ancient cottages). These were rights of common on village 'waste' including pasturage, piscary (fishing), estover and firebote (wood-gathering for house repair and domestic fuel), and turbary (cutting of peat and turf). Other rights covered cutting furze for fuel, bracken (fern) for fuel and animal bedding, and sometimes limited mineral or stone extraction. There were additional rights granted of '*common in gross*' vested in individuals through grant or by prescription (twenty years unchallenged use). Importantly the right was to the *individual* and the *body of the poor inhabitants*.

It is essential to recognize that access to common resources was especially vital for survival for rural poor who in many parts of Britain made up much

[8] Oram, 'Social Inequality'; Oram, 'Arrested Development?'; Oram, 'Fuel Transitions'.

[9] Oram, 'Fuel Transitions', p. 211.

of the population. Nevertheless, such rights and resources were also valued by those in expanding townships such as Sheffield for example. When Sheffield's commons were enclosed in the 1700s, one of the objections was the loss of rights of turbary.[10] At enclosure, lost rights were a huge threat and so the extent and value of land to be enclosed and associated rights were assessed and evaluated. In considering the gathering of fuels there are many factors to consider. These include types of fuels available, accessibility, calorific value and quality (i.e. how much heat they will produce, how easily, and for how long), and cost in terms of effort and energy expended in gathering and processing. Wood, gorse (furze), bracken (fern), heather (ling), and organic soil were all collected along with peat (turf). Even cow-dung might be taken and has high calorific value. However, if materials like dung are used for fuel then there is associated loss of benefits as fertilizer and soil conditioner. This is equally important as fuel in subsistence (closed) economies where exhaustion of productive soils may mean future starvation.

The cost-benefit analysis has to assess the price of buying fuel as opposed to effort expended in gathering and processing which varied widely with location. So the Rev Davies in Barkham, Berkshire,[11] estimated a family could cut a year's fuelwood in one week. To purchase the same fuel would amount to around one tenth of a labourer's annual income of between £1–15s and £4–3s. By 1844, thirty-five years after enclosure, the annual fuel expenditure for a four-room dwelling in Godalming, Surrey, was £3–12s–4d; based on two tons of coal at £1–12s per ton, and a hundred faggots at £1–0s–10d per hundred. The ash was sold-on to a farmer at 12s–6d and so the nutrient content was harvested and used, which was important in a pre-petro-chemical society. Across much of the country coal was available for £1 per ton and faggots at 12s per hundred, giving expected annual fuel bills of £2–10s. Coal prices varied dramatically according to the state of roads, proximity to canals, or later railways, the latter being seventy per cent cheaper than the former in moving coal affected fuel choice decisions.

However, the type of fuel consumed did not merely reflect cost, but also available technology to burn it. Depending on coal quality and type of fireplace, it was not always a favoured fuel. Simple hearths even with a chimney (which most but not all cottages had by the 1700s and 1800s) did not burn coal effectively and generated noxious fumes. Despite having around twice the calories

10 Paulus, *Unpublished Pages*.

11 Davies, *Cases of Labourers*.

of wood weight-for-weight, only with the introduction of metal grates and tall narrow chimneys was coal the better fuel. Wood, cleaner to handle and burn, had the advantage that its ashes could be sold to industry or agriculture and reused. Peat and turf burn long and slow, producing the traditional 'peat reek'; which might be considered disadvantageous but in upland and wet areas served to ward off ubiquitous midges and mosquitoes. Furthermore, soot deposited in thatched roofs could be recycled every few years as fertilizer for in-bye or arable land. Similar in properties to some wood, though faster burning, and giving immediate high temperature, but not long-lasting, furze was commonly used by the poor. It was favoured for communal baking with cloam ovens.

Wood theft and dishonesty of the poor were attributed to loss of traditional fuel rights in a way similar to poaching as a consequence of enclosure and harsh game preservation laws. Overseers of the poor spent more on coal allowances after enclosure, and numerous fuel charities were established. In 1820s' Warwickshire, there were sixty-four endowed charities distributing fuel either free or at discounted prices.[12] Alongside charities there were 'coal clubs' formed from the 1830s onwards mixing self-help and charity. The savings of the poor were topped up by wealthy, charitable parishioners.

Fuel Diversity and Competition

Issues of fuel diversity and usage in the British Isles have been considered by authors such as Hatcher,[13] but with specific emphasis on the evolution of coal usage. Warde examined historic energy use in England and Wales but mostly from the perspective of wood and mineral coal.[14] Warde and Williamson also examine some of the broader matters in relation to fuel availability, use, and competition.[15] One of the few authors to consider the specific issues raised here was David Zylberberg in his 2014 thesis on fuel usage, and we have had long conversations about the importance of peat and turf in the English fuel economy.[16] He raises the matters of peat and turf as fuels and their general neglect by historians considering the history of energy use in Britain. However, nearly twenty-five years ago, my colleagues and I described previously over-

[12] Pinches, 'From Common Rights to Cold Charity'.

[13] Hatcher, *History of the British Coal Industry*.

[14] Warde, *Energy Consumption*.

[15] Warde and Williamson, 'Fuel Supply and Agriculture'.

[16] Zylberberg, 'Plants and Fossils'.

looked effects of medieval and later peat-cutting on British upland landscapes and relationships to fuel for local communities.[17] This issue has been developed further[18] and more recent work has addressed many of these issues in great detail looking at the upland landscapes of northern England and the Scottish Borders.[19] However, it is suggested here that the deeper interactions between the various fuel sources of energy (especially peat, turf, wood, mineral coal and other relatively minor contributors like bracken and gorse) are generally overlooked or misunderstood in terms of the fuel economy of poorer people in both rural and urban situations during medieval and early industrial times. This chapter addresses some of the issues but does not attempt anything more than an overview of historical fuel usage.

Taking an example of ancient woodland management as coppice, for fuelwood and charcoal, is often misinterpreted and the drivers of change that determine contemporary ecology are frequently ignored. Medieval woods, along with heaths, commons, and bogs supplied most people with fuel, building materials, and food over many centuries. Along with energy provision for domestic use, medieval coppice woods and peatland turbaries fuelled much early industry. Different fuels (mineral coal, wood, charcoal, and peat or turf), had varying impacts and yet their landscapes and economics are rarely considered.[20]

Early accounts noted conflicts over resource use; describing for example, medieval iron-masters supposedly destroying woods and competing with other users for possession.[21] But to the untutored eye 'sustainable' coppice management looked devastated. Traditional, managed coppice operated in England for a thousand years, mostly ending by the 1950s; abandoned, grubbed out, or re-planted. Understanding interactions between competing uses and different fuels, helps unravel historic drivers of countryside change and utilization.

Elsewhere, I have discussed woodland-related issues, and those of heaths and commons. Webb considered traditional management of heaths and the impacts and consequences of abandonment.[22]

[17] Rotherham, Egan, and Ardron, 'Fuel Economy and the Uplands'.

[18] Rotherham, 'Fuel and Landscape'.

[19] Winchester, *The Harvest of the Hills*; Rotherham, 'The History of Domestic Peat Fuel Exploitation'; Rotherham, 'History and Heritage in the Bog'.

[20] Rotherham, Egan, and Ardron, 'Fuel Economy and the Uplands'; Rotherham, 'Fuel and Landscape'.

[21] Rotherham and Egan, 'Economics of Fuel Wood'.

[22] Webb, *Heathlands*; Webb, 'Traditional Management of European Heathlands'.

Fuel for People

Fuel is a basic commodity for human existence, particularly in advanced, sophisticated societies. The discovery of fire was a key step in developing human civilization and Pyne has discussed the impact of fire on society.[23] With control and use of fire came ability and technology to use metals and process and prepare foods and other necessities. Fuel for energy is needed to heat people and dwellings, and power machinery and processes of manufacture. Varying demands for energy and heat have influenced and themselves been affected by fuel availability. Resources change, affect, and influence the possible uses, and impact on, the environment which provides them. Transport infrastructure and costs also affect access. Essential for cooking, irrespective of climate, fuel is needed throughout the year. The type of fuel used depends on building construction and the type of fireplace or hearth. Fuel availability is influenced by ownership of land and resources, and in some cases by common rights. Anti-social and environmentally-polluting effects of particular fuels affect usage and acceptability. The impacts of domestic fuel use are discussed by Bond and others,[24] and health effects of cook-stove emissions for example, by Saldiva and Miraglia.[25]

The need for and use of fuel has strong spatial and geographical influences. Higher latitudes and higher altitudes, along with exposed or otherwise extreme environments demand increased fuel availability and use. This is for heating people and spaces (living areas), cooking, and manufacture (part-burning and hardening of implements such as wooden spears), baking clay, and smelting and forging metals. It is important to distinguish between 'fuels' and other non-fuel-based energy sources. Furthermore, energy sources such as fire can be used both as fuel and a tool for land clearance, hunting, and defence. Fuel is distinct from heat energy through passive building design or solar heating.

Fuel Needs

Demand for fuel increases with community sophistication and increased energy expenditure *per capita* by bigger settlements and more complex lifestyles.[26]

[23] Pyne, *Fire*.

[24] Bond, Venkataraman, and Masera, 'Global Atmospheric Impacts'.

[25] Saldiva and Miraglia, 'Health Effects of Cookstove Emissions'.

[26] Rotherham, 'Fuel and Landscape'.

The trend is from essential uses and provision for basic needs, towards complex uses to maintain sophisticated existence and social status. This is paralleled by moves from immediate sourcing from nearby environments, to more complicated acquisition and processing.[27]

Historically, sustainability was a delicate balance and failure devastating. The Western Isles of Scotland provide examples of the collapse of both fuel sources and land fertility, and hence food production. As noted by Kerr, for its existence, life in northern latitudes implies free use of fuel. For wood as fuel and other purposes,[28] Perlin gives detailed insight into issues and demands for wood fuel and resources.[29] As described by Oram, for example,[30] consequence of over-use and exhaustion of particular fuels, or restricted access for socio-political reasons, was the need to find alternatives and maybe use less suitable materials. In some cases, competition or restriction was through interactions of differing and alternative demands such as domestic use versus industrial or military. British iron-masters and the navy competed with demands for fuelwood for rich and poor, but especially the latter. Competition between commoner, peasant, and the lord of the manor; or between industry and domestic use, ran throughout much of the early industrial period. Hayman described eighteenth-century British landowners tightening their control over the countryside, with legislation passed to restrict community customary rights to harvest underwood; a contest between communal resource and private domain.[31] In England, urban populations could resort relatively easily to alternatives such as coal, but rural, and certainly remote rural, areas could not unless supplies were close-by. Indeed, until the 1500s, most mineral coal came from coastal outcrops and coastal transport with limited access and use. Surface-outcropping coal was increasingly utilized but deep-mining did not begin until the 1700s and 1800s. Rural communities especially but urban people too, did as they had always done, and turned to other fuels. Peat, turf, and furze, supplemented by small-wood and heather were often the main fuels of common people. For peasant communities in villages and around towns, fuel was important and often scarce. John Evelyn writing about 'fuel' noted:[32]

[27] Rotherham, 'Fuel and Landscape'.

[28] Kerr, *Peat and its Products.*

[29] Perlin, *Forest Journey.*

[30] Oram, 'Social Inequality'; Oram, 'Arrested Development?'; Oram, 'Fuel Transitions'.

[31] Hayman, *Trees.*

[32] Evelyn, *Silva.*

> But besides the Dung of Beasts, and the Peat and Turf which we may find in our ouzy Lands and heathy Commons) for their Chimneys, Cow-sheds, etc. they make use of Stoves, both portable and standing [...] In many Places (where Fuel is Scarce) poor people spread Fern and Straw inn the Ways and Paths where Cattle dung and tread, and then clap it against a Wall till it be dry [...]).

Tudor writer, the Rev. William Harrison (Withington, 1899; Edelen, 1994) was concerned about the depletion of woods and lack of fuel:[33]

> Howbeit, thus much I dare affirm, that if woods go so fast to decay in the next hundred years of Grace as they have done and are like to do in this [...] it is to be feared that the fenny bote, broom, turf, gall, heath, furze, brakes, whins, ling, dies, hassocks, flags, straw, sedge, reed, rush, and also seascale, will be good merchandise even in the city of London, whereunto some of them even now have gotten ready passage, [...] Of coal-mines we have such plenty in the north and western parts of our island as may suffice for all the realm of England; and so they do hereafter indeed, if wood be not better cherished than it is at this present [...] their greatest trade beginneth now to grow from the forge into the kitchen and hall, as may appear already in most cities and towns that lie about the coast, where they have but little other fuel except it be turf and hassock.

Yet the importance of alternative fuels like peat and turf from the common are generally overlooked. I have noted this significance elsewhere[34] (See Warde for a meticulous study of energy in England and Wales that does not touch on these).[35] A paper by Williamson and Warde provided a more rounded overview and I have explored the consumption of peat fuel in two recent papers.[36]

Fuel Poverty

Dependence on a fuel and loss of access to it has devastated some communities, particularly those in seasonally cold or moist climates where fuel is essential for life. Availability of secure fuel supplies and access to them is important. Over-dependency on one fuel has worried people and politicians over centuries. Kerr noted 'the abject dependence of the nation [Scotland] on coal owners,

[33] Withington, *Elizabethan England*; Harrison, *Description of England.*

[34] Rotherham, 'Peat Cutters and Their Landscapes'; Rotherham, *Peat and Peat Cutting.*

[35] Warde, *Energy Consumption.*

[36] Warde and Williamson, 'Fuel Supply and Agriculture'; Rotherham, 'The History of Domestic Peat Fuel'; Rotherham, 'History and Heritage in the Bog'.

the miners, and the coal factors for the right to live'.[37] This resonates with late twentieth-century debates about access to oil, coal, gas, nuclear power, or alternative energy sources. Kerr argued for peat fuel based on the need to diversify and to have alternatives. Perlin discussed energy issues and the need for wood as fuel and for other purposes.[38] For example, industrial charcoal demand in Elizabethan England combined with bitterly cold weather to create urban fuel shortage. In the long, cold winters of the so-called 'Little Ice Age', conditions for the poor were harsh. 'The meaner sort of people' had to live 'unprovided of fuel' and without fuelwood many of the poor 'perish oft for cold'. They took wood illegally but risked punishment, and they diversified. Mineral coal created major pollution problems, but wood shortage left little choice if people were to survive. In London in 1592, 'the use of coals [...] is of late years greatly augmented for fuel [...]' and prior to 1600, 'London and all other tons near the sea [...] Are mostly driven to burn [...] coals, for most of the woods are consumed [...]'. At this time, mineral coal was useless for metal-smelting, so industry needed wood for charcoal until Abraham Darby the Elder, in 1709, made coke by bringing coal 'to as near a resemblance of charcoal as possible'.[39]

Conflicts can exacerbate fuel-related issues in both town and countryside.[40] For example, during the English Civil War in the 1640s, King Charles I took Newcastle, and consequently Parliament closed the coal trade to impoverish the City of Newcastle. However, the impact on Londoners was severe and, losing their main source of coal, the ordinary people rioted and rampaged through the city in a desperate quest to find wood to burn. To quell unrest Parliament opened the woods and parks of the Royalists but fuel remained critical, affecting commodity availability and prices, and the poor suffered badly. A big problem with the exhaustion of or restricted access to fuel is that it leads to the use of less suitable alternatives. Restrictions to access may be through competing demands such as naval timber or charcoal for smelting, both affecting domestic fuelwood. The varying demands led to competition between peasant and landlord, industrial and domestic use with woods and other countryside resources at the heart of a sometimes bitter struggle. Hayman has described how the notorious English *Black Act* of 1723 restricted woodland access, fuelwood use, and estate timber for building, to bring conflicts between communal resources and

[37] Kerr, *Peat and Its Products*.

[38] Perlin, *Forest Journey*.

[39] Perlin, *Forest Journey*.

[40] Rotherham, 'War and Peat'.

private domains to a head. With widespread enclosure of English commons, peasants frequently became either wage-slaves and/or poachers where they had once taken resources as of right. When the better sorts of fuels were exhausted or otherwise unavailable then communities, both urban and rural might resort to peat, turf, ling, bracken (brake) and the like.

The impacts of such competition for fuel in England were most keenly felt in the lowlands since that was where most people lived and where enclosures were most aggressively pursued. However, we see such competition in the more sparsely populated uplands too, such as the eighteenth-century South Pennine farming communities. Here people used traditional fuels of local wood from the valleys and peat from adjacent moors. With difficult transport conditions making coal expensive, fuel from the countryside was the commoners' resource; farmers burning peat and wood harvested according to traditional rights. However, industrial production of charcoal for metal smelting affected fuelwood availability and the local farmers petitioned for better roads and cheaper coal. Even woodlands of low value since the disestablishment of the Peak Forest during the 1600s, were now needed for charcoal production. So, rising industrial activity meant landowners saw new value in their woods, and this recognition triggered issues of fuel availability, transport, and costs.[41] Access to mineral coal alleviated some problems but the limited technology in poorer cottages made burning it both difficult and potentially unpleasant and dangerous.

Heath, Common, Fen, and Bog

The countryside in which this discussion is set was formerly characterized and often dominated by extensive heathland, wetland, and woodland; all now reduced by agricultural intensification and land 'improvement'. Many lowland heaths and wetlands have been totally obliterated so they and their former importance as providers of basic subsistence are lost from memory.[42] Chadwick suggested that '[…] the particularly British institution of the common may be in greater danger, despite the recent spate of legislation, than at any time since the enclosure acts'.[43] Once the hub of local communities from uplands to lowlands, their diverse uses including providing fuel have ended.

[41] Ardron, 'Peat Cutting in Upland Britain'; Bevan, *The Upper Derwent.*

[42] Rotherham, *The Lost Fens.*

[43] Chadwick, *In Search of Heathland.*

In both uplands and lowlands of Britain, peat was an important fuel source well into post-medieval times and beyond. Frederic Eden reported peat to be 'the usual fuel consumed by labourers' in Lincolnshire, and, for example, the Isle of Axholme.[44] Peat turf was intensively exploited at locations such as south Cambridgeshire in the late 1800s, and at Swaffham, for instance, extraction only ended in the early 1940s. Heathlands (often wooded commons) and lower-lying dry moors could produce large amounts of gorse ('furze') (*Ulex* sp.), broom (*Cytisus scoparius*), heather (*Calluna vulgaris*), heath (*Erica* sp.), and other vegetation such as fern or brake (*Pteridium aquilinum*) that could be burnt as domestic fuel. For many people, both town-dweller and rural peasant, these resources were vital. Thomas Blenerhasset, with regard to Horsford Heath, which was located close to England's second city at that time, stated that 'This heathe is to Norwich and the Countrye heare as Newcastle coales are to London'.[45] Peat fuel and flag were taken from the Norfolk Broads for both the urban centres at Norwich, and for the villages and farmsteads across the East Anglian countryside.

This importance was reflected in the cultivation of lands to provide fuel and also the measures in place to protect from over-exploitation. Furze or gorse for example, might be sown into special enclosures or furze beds and protected from grazing. The crop could be harvested for firing or else bruised as fodder. Turner noted how in 1801 around Sithney, Cornwall, for instance, 'Here are, it is almost literally true, no trees; consequently a considerable part of every estate is under furze, which would frequently, with proper cultivation, produce whatever the cultivated lands now produce'.[46] However, competition for landscape and resources might compromise access to fuel. This was described by John Norden, writing in 1618, in terms of stock grazing destroying gorse fuel, 'Where this was intense plants like gorse and broom could not grow into the large, woody plants suitable for firewood'. Norden described the gorse in the West Country, which grew 'very high, and the stalke great, whereof the people make faggots'. And,

> this kind of Furse groweth also upon the Sea coast of Suffolke: But that the people make not the use of them, as in Devonshire and Cornwalle, for they suffer their sheep and cattell to browse and crop them when they be young, and so they grow too scrubbed and lowe tufts, seldome to that perfection that they might be[47]

[44] Eden, *The State of the Poor*, p. 566.

[45] Barrett-Lennard, 'Two Hundred Years of Estate Management'.

[46] Turner, *Home Office Acreage Returns*, pp. 33–34.

[47] Norden, *The Surveyor's Dialogue*, p. 235.

As discussed by Williamson and Warde, peat, gorse, heather, brake, broom and the other fuels remained important, certainly for poorer folk well into the nineteenth century.[48] This significance was reflected by the detail of many nineteenth-century enclosure awards which allocated 'poor lands'. These were in recognition of customary uses lost and to reduce claims on poor relief. Parliamentary enclosure commissioners set aside land as fuel allotments to be cut by the poor for gorse, turf, or peat, or else rented out to provide income for the purchase of coals for poor-relief. Such awards were widespread across rural England.

Most people in the British countryside were cottagers, labourers, farm servants, and squatters. The cottagers either owned or occupied cottages, which by ancient custom had attached rights of commonage on the wastes. These included pasturing animals, cutting turf and extracting fuel. Massive conversion of heath, moor, waste, and marsh, to arable and enclosed pasture, dramatically ended this utilization by rural populations, especially the poor and poorer commoners. In the Peak District, for example, small-scale peat use lingered in remote, particularly the upland areas, but by the mid-1900s, subsistence exploitation died out. Along with wood, mostly from 'woodland' and wooded common, peat, turf, gorse or furze, broom, bracken, fern or brake, flag, reed, sedge, and ling were harvested from heath, common, and fen. Surface mineral coal might be taken but not favoured because of noxious fumes. Fuels were mixed and combined as need and availability dictated. Indeed, until late in the twentieth century, coal dust might be mixed with peat and clay to form low-cost briquettes for burning in the domestic hearth.[49]

Gorse was often used when wood and peat were in short supply, as kindling, or where high temperature was required (e.g., in baking). This was often the case in villages, but in urban areas too, where the amounts utilized could be massive. So when gorse was brought into and stored in Dublin there were complaints about the hazards of it stacked by the city walls, where they '... doe overtopp the said walles in height'.[50] In rural areas, gorse was important fodder for cattle and horses, particularly in winter when other crops were scarce. In Norfolk, Dorset, or Cornwall, for example, large areas were specifically set-aside for furze-bed cultivation. Many commons yielded underwood and building timber. Bracken

[48] Warde and Williamson, 'Fuel Supply and Agriculture'.

[49] Personal communication with peat workers at Fenn's and Whixall Moss.

[50] Humphries and Shaughnessy, *Gorse*.

'fern' (*Pteridium aquilinum*) was gathered and heather (*Calluna vulgaris*) harvested for fuel and building.

Until the 1900s, gorse and fern were in widespread use across Britain and had considerable value for both commercial and domestic purposes. According to Graham Bathe, in 1734 in Wiltshire, twenty-one loads of fern from Coal Coppice were sold to the brick-makers Hutchins and Hancock, at 2s per load and they took more in 1735.[51] By 1768, a local company was being employed to cut fern and it seems that Hutchins and Hancock mixed this with other fuels purchased in the Forest.[52] These included lop and top, billet wood, faggots and bavins (kindling and small diameter, easily ignited sticks). They consumed many thousands of furze faggots, sold at 20s per thousand. In 1738 alone, they purchased 30,100 furze faggots from a total of 59,600 cut and sold in Furze Coppice.[53] Joseph Hutchins was still buying 37,000 faggots per year in 1765 and was leasing part of Brown's Farm, west of the Salisbury Turnpike Road, to dig brick-clay.[54] His lease mentioned that he would purchase furze or fern at the customary price, and provide all ash to the lessee for manure;[55] a nice demonstration of the interlinking of fuel and food production economies at this local level. For seventy years the price of furze remained constant at 20s per thousand, or 2s per hundred, 'reckoning at six score to the hundred'.[56] The brick-maker at Lye Hill continued purchasing fuel for his operations into the nineteenth century.[57] This research by Graham Bathe demonstrates the potential intensity of fuel demand for an industry such as brick-making and thus the likely competition with domestic users at the lower end of the social scale.

The importance of fuel from the heath and common was demonstrated by Stone (2017) working in East Anglia,[58] and he noted the special protection necessary to conserve or maintain the resource. On most of the Suffolk heaths that he researched, management in the seventeenth and eighteenth centu-

[51] Thanks to Graham Bathe (personal communication) for information on gorse use in Wiltshire with material from the Wiltshire and Swindon Archives. Wiltshire and Swindon Archives, WSA 9–1-254, WSA 9–1-125.

[52] WSA 9–1-285.

[53] WSA 9–1-258.

[54] WSA 9–1-283.

[55] WSA 1300–1850.

[56] WSA 9–1-269.

[57] WSA 9–1-318.

[58] Stone, 'Heaths, Commons and Wastes'.

ries was focused on the maintenance of both heather and gorse. Tenants were allowed to gather these but only for use within the manor and certainly not for sale. This was confirmed by inspection of the court roles where, in 1644, a man was fined three pence at the manor court for 'cutting takeing & Carrying away Ling growing upon the Comon heaths of said town'. Two others were fined the same amount for selling heather cut from the heath contrary to local bye-laws.[59] It was also discovered that the rights of commoners to take gorse as fuel in the 1600s and 1700s, depended on the state of the heath and its vegetation. If the common was mismanaged and so in poor condition, the rights could be suspended to allow recovery. Furthermore, perhaps reflecting reduced access, deteriorating sites, and growing local populations, the punishments for offenders over-exploiting or damaging the heather and gorse became more severe as the centuries progressed. In 1698, four people were fined ten shillings for over-cutting. Courts began to list not only those allowed to take fuel but also those barred from doing so. Stone found references for Wrampit Heath allowing the taking of furze, ling, and stubbed-up turves of peat (known as 'flags') as allowed, or the 'poor's firing'.[60] The importance of these resources for the very poor in both town and country are made clear but controlling over-exploitation was also significant. Conflicts with commoners grazing cattle or sheep and thus damaging the ling and gorse were also troublesome.

Countryside Utilization

Subsistence use of countryside resources over centuries causes dramatic change with highly regional impacts of utilization for fuel or foods intertwined. British upland landscapes were shaped and manipulated by fuel-use, grazing, and later game preservation. Lowland areas like the Humberhead Levels, Cambridgeshire Fens, Norfolk Broads, or Somerset Levels, had extensive peat removal, causing massive change in ecology and landscape. Medieval peat-fuel use created the East Anglian 'Broads' landscape and provided fuel for common people, major towns and cities, including Norwich. The scale of peat and turf exploitation was massive, with the Norwich Cathedral Priory chamberlains purchasing *c.* 5000 turves per year during the 1300s and 1400s. Purchases by the kitchen cellarer were colossal with 300,000 to 400,000 turves per year.[61]

59 Stone, 'Heaths, Commons and Wastes'.

60 Stone, 'Heaths, Commons and Wastes'.

61 Lambert and others, *The Making of the Broads*.

However, the rural community was also a major part of the consumption but perhaps less well documented. At Martham in East Anglia, households used 5000 to 8000 turves per year and at Scratby around 10,000. Vast areas were maintained as peat-moor and fen, but climate change and sea-level rise led to abandonment and today's farming and tourism landscape.

In Northwold, Norfolk, a 1.75-acre fuel allotment produced 12,000 turves annually:[62] the calculated consumption of one hearth. Cottagers of twenty years' standing without common rights were treated meanly, being permitted by the fen reeves only 800 turves each a year, insufficient to keep a hearth alight. Fuelwood was bundled as faggots, but variable in size, with great faggots quoted for Castle Rising in Norfolk. The price varied from year to year and with location.

Faggots cut from coppice were generally around 2s–6d per hundred, varying from 8d to 8s. Tallwood prices were similar to faggots, but probably refer to lop-and-top of tall trees or pollards. Coppice woodland production comparisons can be made with underwood (*boscum*) price per acre. Rogers suggested that whilst relevant information is scarce, thirteenth- and fourteenth-century prices were around 6s an acre up to 13s–8d.[63] Underwood produced from 300 to 600 faggots per acre depending on the quality of the coppice woodland, selling at 20s per 100. Exceptionally, sixteenth-century woods produced 1700 faggots per acre. Underwood producing 400 faggots might sell for £5–5s per acre with prices rising eight times from the medieval period and raw materials about twenty times. Seventeenth-century wood prices rose steadily. Rogers questioned whether commodity price included labour as it can be the principal cost. In fifteenth- and sixteenth-century England, no land was without an owner, and the privilege of cutting even the cheapest type of fuel was either as a common right or granted on payment for a licence. Underwood was selling at 20s to 50s an acre; and faggots at around 7s–11d per hundred delivered. English eastern counties also used sedge as fuel, continuing well into the 1800s.

When wood was scarce or unavailable in Britain:

> Peat was the only alternative fuel to be had in any quantity. There is even more of it in Britain than in Ireland, despite the poets and the travel brochures; but it was too bulky to be carried far from its source. Peat burns readily; its merit is to smoulder without a blaze, though this makes for a smoky fire. The 'peat-reek' is pleasant at a distance, and a whiff of it is not unwelcome in one's Harris tweeds; but in a 'lum'

[62] Young, *General View*.

[63] Rogers, *History of Agriculture*.

> cottage its pungency is dreadful. Peat has been 'coked' to destroy the reek, powdered, mixed with pitch or rosin, and compressed into bricks that were claimed to be better than coal. Lacking true peat, some burn turf, or parings of peaty soil with roots of heather and gorse. (Peat and turf, accidentally ignited, have caused slow, widespread and unquenchable fires, which have even consumed whole villages.) Dried cowdung is a good fuel, and the scent of its fire sweeter than might be supposed; it found some favour during the wood famine, but to burn dung that ought to enrich the soil is bad husbandry.[64]

These issues have been discussed in some detail by Hammond and Hammond,[65] addressing the plight of poorer village labourers. Loss of commons affected many aspects of life and particularly the production and preparation of food. For example, a Cumberland housewife could bake barley bread in her oven 'heated with heath, furze, or brush-wood, the expense of which is inconsiderable'. With stretches of 'waste land' at her door, children could be sent to fetch fuel. It was said that there was no doubt that brakes, goss, furz, broom, and heath off the common, were more valuable to the community than good wheat, rye, and turnips from the improvement agriculture. Yet the inexorable enclosure of commons rolled on and in the south and midlands of England, the fuel of the poor grew ever 'scantier'.

> When the common where he had gleaned his firing was fenced off, the poor man could only trust for his fuel to pilferings from the hedgerows. To the spectator, furze from the common might seem gathered with more loss of time than it appears to be worth; to the labourer whose scanty earnings left little margin over the expense of bread alone, the loss of firing was not balanced by the economy of time.[66]

Hammond and Hammond explain how lack of firing added to problems of insufficient clothing and food;[67] the lot of the poor commoner was pretty miserable. A writer to the *Annals of Agriculture* even suggested the poor should spend wintertime with animals in the stables as it would cost them nothing to enjoy warmth from the cattle. Indeed, fewer would suffer death from want of fire. Another correspondent noted a connection between fuel and food in that where the former was scarce and expensive the poorer people found it cheaper to buy bread rather than bake it themselves. However, '... where fuel abounds,

[64] Wright, *Home Fires Burning*.

[65] Hammond and Hammond, *The Village Labourer*.

[66] Hammond and Hammond, *The Village Labourer*.

[67] Hammond and Hammond, *The Village Labourer*.

and costs only the trouble of cutting and carrying home, there they may save something by baking their own bread'. With loss of commonland there were complaints of pilfering from hedgerows; and some families avoided paying for fuel by gathering and drying cow-dung. In some cases, it was found that even if the peasants had money to buy fuel, none was to be had locally. In one example at least, they clubbed together to fetch pit-coal to bring to their village and to sell to the needy. Sometimes the parish helped out but in other situations treatment was harsh; as in the case of a peasant whose master imprisoned him for stealing wood from his hedges.

Fuel Process, Usage Rights, and Cost

Common rights to specific fuels varied through time and across regions with occurrence, competition, and economics. Availability and transport, mode and efficiency of use, and social or other controls or barriers all had effects, and these have been discussed with regard to Scotland by Oram.[68] Geographical location influenced demographic trends, and development of transportation infrastructure.[69] Detailed assessment of interactions is difficult and presents basic challenges in 1) understanding and transference of nomenclature of commodities and 2) the comparability of monetary figures. Texts such as Thorold Rogers', whilst not comprehensive, are helpful.[70] However, since many estates manufactured or harvested their own fuel, particularly charcoal or fuelwood, it may not be accounted in external trading payments. Furthermore, fuel extraction and usage by common right or through charitable fuel allotments are also disguised in such analyses and lead to oversight of common fuels like peat.

Thorold Rogers provides context for wider discussion. He noted that in attempting a money estimate of various utilities it is of great interest to determine, as accurately as possible, the proportionate price of fuel. In rural areas of low population density, he suggested people near forests and chases, and similar open grounds, sourced their fuel from such areas. He noted sale and value of these fuels were important to landowners and countryside dwellers. Rogers stated that until mineral coal use became general and transport over-land relatively easy and cheap, all inland places depended on other energy sources, including local wood, turf, or peat. However, it must be noted that for an illit-

68 Oram, 'Social Inequality'; Oram, 'Arrested Development?'; Oram, 'Fuel Transitions'.

69 Rotherham and Egan, 'The Economics of Fuel Wood, Charcoal and Coal'.

70 Rogers, *History of Agriculture*.

erate, poor community the 'free' usage of lower grade fuels from the commons will be undocumented and under-recorded; but absence of evidence should not be taken as evidence for absence.

Rogers considered interactions between availability and prices of fuelwood and sea-coal, and modes of transport and supply to centres like London. For the latter, the River Thames was the main transport route, upstream from the sea, or downstream from inland. Rogers felt he could reasonably assess comparative prices of charcoal, sea-coal, and turf. Nevertheless, he considered that measures and values for wood, for example, present difficulties. In order to achieve reliable comparative analysis, standard measures, such as the thousand or the hundred of a particular item, are required. With defined names for sub-units and accepted measures, analysis should be straightforward. However, for fuelwood alone, Rogers identified terms like fagot and fardel, spelden, tosards, kiddles, bavins, billets, bobbelyns, Shrof faggots, ascels, astells, kydes, kedys, brushwood, and tallwood (or tall wood). Furthermore, faggots (fagots) may be varying sized bundles of fuelwood, sold by the load (quarter of a hundred) whereas fuel or *focalia* was sold by the hundred, and so were standards, shides, sheddings, pole wood, great wood, and furze.

To understand fuel usage, Rogers addressed building status and design related to consumption for heating and cooking. Considering medieval times, he noted that chimneys were only used in castles and great houses. Typical manor houses of lesser gentry, or poor cottages, had fires against a hob of clay with smoke escaping via holes in the roof, or through doorways. With no chimney, then turf or charcoal was preferred though cost restricted the latter to wealthier households. Charcoal was used in most college halls until relatively modern times (in Cambridge colleges into the 1800s). The fuel was laid in an iron frame in the centre of the hall and fumes escaped through a lantern in the roof. For earlier periods, most information was for wood, faggots, firewood, and charcoal, and later for sea-coal, pit-coal, and charcoal. Rogers suggested charcoal was manufactured and sold by the quarter with no great variation in cost for domestic heating and cooking.

Sea-coal was used by smiths and for burning in greater houses, and Rogers described sea-coal from dates as early as 1279 in Dover; carried by coastal traffic, purchased for the Castle and burnt in a fire with a chimney. It was similarly used in Suffolk, Hertfordshire, Kent, and in Southampton, and later recorded for Kent, Essex, and by 1378, in London. Sea-coal was sold by the chaldron at Southampton, being half the price as in London, and prices rose enormously because of plague. There are additional complications with mineral coal becoming increasingly available where it was mined inland in counties like Derbyshire and yet was still sometimes called 'seacole'. The first Derbyshire reference to coal-mining was 1285, when Hugh de Morley 'granted and confirmed to Simon, Abbott of Chester, free entrance and exit to his mines of sea coal wheresoever they may be found in his lands of Morley and Smalley [...]'. Another was in 1322, when Emma Culhane was killed 'by the damp while drawing water from a colepit'.[71] By the fifteenth and sixteenth centuries, sea-coal was more frequent and sold by chalder, chaldron, fother, quarter, and bushel. The prices varied according to proximity to, or ease of transport from, the coalfield. Sea-coal was always more expensive than charcoal at this time, except close to coalmines. It is perhaps the greater awareness of this transition from wood to coal that has masked the wider importance of peat and turf as common fuels.

In the 1600s and 1700s, Cambridge sea-coal prices were raised by poor weather affecting both land and water transport. By 1652, Eton College began to purchase sea-coal; thereafter using little else. Sometimes transport costs are specified or else included and means transport from wharf to college coal-yard. By the 1700s, there were coalmines in Newcastle-upon-Tyne, Durham, Lancashire, Derbyshire, the Midlands (especially around Leicester), the Forest of Dean, South Wales, Somerset, and Southern Scotland. Fuel demand could be immense and often involved mixed fuels for different purposes. In 1403 for example, a thousand cart-loads of dry wood and a large barnful of coal were used for a two-day feast in Savoy.[72] Rogers found turf recorded in 1337 with amounts given as the thousand (1200) or the last (12,000). In Cambridge in 1334, turf was specified as heather with average price around 9d per thousand. Turf was frequent in Southampton and Kent. Between 1415 and 1450, Cambridge colleges used turves extensively, bought by the thousand at around 2s; but they were not used after this. So again here we see the situations where turf is purchased rather than where peat and turf are burnt as fuels freely sourced from the local common.

Hartley and Ingilby stated that around the North Yorkshire Moors:

> Rights of turbary appear frequently in monastic charters and peat and turf were dug in more places than would appear possible now, including the Carrs of the Vale of Pickering almost to the coast near Filey. Until the seventeenth century, fuel used at Scarborough was mostly got from the moorlands north of it and from the extensive Allerston Moor near Pickering. Special areas were sometimes allotted on the moors for the use of the tenants of a manor, and the sale of peat and turf, regu-

[71] Heath, *Illustrated History of Derbyshire*.

[72] Porter, *Yesterday's Countryside*.

> lated by manor courts often caused trouble and ill feeling. From the fourteenth to the nineteenth century, records show that this occasionally took the form of the throwing down and burning of the stacks drying remote from the Company in the wide Moors.[73]

Competition between fuels was often from ones cheaper, more accessible, or better quality. *The Statistical Account of the Scotland for the Parishes of Anworth and Girthon, 1792,*[74] noted for coastal south-west Scotland,

> Peats, the fuel used by the farmers and cottars, are dear, owing to the distance of the mosses, and, the bad roads which lead to them [...] Coals are the general fuel here. They are imported from Whitehaven, Newcastle, etc. and run from 30s to 40s a ton.

> Peats are not used so widely as in former years. Prices are high normally £2 and over per cartload. Coal is brought to the door by motor lorry from Aberdeen and sells at the high price of £5 a ton for English coal and £4–15 for Scottish coal.

Smith stated that in north-east Scotland,

> Peats are not used so widely as in former years. Prices are high normally £2 and over per cartload. Coal is brought to the door by motor lorry from Aberdeen and sells at the high price of £5 a ton for English coal and £4–15 for Scottish coal.[75]

Accessibility to fuel transport was the key and locations difficult for transport or remote from coastal distribution or local coal supplies, maintained peat-use for longer. Limited availability of higher grade fuels meant reliance on peat, turf and associated resources.

Around Rhayadder in mid-Wales, coal arriving by train in the 1860s to 1870s was the main reason for abandonment of hilltop turbaries, valley-bottom bogs having been long-since exhausted. In the Yorkshire Dales, and North York Moors, extensive cutting probably finished by the 1930s to 1940s. Such use was often not sustainable in the longer-term and when the fuel ran out either alternatives were found, or the community moved on. Upland landscapes like the English Pennines, the Peak District, the Yorkshire Dales, the North Yorkshire Moors, much of Wales and large areas of Scotland have changed dramatically through fuel exploitation, mostly for peat and turf. For parts of the Peak District

[73] Hartley and Ingilby, *Life and Tradition*.

[74] Mann, *Cattle, Cotton and Commerce*.

[75] Smith, ed., *The Third Statistical Account of Scotland*.

and South Pennines as demonstrated by Ardron, Rotherham and Gilbert,[76] there was more medieval peat cut (*c.* thirty-four million cubic metres), than from the Norfolk Broads at the same time, most in the early medieval period.[77] Such usage has been largely overlooked.

The Impacts of Enclosures and Improvement

Agrarian and early industrial communities maintained vital environmental resources through complex social, economic, and legal mechanisms. The productive landscapes that were created reflected their social needs. The mechanisms in place to manage exploitation generated continuity over time. However, with parliamentary and private 'enclosures' of heath, moor, common, bog, and 'waste', lower-lying sites were exploited and progressively destroyed during the sixteenth, seventeenth, and eighteenth centuries. With increasing technological industrialization, urbanization, and rural de-population, these essentially conservative cultural landscapes were swept away or abandoned. Industrial and urban areas became more techno-centric and agriculture industrial. With social and technological innovation during the 1700s and 1800s, fuelled by coal then oil, the countryside was largely freed from dependence on local sustainability and need for local subsistence. By the late twentieth century, the cultural landscape was a largely forgotten, archaic remnant of a lost way of life; a mere shadow etched in landscapes across upland hill and valley, and lowland plain. Traditional exploitation lingered longer in the British uplands and economically peripheral areas. Even here though, it was in decline by the late 1800s. Edward Bradbury noted how 2000 acres of common land around Kinder Scout had passed to private ownership, and that the poor of Hayfield had been robbed of their forty acres of turbary at Poor Man's Piece.[78] This was near the end of common subsistence usage.

[76] Ardron, Rotherham, and Gilbert, 'Peat-Cutting and Upland Landscapes'.

[77] Ardron, 'Peat Cutting in Upland Britain'.

[78] Bradbury, *In the Derbyshire Highlands*.

Some Case-Study Examples of Fuel Allotments

Fuel Charity and the North Devon Turbaries

Stockland Turbaries are located near Honiton in North Devon and now maintained as nature conservation areas.[79] The Turbaries were established by an early 1800s Act of Parliament to enclose the local commons; probably enacted in 1807 with enclosure of the waste somewhat later, as the accompanying map was dated 1824. The Turbaries were described in Reports to the Charity Commissioners included in the Return to an Order of the Honourable The House of Commons 26 July 1905, which includes Reports on Stockland and Dalwood Parishes dated 19 July 1834. The lands and other properties generated income to be used for '[...] any other pious and charitable use or uses within the said parish as the feoffees for the time being should think fit'. These included the Stockton Ecclesiastical Charity, Stockland Educational Foundation, and Stockton Parish Lands (General Charity); collection and distribution overseen by appointed trustees. In 1908, total annual income was just over £44, of which £4 was given to the parish Christmas coal-fund.

Along with the coal-fund there was specific mention of turbary lands:

> By the Inclosure Award already mentioned there were allotted and awarded to the churchwardens and overseers of the poor of the parish of Stockland for the time being, and their successors for ever the twelve several parcels of land hereinafter described (that is to say).[80]

The document details individual turbaries at Bucehays Turbary, Quantock Turbary, Short Moor Turbary, Huntshays Turbary, Hamer Hill Turbary, and Shore Turbary.

The land awarded in 1807 and marked on the award map was,

> one hundred and sixty-five acres and fourteen perches, and were set out deemed sufficient for the poor settled inhabitants of the parish of Stockland aforesaid residing therein, to dig, or cut turf or furze for their own use, but not for sale, subject to the several provisos, powers, and restrictions in the Act of the forty-seventh year of His late Majesty's reign mentioned and contained, and also subject to certain rights of way in, over and across the Same, therein set out, allotted and awarded.[81]

[79] *The Stockland Turbaries. Information Leaflet.*

[80] *The Stockland Turbaries. Information Leaflet.*

[81] *The Stockland Turbaries. Information Leaflet.*

In the tithe apportionment agreement of 21 December 1844, the area was less, 130 acres with a tithe value of £1–10s. It was noted that the turbary land '[...] is occupied and used as stated in the Award, and no yearly rent or profit is received for it'. Interestingly by the late 1990s, the turbaries were described as seven parcels of land though the maps suggest that these included all twelve original areas. The knowledge and memory of their use as working fuel turbaries had long gone, abandoned between 1910 and 1930, as individuals interviewed had personal knowledge pre-1940s. Allen shows a photograph of turf-cutting in 1914;[82] assuming this was local it could have been at the very end of their use. Perhaps with post-WW1 loss of rural labour and the 1920s 'Depression', the turbaries fell into disuse. This combined with improved road and rail transport, and availability of coal and coal-burning stoves, to end their use.

Examples from Norfolk Such as Lingwood Poor Lands Fuel Allotment, and Others

Lingwood is a scattered village, eight miles east of Norwich. According to William White's *History, Gazetteer, and Directory of Norfolk*, in 1841 the parish had 643 acres of fertile land and 473 inhabitants.[83] The 1821 Census stated that 'An inclosure of land, and small tenements built thereon, is mentioned, in the Return of Lingwood, as having caused an increase of Population'. However, poverty was an issue with 102 persons in 1841, 135 in 1851, and ninety-five as late as 1911, in the Blofield Union Workhouse. Fuel was an issue for the poor and the *Fuel Allotment* awarded at the 1803 enclosure was let for £20; the amount distributed in coals.

The Lingwood Poor Land Trustees are empowered under the Charities Act 1960 to provide funds to support the poor. To this end they advertise locally asking for information on those experiencing hard times and who might be helped. The Lingwood Poor Lands Fuel Allotment Deed dates from 3 November 1803 and presumably followed the enclosure award for the parish. Plots in Buckingham Road, Lingwood are let to parishioners with money raised from annual rents. Its original purpose was to purchase coal for the poor and needy families of the parish. In recent times, it has apparently been quite difficult to distribute the money to the now larger and presumably more affluent population.

82 Allen, *Heathland in East Devon*.

83 White, *History, Gazetteer, and Directory of Suffolk*.

At Merrow, Surrey, community use and dependence on free fuel off the common waste was in-part commuted to allocations of coal funded by rental from vegetable allotments. Shipdham near East Dereham in Norfolk had fuel allotments, awarded in 1809, with 126 acres in 1883 generating £140 annually. This was distributed among the poor in coals with £4–5s from other charities. Brome in the Hartismere Hundred in Suffolk (enclosure act, 1808) had a rental of £15 per year charged on an allotment for providing '[...] fuel for the poor, in lieu of the right they had of cutting firing on the commons'.[84] Bunwell, a parish of 2423 acres in Norfolk, had a fuel allotment of eighteen acres, one rod, and twenty-two perches awarded at enclosure in 1812 and let for £19 a year in 1854.

Thetford in Norfolk had extensive common waste with enclosure awards from 1806 to 1810 affecting 5616 acres of heath and grassland and some was reserved as fuel allotments for the town poor. Each of the surrounding parishes had heath retained for grazing, and gathering furze, heather, or bracken. Enclosure left scattered pockets of unenclosed land for communal use, and some survive today as Public Open Spaces. The main fuel allotments were: St Peter's Parish, thirty acres; St Cuthbert's Parish, seven acres; and St Mary's Parish, 163 acres.

South Walsham Parish in Norfolk appointed a Poor Overseer and Assistant Overseer for forty-four acres of Poors Allotments on South Walsham Marshes with a reed-boat and the right to use an adjoining staithe. Reed-cutting and sporting rights generated income used to support the poor with food, clothing, and fuel. Payment was as an order to local tradesman which kept money circulating within the village economy. At enclosure in 1820, one acre, three rods, and eight perches was awarded as poor allotment.

The Norfolk Records Office provides has abundant archives relating to such lands, including: East Ruston Poor's Allotment, 1810; Tharston Poor's Allotment Charity, 1781–1952; Rockland St Mary Poor's Fuel Allotment Charity Trustees, 1789; Great Plumstead Fuel Allotment Trustees, 1907–1987; Marham Poor Allotment Trustees, 1796–1970; and Sprowston Fuel Fund and Warner Charity Trustees, 1830–1960. These are just a few of the many fuel allotments awarded in Norfolk during enclosures. Some charities still make fuel allotment awards in the Kings Lynn area for example: a) The Great Hockham Fuel and Furze Trust. In this case, people in parish of Gt. Hockham can apply to the parish clerk to help with cost of fuel. b) Harling Fuel Allotment Trust makes grants to people in need for fuel, for services or household items. c)

[84] White, *History, Gazetteer, and Directory of Suffolk*.

The Horstead's Poor Land gives amounts according to need. d) Lyng Heath Charity dispenses £20 to £50 mainly for fuel. d) Shipdham Parochial and Fuel Allotment provides grants of £60.

Frimley Fuel Allotments, Surrey

Throughout the late 1700s and early 1800s, funded by a levy of a Poor Rate on the parishioners, the Frimley overseers dealt with helping the parish poor. Assistance included money, clothing, food, and fuel; and in some cases, if not infirm, the poor might be employed. According to Wellard, work included physical labour such as cutting turves for fuel, digging graves, or extracting building stone. Turf was cut on the common waste which was largely unenclosed peat-moor.[85]

Frimley Fuel Allotments Charity (founded 1801), has been particularly well-documented to provide unique insight into provision for the upkeep of the poor at the time of enclosure. The Fuel Allotment Charity owns land on which the Pine Ridge Golf Centre is built, along with around one hundred acres of open access heath. The Charity was established when Parliament under George III passed the 1801 Frimley Enclosure Act. The common was enclosed in 1826 with land set aside for '*Fuel for Firing*' of the poor of Frimley.[86] In 1793, the extensive 'wastes' of Frimley served a small rural population until almost ninety per cent of the modern settlements of Camberley, Frimley, and Frimley Green were built on the expansive, open heaths, including Frimley Heath, Cow Moor, Bisley Common, Pirbright Common, and Chobham Common. The area was covered by gorse, heather, scrub, and sparse grass and had abundant deer but only poor grazing for sheep. Commoners held rights to cut turf or wood, to fish, and to pasture cattle. The 1801 Act allowed for dividing, allotting, and inclosing of waste grounds and commons and commonable lands within the Manor of Frimley in the Parish of Ash, in the County of Surrey. In the lead-in to physical enclosure by 1820, the parish workhouse housing nineteen paupers was established on part of the fuel allotments.

At the time of enclosure, the Act stipulated that

> such part of the waste Lands of Frimley as in the judgement of the Commissioners was adequate to provide a reasonable supply of fuel for those inhabitants of the

[85] Wellard, *200 Years of Frimley's History*.

[86] Wellard, *200 Years of Frimley's History*.

> Hamlet who did not occupy lands or a dwelling of an annual value of more than Five Pounds.

In effect, all agricultural labourers, cottagers, and small tradesmen qualified for fuel from the 'Firing' Allotments. Many of these people were poor due to changing urban and rural economics of the time and the demise of small rural crafts and industries. This was in the aftermath of the Napoleonic War and in the context of rising grain prices. Land was specifically set aside to provide fuel. Along with the right to take fuel away for those qualified under the Act, the Trustees were empowered to lease the whole or part of the allotments to any person they thought suitable as a tenant. This was for a term not exceeding twenty-one years with rent paid quarterly. On expiration, the tenant would have to leave the land in good condition. The Trustees had to ensure the money raised was spent to purchase 'fuel for firing' under the £5 qualification.

The cost of providing fuel was not too great when the bulk was wood or peat. However, costs increased as these declined and coal became the more commonplace fuel. By the early 1860s, the situation became critical but was offset by an offer to purchase land from the Trust. With permission from the Charity Commissioners, sale was allowed, and money invested. The Charity Commissioners allowed a waiver in the interpretation of the original Act to provide fuel.

> When not required for the purchase of fuel shall be laid out by the Trustees in the purchase of warm clothing and blankets to be distributed by the Trustees at their discretion to the deserving poor resident in the parish of Frimley.

So, whilst fuel was the first priority, warm clothing and blankets were also allowed. Those residing in properties valued under £5 qualified automatically but there was discretion to allow others too.

There were disputes over the allocation and interpretation of rights allowed under the 1801 enclosure, and by the 1860s the poor burnt coal along with wood and peat, so money was required. With a total income of around £30 per year from rents and investments, the outlay on coal was about £1 a ton, with a quarter of a ton per person per year. For clothing expenditure, a blanket was about 3s, a pair of shoes 12s, a petticoat 5s, and a pair of trousers and a man's coat a pound.

Frimley Urban District Council was formed in 1894, and after four years, responsibility for the fuel allotments was transferred from the Fuel Allotment Charity's Trustees. Soon after, the Council sought to sell the land, extinguish the rights, and purchase land elsewhere for a recreation ground. Whilst the first proposals were rejected by the Charity Commission, a scheme that allowed military purchase and use of the recently enclosed common was found acceptable and was to generate an income of £140 per year. Agreement allowed the maintenance of heath and the cutting of furze fuel, and the fuel charity was administered by the Council. In 1904, for example, costing £50, 112 people or families received Christmas coal allocations in Camberley and Yorktown (thirty tons), and another 120 recipients in Frimley, Frimley Green, and Mytchett (nineteen tons). By 1909 the annual income was around £200 or around £20,000 at 1990s values. Coal allocations varied from 2 cwt to 6 cwt per recipient. By 1914, there were 327 recipients with £360 paid at 25s per ton. Subsequent records until the 1940s have been lost.

In 1939, the income and expenditure of the Trust were unchanged from 1909, though coal had increased in price. By the 1960s, with an annual income of £450, there were continuing problems of maintenance of the allotments and the Trust's other assets so in 1967 permission was given to sell land to build a County Council school. This brought increased revenue and the Council began to establish partnerships with other local charities to further the aims of the Trust and dispense awards. Around £1255 per annum was given as £5 fuel vouchers, £2000 to support old people, and £750 for other charitable purposes. By the 1970s, increasing urbanization meant the allotments were one of the few remaining open spaces in the area. However, alongside other pressures, a golf course was proposed on the land, so the Charity Commission reminded the Council their specific charitable purpose to benefit the Frimley Parish poor and not 'all residents'. Any capital from land sales was for this specific function; ultimately generating around £65,000 per year (1990s) dispensed for Frimley Fuel Allotments Charity to help the elderly poor with fuel bills. Frimley provides a uniquely well-documented case-study of the relationship between commonland, allotments, and the alleviation of poverty.

Conclusions

The Impact of the Enclosures

Enclosures and the loss or modification of common rights occurred over a long period, perhaps several centuries, and the process and impacts vary considerably from region to region. However, there are key trends that can be teased out from general observations. Firstly, the right to fuel sources was hugely important to everyone, and that to most people in a rural society, the common rights to fuels were vital for survival. This was very acutely the case during the very cold 'Little Ice Age' from around CE 1300 to the 1800s and was most significant to the abundant poor. Without alternative fuels, limited by availability, transport or cost, free access to the commons was essential. Particularly during the parliamentary enclosures in Britain, there was the serious problem of the loss of such common rights that allowed many poor people to subsist.

Recipients of larger allotments of enclosed land would have two advantages over the small commoner or the poor of the parish. Firstly, their land holding might include such fuel sources that they might require. Secondly, their level of income would facilitate the purchase of alternative fuels, often wood or, increasingly, mineral coal. For the poor, this was not possible and so for their survival it was generally recognized at enclosure that special provision might be needed. The result was generally the setting aside of specific small areas of unenclosed land as 'fuel allotments'. These might be turbaries for the digging of peat, wooded common for taking fuelwood, or furze fields for gorse. In many cases, it seems that fuel allotments were actually land rented out for cultivation and the income so generated for the parish was invested in coal. This coal was then distributed to the poor and the needy.

Cultural Severance

Many but not all fuel allotments have long-been abandoned; some small areas remain, though not managed in their traditional way. Like many cultural relicts in the landscape the presence of this use remains like a shadow, half-forgotten but hinted at in place-names, field-names and road-names. With the agricultural and industrial revolutions, the long process of severance began in earnest, and people, food, building materials, and fuel supplies have been increasingly separated. The timing and impacts vary across Britain, but the inexorable trends are the same. As industry and agriculture have moved more to technological processes, resources, and solutions, nature and local countryside have become of secondary importance. Following these changes has been a dramatic shift from rural to urban populations which continues to accelerate into the twenty-first century. In many regions, the rural working community in often subsistence countryside are displaced to become the urban poor. Traditional lands become disputed spaces from which communities are squeezed, with inexorable change from subsistence and tradition. Landscapes are abandoned to become backdrops to tourism, and the weekend recreation of the affluent; traditional rural economies replaced by leisure visitors.

Progress in the modern economy generally equates to socio-economic development with rural depopulation and urban growth. There is technological provision for needs, and a separation of people from nature. This process is an on-going part of human cultural evolution, but with major environmental consequences. In terms of the severance of people and landscape, there has been a rapid de-coupling of communities from their local environments. The creation and then loss of fuel allotments is one such example of these changes, and there are consequences of the cessation of traditional land uses.

Even a superficial examination of the case-studies raises interesting issues and concerns for future conservation of nature and heritage. There are matters of competition, of displaced rights and of misunderstanding of status when fuel allotments were established. Some allotments were originally awarded in order to generate income to purchase fuel for the poor; others began life as primary sources of fuel for the same. They were later either abandoned or the right converted to a cash allowance for fuel if there was income from rental or interest on capital sales.

Fuel and the Poor

For the poor in any society, fuel and energy are of major importance and can mean the difference between life and death. Competition for resources between social strata within a community can become acute when these are scarce, or accessibility is restricted. Climate change and conflicts such as war can further push societies into disputes over fuel access and the poor into personal tragedy. Whilst this short account considers the British Isles in the medieval and early modern periods through to the early twentieth century, fuel shortage and poverty remain today even in modern-day Europe. With bitterly cold winters and shortage of fuel, the poor still suffer and some die.

Acknowledgements

I am grateful particularly to Richard Oram, Chris Smout, and Tom Williamson for encouragement, discussions, and shared enthusiasm for the topic, and to Alasdair Ross for his generous sharing of insights and knowledge of fuel usage. I also wish to thank David Zylberberg for stimulating conversations and Graham Bathe for permission to cite his unpublished research.

Works Cited

Manuscripts

National Library of Scotland

NLS Adv. MS 34.5 (Cartulary of Kelso Abbey)
NLS, MS 2131, f. 69

National Records of Scotland

Church Records: Cartularies (Coupar Angus Abbey), CH 6/2/4
Fraser-Mackinosh Collection (Chisholm Papers), GD128/64/4 (2)
Papers of the Mackintosh Family of Mackintosh, GD176/7
Papers of the Society of Antiquaries of Scotland: Other papers, GD103/2/12–14
Register House Plans, RHP142397, Sheet 6
Register of Deeds Second Series, Dalrymple's Office (Dal), RD2/230/2

University of Western Australia, Special Collections Department

MS Med. 2

Wiltshire and Swindon Record Office/Wiltshire and Swindon Archives

WSRO 9–1-254
WSRO 9–1-125
WSRO 9–1-285
WSRO 9–1-258
WSRO 9–1-283
WSRO 1300–1850
WSRO 9–1-269
WSRO 9–1-318

Primary Sources

Acts of the Lords of Council, II: 1496–1501, ed. by George Neilson and Henry Paton (Edinburgh: H.M.S.O., 1918)

Acts of the Lords of Council, III: 1501–1503, ed. by Alma B. Calderwood (Edinburgh: H.M.S.O., 1993)

Adomnán, *Vita S. Columbae*, ed. by Joseph Thomas Fowler from the text of William Reeves (Oxford: Clarendon, 1894)

——, *Life of Columba*, ed. and trans. by Alan Orr Anderson and Marjorie Ogilvie Anderson (London: Nelson, 1961)

——, *The Life of Saint Columba*, trans. by Richard Sharp (London: Penguin, 1995)

Agnellus of Ravenna, 'Liber Pontificalis Ecclesiae Ravennatis', in *Scriptores rerum Langobardicarum et Italicarum: saec. VI–IX*, ed. by Oswald Holder-Egger, MGH, Scriptores rerum Langobardicarum et Italicarum, 1 (Hanover: Hahn, 1878), pp. 265–391

——, *The Book of Pontiffs of the Church of Ravenna*, trans. by Deborah Mauskopf Deliyannis, Medieval Texts in Translation (Washington: Catholic University of America Press, 2004)

Al-Ṭabarī, *The History of al-Ṭabarī V: The Sāsānids, the Byzantines, the Lakmids, and Yemen*, trans. by Clifford Edmund Bosworth, Bibliotheca Persica, SUNY Series in Near Eastern Studies (Albany: State University of New York Press, 1999)

Anderson, Alan Orr, ed. and trans., *Early Sources of Scottish History AD 500 to 1286*, 2 vols (Edinburgh: Oliver and Boyd, 1922)

Andrew of Wyntoun, *Orygynale Cronykil of Scotland*, ed. by David Laing, Historians of Scotland, 2; 3; 9, 3 vols (Edinburgh: Edmonston and Douglas, 1872–1879)

——, *The Original Chronice of Andrew of Wyntoun*, ed. by François Joseph Amours, 6 vols (Edinburgh: Blackwood, 1903–1914)

Annala Uladh: Annals of Ulster, otherwise Annala Senait, Annals of Senat: A Chronicle of Irish Affairs from AD 431 to AD 1540, ed. and trans. by William Maunsell Hennessy and Bartholomew MacCarthy, 4 vols (Dublin: Thom & Co., 1887–1901), I (1887)

'Annala Uladh: Annals of Ulster otherwise Annala Senait, Annals of Senat', *CELT* <http://www.ucc.ie/celt/published/T100001B/index.html> [accessed 7 July 2022]

'Annales Inisfalenses, ex Codice Bodleiano Rawlinson, no. 503', in *Rerum Hibernicarum Scriptores*, ed. by Charles O'Conor, 4 vols (London: Payne, 1814–26), II (1825)

The Annals of Clonmacnoise, trans. by Conell Mageoghagan, ed. by Denis Murphy (Dublin: University Press, 1896)

'Annals of Connacht', *CELT* <http://www.ucc.ie/celt/published/T100011/index.html> [accessed 7 July 2022]

'Annals of the Four Masters', *CELT* <http://celt.ucc.ie/published/T100005D/index.html> [accessed 7 July 2022]

'Annals of Innisfallen', *CELT* <http://www.ucc.ie/celt/published/T100004/index.html> [accessed 7 July 2022]

'Annals of Loch Cé', *CELT* <https://www.ucc.ie/celt/published/T100010A/index.html> [accessed 4 August 2022]

The Annals of St-Bertin, trans. by Janet Laughland Nelson, Machester Medieval Sources, Ninth-Century Histories, 1 (Manchester: Manchester University Press, 1991)

'The Annals of Tigernach: Third Fragment, AD 489–766', ed. and trans. by Whitley Stokes, *Revue Celtique*, 17 (1896), 6–33; 116–263; 337–420

Barbour, John, *The Bruce*, ed. and trans. by Archibald Alexander McBeth Duncan, Canongate Classics, 78 (Edinburgh: Canongate, 1997)

Barrow, Geoffrey Wallis Steuart, ed., *The Acts of Malcolm IV, King of Scots, 1153–1165: Together with Scottish Royal Acts Prior to 1153 Not Included in Sir Archibald Lawrie's Early Scottish Charters*, Regesta Regum Scottorum, 1 (Edinburgh: Edinburgh University Press, 1960)

——, ed., *The Acts of William I, King of Scots, 1165–1214*, Regesta Regum Scottorum, 2 (Edinburgh: Edinburgh University Press 1971)

Bede, 'Life of Cuthbert', trans. by James Fracis Webb, in *The Age of Bede*, ed. by David Hugh Farmer (London: Penguin, 2004), pp. 43–104

The Berwickshire Place-Name Resource <https://berwickshire-placenames.glasgow.ac.uk/>

Bower, Walter, *Scotichronicon*, ed. by Donald Elmslie Robertson Watt, 9 vols (Aberdeen: Aberdeen University Press, 1987–1998)

Calendar of Close Rolls, 47 vols (London: H.M.S.O., 1900–1963)

Calendar of Documents Relating to Scotland, ed. by Joseph Bain, Grant G. Simpson, and James D. Galbraith, 5 vols (Edinburgh: H.M. General Register House, 1881–1985)

Calendar of Writs Preserved at Yester House, 1166–1503, ed. by Charles Cleland H. Harvey (Edinburgh: Scottish Record Society, 1916)

CELT: The Corpus of Electronic Texts <www.ucc.ie/celt>

Charters of the Abbey of Coupar Angus, ed. by David Edward Easson, Publications of the Scottish History Society, Series 3, 40–41, 2 vols (Edinburgh: Constable, 1947)

The Charters of David I: The Written Acts of David I King of Scots, 1124–53, and of His Son Henry, Earl of Northumberland, 1139–52, ed. by Geoffrey Wallace Steuart Barrow (Woodbridge: Boydell, 1999)

The Chronicle of Melrose: From the Cottonian Manuscript, Faustina B. IX in the British Museum, ed. by Alan Orr Anderson and Marjorie Ogilvie Anderson, with William Croft Dickinson (London: Lund, 1936)

Chronicon de Lanercost, ed. by Joseph Stevenson, Maitland Club Publications, 46 (Edinburgh: Maitland Club, 1839)

Constantine the African, *Constantini Africani post Hippocratem et Galenum* (Basel: Petri, 1536)

A Corpus of Scottish Medieval Parish Churches <https://arts.st-andrews.ac.uk/corpusofscottishchurches/> [accessed 7 July 2022]

The Correspondence, Inventories, Account Rolls and Law Proceedings of the Priory of Coldingham, ed. by James Raine, Publications of the Surtees Society, 12 (London: Nichols, 1841)

Crossley-Holland, Kevin, ed. and trans., *The Anglo-Saxon World: An Anthology* (Oxford: Oxford University Press, 1999)

Dado of Rouen, 'Life of St Eligius of Rouen', trans. by Jo Ann McNamara in *Medieval Hagiography: An Anthology*, ed. by Thomas Head, repr. (London: Routledge, 2020), pp. 137–68

Dentrecolles, François-Xavier, 'Lettre du F. Dentrecolles, Missionnaire de la Compagnie de Jesus', in *Lettres Édifiantes et Curieuses Ecrites des Missions etrangeres par quelques missionnaires de la Compagnie de Jesus*, ed. by Jean Baptiste Du Halde, 34 vols (Paris: Le Clerc, 1703–1776), xx, pp. 304–61

A Dictionary of the Older Scots Tongue to 1707 <http://www.dsl.ac.uk/>

Dio Cassius, *Roman History IX: Books 71–80*, trans. by Earnest Cary and Herbert B. Foster, Loeb Classical Library, 177 (Cambridge: Harvard University Press, 1927)

DMLBS – Dictionary of Medieval Latin from British Sources, ed. by Ronald Edward Latham, David R. Howlett, and Richard Ashdowne, online at: <http://www.dmlbs.ox.ac.uk/web/index.html>

Duncan, Archibald Alexander McBeth, ed., *Formulary E: Scottish Letters and Brieves, 1286–1424*, Scottish History Department Occasional Papers (Glasgow: University of Glasgow, 1976)

——, ed., *The Acts of Robert I, King of Scots, 1306–1329*, Regesta Regum Scottorum, 5 (Edinburgh: Edinburgh University Press, 1988)

eDIL: An Electronic Dictionary of the Irish Language <www.dil.ie>

Eusebius, *Church History, Life of Constantine the Great, and Oration in Praise of Constantine*, ed. and trans. Henry Wallace and Philip Schaff, repr. (Cosimo Classics: New York, 2007)

Evagrius Scholasticus, *Ecclesiastical History*, trans. by Michael Whitby, Translated Texts for Historians, 33 (Liverpool: Liverpool University Press, 2001)

Flodoard of Rheims, 'Historia Remensis ecclesiae', in *Supplementa Tomorum, I-XII.1*, ed. by Johannes Heller and Georg Waitz, MGH, Scriptores, 13 (Hanover: Hahn, 1881), pp. 405–599

Franck, Richard, *Northern Memoirs Calculated for the Meridian of Scotland to Which is Added, the Contemplative and Practical Angler Writ in the Year 1658* ([London]: Henry Mortclock[sic] for the author, 1694)

Godman, Peter, *Poetry of the Carolingian Renaissance*, Duckworth Classical, Medieval, and Renaissance Editions (London: Duckworth, 1985)

Gregory of Tours, *Libri historiarum decem*, ed. by Bruno Krusch and Wilhelm Levison, MGH, Scriptores rerum Merovingicarum, 1.1 (Hanover: Hahn, 1851)

——, 'De cursu stellarum ratio', in *Miracula et opera minora*, ed. by Bruno Krusch, MGH, Scriptores rerum Merovingicarum, 1.2 (Hanover: Hahn, 1885), pp. 404–22

——, *The History of the Franks*, trans. by Lewis Thorpe (London: Penguin, 1975)

Harrison, William, *The Description of England: the Classic Contemporary Account of Tudor Social Life*, ed. by Georges Edelen (Washington: The Folger Shakespeare Library, 1994)

High Life Highland, 'Reference and Local History: Special Collections', <https://www.highlifehighland.com/reference-and-local-history/special-collections/> [accessed 22 August 2022]

Ibn Isḥāq, Muḥammad, *The Life of Muhammad*, trans. by Alfred Guillaume (London: Oxford University Press, 1955)

Ibn Wāḍiḥ al-Yaʿqūbī, 'The History (Taʾrīkh): The Rise of Islam to the Reign of al-Muʿtamid', in *The Works of Ibn Wāḍiḥ al-Yaʿqūbī: An English Translation*, ed. and trans. by Matthew Gordon, Chase F. Robinson, Everett K. Rowson, and Michael Fishbein, Islamic History and Civilization, 152, 3 vols (Leiden: Brill, 2018), III, pp. 595–1293

John of Fordun, *Chronica Gentis Scotorum*, ed. by William F. Skene, 2 vols (Edinburgh: Edmonston and Douglas, 1871–1872)

John of Nikiu, *Chronicle*, trans. by Robert Henry Charles, Text and Translation Society Publications (London: William and Norgate, 1916)

Lawrie, Archibald Campbell, ed., *Early Scottish Charters Prior to AD 1153* (Glasgow: Maclehose, 1905)

Liber Cartarum Prioratus Sancti Andree in Scotia, E Registro Ipso In Archivis Baronum De Panmure Hodie Asservato, ed. by Thomas Thomson, Bannatyne Club Publications, 69 (Edinburgh: Bannatyne Club, 1841)

Liber Cartarum Sancte Crucis: munimenta Ecclesie Sancte Crucis de Edwinesburg, ed. by Cosmo Nelson Innes, Bannatyne Club Publications, 70 (Edinburgh: Bannatyne Club, 1840)

Liber Ecclesie de Scon, ed. by Cosmo Nelson Innes, Bannatyne Club Publications, 78 (Edinburgh: Bannatyne Club, 1843)

Liber S. Marie de Calchou: registrum cartarum Abbacie Tironensis de Kelso, 1113–1567, ed. by Cosmo Nelson Innes, Bannatyne Club Publications, 82, 2 vols (Edinburgh: Bannatyne Club, 1846)

Liber S. Thome de Aberbrothoc, ed. by Cosmo Nelson Innes, Bannatyne Club Publications, 86, 2 vols (Edinburgh: Bannatyne Club, 1848–1856)

Macfarlane, Walter, *Genealogical Collections Concerning Families in Scotland Made by Walter Macfarlane, 1750–1751*, ed. by James Toshach Clark, Publications of the Scottish History Society, 33–34, 2 vols (Edinburgh: Scottish Historical Society, 1900)

Marius of Avenches, 'Chronica', in *Chronica minora saec. IV, V, VI, VII*, ed. by Theodore Mommsen, MGH, Auctores antiquissimi, 9, 11, 2 vols (Berlin: Weidmann, 1894), II, pp. 225–39

Maule, Harry, *Registrum de Panmure: Records of the families of Maule, De Valoniis, Brechin, and Brechin-Barclay, United in the Line of the Barons and Earls of Panmure*, ed. by John Stuart, 2 vols (Edinburgh: [n. pub.], 1874)

Maunoir, C., 'Lettre aux Rédacteurs', *Bibliotheque Britannique ou Recueil*, 18 (Geneva, 1801), 102–04

Munro, Jean, and Robert William Munro, eds, *Acts of the Lords of the Isles 1336–1493*, Scottish History Society, 22 (Edinburgh: Scottish History Society, 1986)

Neville, Cynthia J., and Grant G. Simpson, eds., *The Acts of Alexander III, King of Scots, 1249–1286*, Regesta Regum Scottorum, 4, pt 1 (Edinburgh: Edinburgh University Press, 2013)

Norden, John, *The Surveyor's Dialogue (1618): A Critical Edition*, ed. by Mark Netzloff (Farnham: Ashgate, 2010)

NRS, 'Fraser-Mackintosh Collection', <http://catalogue.nrscotland.gov.uk/nrsonline-catalogue/details.aspx?reference=GD128&st=1&tc=y&tl=n&tn=n&tp=n&k=Fraser+Mackintosh+collection&ko=a&r=&ro=s&df=&dt=&di=y> [accessed 22 August 2022]

Paris, Matthew, *Chronica Majora*, ed. by Henry Richards Luard, Rolls Series, 57, 7 vols (London: Longman, 1872–1883), v (1880)

Paul the Deacon, 'Historia Langobardorum', in *Scriptores rerum Langobardicum et Italicarum: saec.VI–IX*, ed. by L. Bethmann and G. Waitz, MGH, Scriptores rerum Langobardicum et Italicarum, 1 (Hanover: Hahn, 1878), pp. 12–187

Pegolotti, Francesco Balducci, *La Practica della Mercurata*, ed. by Allan Evans (Cambridge: Medieval Academy of America, 1936)

People of Medieval Scotland <https://www.poms.ac.uk/>

Procopius, *History of the Wars, I: Books 1–2 (Persian War)*, trans. by Henry Bronson Dewing, Loeb Classical Library, 48 (Cambridge: Harvard University Press, 1914)

Pseudo-Joshua the Stylite, *Chronicle*, trans. by Frank R. Trombley and John W. Watt, Translated Texts for Historians, 32 (Liverpool: Liverpool University Press, 2000)

Raine, James, ed., *The History and Antiquities of North Durham* (London: Bowyer Nichols & Son, 1852)

Raine, James, ed., *The Correspondence, Inventories, Account Rolls, and Law Proceedings, of the Priory of Coldingham*, Publications of the Surtees Society, 12 (London: Nichols, 1841)

Ramsay, James Henry, ed., *Bamff Charters AD 1232–1703* (London: Oxford University Press, 1915)

The Records of the Parliaments of Scotland to 1707 <http://www.rps.ac.uk/>

'The Register: Bishoprics of Scotland; Memorandum on writs; Caldbeck (continued)', in *Register and Records of Holm Cultram*, ed. by Francis Grainger and William Gersholm Collingwood (Kendal: Wilson, 1929), pp. 105–06 online at: <http://www.british-history.ac.uk/n-westmorland-records/vol7/pp105-106> [accessed 7 July 2022]

The Register of John De Halton, Bishop of Carlisle AD 1292–1324, transcribed by W. N. Thompson, with introduction by Thomas Frederick Tout, Canterbury and York Series, 12–13, 2 vols (London: Canterbury and York Society, 1913)

Registrum de Dunfermelyn, ed. by Cosmo Nelson Innes, Bannatyne Club Publications, 74 (Edinburgh: Bannatyne Club, 1842)

Registrum Domus de Soltre, necnon Ecclesie Collegiate S. Trinitatis prope Edinburgh/ Charters of the Hospital of Soltre, of Trinity College, Edinburgh, and other Collegiate Churches in Mid-lothian, Bannatyne Club Publications (Edinburgh: Bannatyne Club, 1861)

Registrum Episcopatus Aberdonensis: ecclesia cathedralis Aberdonensis: regesta que extant inunum collecta, ed. by Cosmo Nelson Innes, Maitland Club Publications, 63, 2 vols (Edinburgh: Maitland Club, 1845)

Registrum Episcopatus Moraviensis, ed. by Cosmo Nelson Innes, Bannatyne Club Publications, 58 (Edinburgh: Bannatyne Club, 1837)

Registrum Monasterii de Passelet, 1163–1529, ed. by Cosmo Nelson Innes, Maitland Club Publications, 17 (Edinburgh: Maitland Club, 1832)

Registrum Sancte Marie de Neubotle: Abbacie Cisterciensis Beate de Virginis de Neubotle chartarium vetus: accedit appendix cartarum originalium, 1140–1528, ed. by Cosmo Nelson Innes, Bannatyne Club Publications, 89 (Edinburgh: Bannatyne Club, 1849)

Rental Book of the Cistercian Abbey of Cupar-Angus, with the Breviary of the Register, ed. by Charles Rogers, 2 vols (London: Grampian Club, 1879)

RMS – Registrum Magni Sigilli Regum Scottorum (Register of the Great Seal), 11 vols, ed. John Maitland Thomson, rpt (Edinburgh: H. M. S. O., 1984)

Ross, Alasdair, 'Two 1585×1612 Surveys of Vernacular Buildings and Tree Usage in the Lordship of Strathavon, Banffshire', *Miscellany of the Scottish History Society*, 14 (2012), 1–52

Rotuli Scaccarii regum Scotorum = The Exchequer Rolls of Scotland, ed. by John Stuart, George Burnett, Aeneas James George Mackay, and George Powell MacNeill, 23 vols (Edinburgh: H.M. General Register House, 1878–1908)

Schoenberg Database of Manuscripts <https://sdbm.library.upenn.edu/>

ScotlandsPlaces <https://scotlandsplaces.gov.uk/>

Shead, Norman F., ed., *Scottish Episcopal Acta*, 2 vols (Woodbridge: Boydell, 2015)

Somerville, Robert, ed., *Scotia Pontificia: Papal Letters to Scotland before the Pontificate of Innocent III* (Oxford: Clarendon Press, 1982)

Statuta Capitulorum Generalium Ordinis Cisterciensis ab Anno 1116 ad Annum 1786, ed. by Joseph Marie Canivez, with Auguste Trilhe, 8 vols (Louvain: Bureaux de la Revue, 1933–1941)

Theiner, Augustin, ed., *Vetera Monumenta Hibernorum et Scotorum Historiam Illustrantia* (Rome: Vaticania, 1864)

Turnbull, William Barclay D. D., ed., *The Chartularies of Balmerino and Lindores / Liber Sancte Marie de Balmorinach*, Abbotsford Club Series, 22 (Edinburgh: Abbotsford Club, 1841)

Venantius Fortunatus, *Poems*, ed. and trans. by Michael Roberts, Dumbarton Oaks Medieval Library, 46 (Cambridge: Harvard University Press, 2017)

'Vita Audomari Altera', in *AASS*, Septembri, 3 (Antwerp: Société des Bollandistes, 1750), pp. 402–06

'Vita S. Condedi', in *AASS*, Octobri, 9 (Antwerp: Société des Bollandistes, 1858), pp. 354–55

'Vita Filiberti', in *Passiones Vitaeque Sanctorum Aevi Merovingci III*, ed. by Bruno Krusch and Wilhelm Levison, MGH, Scriptores rerum Merovingicarum, 5 (Hanover: Hahn, 1920), pp. 568–606

'Vita Geretrudis', in *Fredegarii at Aliorum Chronica. Vitae Sanctorum*, ed. by Bruno Krusch, MGH, Scriptores rerum Merovingicarum, 2 (Hanover: Hahn, 1888), pp. 447–74

Voltaire, *Lettres Philosophiques par M de Voltaire* (Rouen: Jore, 1734)

Waddell, Chrysogonus, ed., *Twelfth-Century Statutes from the Cistercian General Chapter*, Cîteaux: Commentarii Cistercienses, Studia et Documenta, 12 (Brecht: Cîteaux, 2002)

Wedderburn, John, *Ane compendious buik of godlie psalmes and spirituall sangis* (Edinburgh: [n. pub.], 1578)

Wikipedia: The Free Encyclopedia <https://www.wikipedia.org/>

William of Newburgh, 'Continuation of the Chronicle of William of Newburgh', in *Chronicles of the Reigns of Stephen, Henry II and Richard I*, ed. by Richard Howlett, Rolls Series, 82, 4 vols (London: Longman, 1884–1889), II (1884)

Secondary Works

Adam, Dillon C., Matthew Scotch, and Chandini Raina MacIntyre, 'Bayesian Phylogeography and Pathogenic Characterization of Smallpox Based on HA, ATI, and CrmB Genes', *Molecular Biology and Evolution*, 35 (2018), 2607–17

Adam, Frank, *The Clans, Septs and Regiments of the Scottish Highlands* (Edinburgh: Johnston, 1908)

Adderley, W. Paul, and Ian A. Simpson, 'Soils and Palaeo-climate based evidence for past irrigation requirements in Norse Greenland', *Journal of Archaeological Science*, 33 (2006), 1666–1679

Ainslie, Whitelaw, 'Observations Respecting the Small-Pox and Inoculation in Eastern Countries', *Transactions of the Royal Asiatic Society of Great Britain and Ireland*, 2 (1829), 52

Allen, David J., *Heathland in East Devon and the Blackdown Hills* (London: Heritage Lottery Fund, 2004)

Andam, Cheryl P., Colin J. Worby, Qiuzhi Chang, and Michael G. Campana, 'Microbial Genomics of Ancient Plagues and Outbreaks', *Trends in Microbiology*, 24 (2016), 978–99

App, Urs, *The Birth of Orientalism* (Pennsylvania: University of Pennsylvania Press, 2010)

Ardron, Paul A., 'Peat Cutting in Upland Britain, with Special Reference to the Peak District' (unpublished doctoral thesis, University of Sheffield, 1999)

——, Ian D. Rotherham, and Oliver L. Gilbert, 'Factors Determining Contemporary Upland Landscapes: A Re-evaluation of the Importance of Peat-Cutting and Associated Drainage, and the Implications for Mire Restoration and Remediation', in *Blanket Mire Degradation. Causes, Consequences and Challenges. Proceedings of the British Ecological Society Conference in Manchester, 1997*, ed. by John H. Tallis, Roger Meade, and P. D. Hulme (Aberdeen: British Ecological Society and the Macaulay Land Use Research Institute, 1997), pp. 38–41

——, Ian D. Rotherham, and Oliver L. Gilbert, 'Peat-Cutting and Upland Landscapes: Case-Studies from the South Pennines', in *Landscapes – Perception, Recognition and Management: Reconciling the Impossible? Proceedings of the Conference Held in Sheffield, UK, 2–4 April, 1996*, ed. by Mel Jones, Landscape Archaeology and Ecology, 3 (Sheffield: Sheffield Hallam University, 1998), pp. 65–69

Armstrong, Lawrin, Ivana Elbl, and Martin M. Elbl, eds, *Money, Markets and Trade in Late Medieval Europe: Essays in Honour of John H. A. Munro*, Later Medieval Europe, 1 (Leiden: Brill, 2007)

Arnold, Ellen F., 'Hagiography and Environmental History', in *Hagiography and the History of Latin Christendom, 500–1500*, ed. by Samantha Herrick (Brill: Leiden, 2019), pp. 351–72

——, 'Washed Ashore: Marine Mammals from Medieval Times to Today' <http://origins.osu.edu/connecting-history/washed-ashore-marine-mammals-medieval-times-today> [accessed 7 July 2022]

Arrizabalaga, Jon, 'Problematizing Retrospective Diagnosis in the History of Disease', *Asclepio*, 54 (2002), 51–70

Assemani, Giuseppe Simone, *Bibliotheca Orientalis Clementino-Vaticana*, 3 vols (Rome: Sacræ Congregationis de Propaganda Fide, 1719–28)

Baas, Johann Hermann, *Grundriss der geschichte der medicin und des heilenden standes* (Stuttgart: Enke, 1876)

Babkin, Igor V., and Irina N. Babkina, 'A Retrospective Study of the Orthopoxvirus Molecular Evolution', *Infection, Genetics and Evolution*, 12 (2012), 1597–604

——, and Irina N. Babkina, 'Origin of the Variola Virus', *Viruses*, 7 (2015), 1100–12

——, and Sergei N. Shchelkunov, 'Time Scale of Poxvirus Evolution', *Molecular Biology*, 40 (2006), 16–19

——, and Sergei N. Shchelkunov, 'Molecular Evolution of Poxviruses', *Russian Journal of Genetics*, 44 (2008), 895–908

Bachrach, Bernard S., 'Plague, Population, and Economy in Merovingian Gaul', *Journal of the Australian Early Medieval Association*, 3 (2007), 29–57

Baillie, Michael G. L., and David M. Brown, 'Dendrochronology and the Reconstruction of Fine Resolution Environmental Change in the Holocene', in *Global Change in the Holocene*, ed. by Anson W. Mackay, John Birks, and Rick Battarbee (London: Hodder, 2005), pp. 75–91

Baker, Andy, John C. Hellstrom, Bryce F. J. Kelly, Gregoire Mariethoz, and Valerie Trouet, 'A Composite Annual-Resolution Stalagmite Record of North Atlantic Climate over the Last Three Millennia', *Scientific Reports*, 5 (2015), 103–07

Bannerman, John, *Studies in the History of Dalriada* (Edinburgh: Scottish Academic Press, 1974)

Baril, Laurence, Xavier Vallès, Nils Christian Stenseth, Minoarisoa Rajerison, Maherisoa Ratsitorahina, Javier Pizarro-Cerdá, Christian Demeure, Steve Belmain, Holger Scholz, Romain Girod, Joseph Hinnebusch, Ines Vigan-Womas, Eric Bertherat, Arnaud Fontanet, Yazdan Yazadanpanah, Guia Carrara, Jane Deuve, Eric D'ortenzio, Jose Oswaldo Cabanillas Angulo, Paul Mead, and Peter W. Horby, 'Can We Make Human Plague History? A Call to Action', *BMJ Global Health*, 4 (2019), e001984

Baron, John, *The Life of Edward Jenner*, 2 vols (London: Colburn, 1827–38)

Barrett, James H., 'Fish Trade in Norse Orkney and Shetland: A Zooarchaeological Approach', *Antiquity*, 71 (1997), 616–38

——, Alison M. Locker, and Callum M. Roberts, '"Dark Age Economics" Revisited: The English Fish Bone Evidence AD 600–1600', *Antiquity*, 78 (2004), 618–36

——, Cluny Johnstone, Jennifer Harland, Wim Van Neer, Anton Ervynck, Daniel Makowiecki, Dirk Heinrich, Anne Karin Hufthammer, Inge Bødker Enghoff, Colin Amundsen, Jørgen Schou Christiansen, Andrew K. G. Jones, Alison Locker, Sheila Hamilton-Dyer, Leif Jonsson, Lembi Lõugas, Callum Roberts, and Michael Richards, 'Detecting the Medieval Cod Trade: A New Method and First Results', *Journal of Archaeological Science*, 35 (2008), 850–61

——, David Orton, Cluny Johnstone, Jennifer Harland, Wim Van Neer, Anton Ervynck, Callum Roberts, Alison Locker, Colin Amundsen, Inge Bødker Enghoff, Sheila

Hamilton-Dyer, Dirk Heinrich, Anne Karin Hufthammer, Andrew K. G. Jones, Leif Jonsson, Daniel Makowiecki, Peter Pope, Tamsin C. O'Connell, Tessa de Roo, and Michael Richards, 'Interpreting the Expansion of Sea Fishing in Medieval Europe Using Stable Isotope Analysis of Archaeological Cod Bones', *Journal of Archaeological Science*, 38 (2011), 1516–24

——, Rebecca A. Nicholson, and Ruby Cerron-Carrasco, 'Archaeo-ichthyological Evidence for Long-Term Socioeconomic Trends in Northern Scotland: 3500 BC to AD 1500', *Journal of Archaeological Science*, 26 (1999), 353–88

Barrett-Lennard, Thomas, 'Two Hundred Years of Estate Management at Horsford During the 17th and 18th Centuries', *Norfolk Archaeology*, 20 (1921), 57–139

Barrow, Geoffrey Wallis Steuart, *The Anglo-Norman Era in Scottish History*, Ford Lectures, 1977 (Oxford: Clarendon, 1980)

Behringer, Wolfgang, *A Cultural History of Climate*, trans. by Patrick Camilller (Cambridge: Polity, 2010)

Bell, Adrian Robert, Chris Brooks, and Paul R. Dryburgh, *The English Wool Market, c. 1230–1327* (Cambridge: Cambridge University Press, 2007)

Bevan, William, *The Upper Derwent: 10,000 Years in a Peak District Valley* (Stroud: Tempus, 2004)

Beglane, Fiona, *Anglo-Norman Parks in Medieval Ireland* (Dublin: Four Courts Press, 2015)

Bibby, John S., H. A. Douglas, Arthur Jackson Thomasson, and James Ian Summers Robertson, *Land Capability Classification for Agriculture* (Aberdeen: The Macaulay Institute for Soil Research, 1982)

Biraben, Jean Noël, and Jacques Le Goff, 'La peste dans le Haut Moyen Age', *Annales ESC*, 26 (1969), 1484–510

Birtles, Sara, 'Common Land, Poor Relief and Enclosure: The Use of Manorial Resources in Fulfilling Parish Obligations 1601–1834', *Past and Present*, 165 (1999), 74–106

Blanchard, Ian, 'Northern Wools and Netherlands Markets at the Close of the Middle Ages', in *Scotland and the Low Countries, 1124–1994*, ed. by Grant G. Simpson, Mackie Monographs, 3 (East Linton: Tuckwell, 1996), pp. 76–88

Boardman, Stephen, 'Chronicle Propaganda in Fourteenth-Century Scotland: Robert the Steward, John of Fordun and the "Anonymous Chronicle"', *Scottish Historical Review*, 76 (1997), 23–43

——, and Ross, Alasdair, eds, *The Exercise of Power in Medieval Scotland c. 1200–1500* (Dublin: Four Courts Press, 2003)

Bond, C. James, 'Monastic Fisheries', in *Medieval Fish, Fisheries and Fishponds in England*, ed. by Michael Aston, B.A.R. British Series, 182 (Oxford: B.A.R., 1988), pp. 69–112

Bond, Tami, Chandra Venkataraman, and Omar Masera, 'Global Atmospheric Impacts of Residential Fuels', *Energy for Sustainable Development*, 8 (2004), 20–32

Bonser, Wilfrid, *The Medical Background of Anglo-Saxon England: A Study in History, Psychology and Folklore* (London: Wellcome Historical Medical Library, 1963)

Borsay, Anne, and Peter Shapely, eds, *Medicine, Charity and Mutual Aid: The Consumption of Health and Welfare in Britain, c. 1550–1950*, Historical Urban Studies (Aldershot: Ashgate, 2004)

Bradbury, Edward, *In the Derbyshire Highlands: Highways, Byeways, and My Ways in the Peake Countrie* (Buxton: Bates, 1881)

Brander, Keith, 'Impacts of Climate Change on Fisheries', *Journal of Marine Systems*, 79 (2010), 389–402

Briffa, Keith Raphael, 'Annual Climate Variability in the Holocene: Interpreting the Message of Ancient Trees', *Quaternary Science Reviews*, 19 (2000), 87–105

——, Philip D. Jones, Fritz Hans Schweingruber, and Timothy J. Osborn, 'Influences of Volcanic Eruptions on Northern Hemisphere Summer Temperatures over the Past 600 Years', *Nature*, 393 (1998), 450–55

——, Philip D. Jones, R. Brennan Vogel, Fritz Hans Schweingruber, Michael G. L. Baillie, Stepan G. Shiyatov, and Eugene A. Voganov, 'European Tree Rings and Climate Change in the Sixteenth Century', *Climate Change*, 43 (1999), 151–68

——, Timothy J. Osborn, Fritz Hans Schweingruber, Ian C. Harris, Philip D. Jones, Stepan G. Shiyatov, and Eugene A. Vaganov, 'Low-Frequency Temperature Variations from a Northern Tree Ring Density Network', *Journal of Geophysical Research*, 106. D3 (2001), 2929–41

Brown, Ian Michael, Willie Towers, Mike Rivington, and Helaina I. J. Black, 'Influence of Climate Change on Agricultural Land-Use Potential: Adapting and Updating the Land Capability System for Scotland', *Climate Research*, 37 (2008), 43–57

Brown, Michael, *The Wars of Scotland, 1214–1371*, New Edinburgh History of Scotland, 4 (Edinburgh: Edinburgh University Press, 2004)

——, *James I*, 2nd edn (Edinburgh: Birlinn, 2015)

Bruce, James, *Travels to Discover the Source of the Nile, in the Years 1768, 1769, 1770, 1771, 1772, and 1773*, 5 vols (Edinburgh: Ruthven, 1790)

Brüning, Gertrud, 'Adamnan's Vita Columbae und ihre ableitungen', *Zeitschrift für Celtische Philologie*, 11 (1917), 227–29

Bruun, Christer, 'La mancanza di prove di un effetto catastrofico della "peste antonina" (dal 166 d.C. in poi)', in *L'impatto della 'peste antonina'*, ed. by Elio Lo Cascio, Pragmateiai, 22 (Bari: Edipuglia, 2012), pp. 123–65

Cage, Alix G. and William E. N. Austin, 'Marine Climate Variability During the Last Millennium: The Loch Sunart Record, Scotland, UK', *Quaternary Science Reviews*, 29 (2010), 1633–47

Caldwell, David H., 'The Lordship of the Isles. Identity Through Materiality', in *The Lordship of the Isles*, ed. by Richard D. Oram, The Northern World, 68 (Leiden: Brill, 2014), pp. 227–53

Camenisch, Chantal, Kathrin M. Keller, Melanie Salvisberg, Benjamin Amann, Martin Bauch, Sandro Blumer, Rudolf Brázdil, Stefan Brönnimann, Ulf Büntgen, Bruce M. S. Campbell, Laura Fernández-Donado, Dominik Fleitmann, Rüdiger Glaser, Fidel González-Rouco, Martin Grosjean, Richard C. Hoffmann, Heli Huhtamaa, Fortunat Joos, Andrea Kiss, Oldřich Kotyza, Flavio Lehner, Jürg Luterbacher, Nicolas Maughan, Raphael Neukom, Theresa Novy, Kathleen Pribyl, Christoph C. Raible, Dirk Riemann, Maximilian Schuh, Philip Slavin, Johannes Werner, and Oliver Wetter, 'The 1430s: A Cold Period of Extraordinary Climate Variability during the Early Spörer

Minimum with Social and Economic Impacts in North-Western and Central Europe', *Climate of the Past*, 12 (2016), 2017–26

Cameron, Sonja, 'Contumaciously Absent'? The Lords of the Isles and the Scottish Crown', in *The Lordship of the Isles*, ed. by Richard D. Oram, The Northern World, 68 (Leiden: Brill, 2014), pp. 161–67

——, and Alasdair Ross, 'The Treaty of Edinburgh and the Disinherited (1328–1332)', *History*, 84 (1999), 237–56

Campbell, Bruce M. S., 'Benchmarking Medieval Economic Development: England, Wales, Scotland, and Ireland, c. 1290', *The Economic History Review*, 61 (2008), 896–945

——, 'Nature as Historical Protagonist: Environment and Society in Pre-Industrial England', *Economic History Review*, 63 (2010), 287–93

——, 'Physical Shocks, Biological Hazards, and Human Impacts: The Crisis of the Fourteenth Century Revisited', in *Le Interazioni Fra Economia E Ambiente Biologico Nell'Europa Preindustriale Secc.XIII–XVIII = Economic and biological interactions in pre-industrial Europe, from the 13th to the 18th century: atti della Quarantunesima Settimana di studi, 26–30 aprile 2009*, ed. by Simonetta Cavaciocchi, Pubblicazioni, 41 (Florence: Firenze University Press, 2010), pp. 13–32

——, *The Great Transition: Climate, Disease and Society in the Late-Medieval World*, Ellen McArthur Lectures, 2013 (Cambridge: Cambridge University Press, 2016)

Carlson, Catherine C., 'Where's the Salmon? A Reevaluation of the Role of Anadromous Fisheries to Aboriginal New England', in *Human Holocene Ecology in Northeastern North America*, ed. by George P. Nicholas (New York: Plenum Press, 1988), pp. 47–80

——, 'The (In)Significance of Atlantic Salmon', *Federal Archeology*, 8.3/4 (1996), 22–30

Carmichael, Ann G., and Arthur M. Silverstein, 'Smallpox in Europe before the Seventeenth Century: Virulent Killer or Benign Disease?', *Journal of the History of Medicine and Allied Sciences*, 42 (1987), 147–68

Carrell, Jennifer Lee, *The Speckled Monster: A Historical Tale of Battling Smallpox* (New York: Dutton, 2004)

Cerron-Carrasco, Ruby, *'Of Fish and Men' (De iasg agus dhaione): A Study of the Utilization of Marine Resources as Recovered from Selected Hebridean Archaeological Sites*, B.A.R. British Series, 400 (Oxford: Archaeopress, 2005)

Chadwick, Lee, *In Search of Heathland* (Durham: Dobson Books, 1982)

Chais, Charles, *Essai apologétique sur la méthode de communiquer la petite vérole par inoculation, où, l'on tâche de faire voir que la conscience ne sauroit en être blessée, ni la religion offensée* (The Hague: de Hondt, 1754)

Charles, Michael, 'The Elephants of Aksum: In Search of the Bush Elephant of Late Antiquity', *Journal of Late Antiquity*, 11 (2018), 166–92

Charles-Edwards, Thomas, *The Chronicle of Ireland*, Translated Texts for Historians, 44, 2 vols (Liverpool: Liverpool University Press, 2006)

Charman, Dan J., Chris Caseldine, Andy Baker, Ben Gearey, Jackie Hatton, and Chris Proctor, 'Paleohydrological Records from Peat Profiles and Speleothems in Sutherland, Northwest Scotland', *Quaternary Research*, 55.2 (2001), 223–34

Checkley, David M., Jürgen Alheit, Yoshioki Oozeki, and Claude Roy, eds, *Climate Change and Small Pelagic Fish* (Cambridge: Cambridge University Press, 2009)

Chorley, Patrick, 'The Cloth Exports of Flanders and Northern France during the Thirteenth Century: A Luxury Trade?', The Economic History Review, 40 (1987), 349–79

Cibot, Pierre-Martial, 'De La Petite Vérole', in *Mémoires concernant l'histoire, les sciences, les arts, les moeurs, les usages, etc. des Chinois par les missionaires de Pe-Kin*, 16 vols (Paris: Nyon, 1776–91), IV (1779)

Cioc, Mark, *The Rhine: An Eco-Biography, 1815–2000*, Weyerhaeuser Environmental Books (Seattle: University of Washington Press, 2002)

Coates, Peter, *Salmon* (London: Reaktion, 2006)

Cockburn, T. Aidan, *The Evolution and Eradication of Infectious Diseases* (Baltimore: Johns Hopkins University Press, 1963)

——, 'Infectious Diseases in Ancient Populations', *Current Anthropology*, 12 (1971), 45–62

Conrad, Lawrence I., 'Tāʿūn and Wabāʾ: Conceptions of Plague and Pestilience in Early Islam', *Journal of the Economic and Social History of the Orient*, 25 (1982), 268–307

——, 'Abraha and Muhammad: Some Observations Apropos of Chronology and Literary "topoi" in Early Arabic Historical Tradition', *Bulletin of the School of Oriental and African Studies*, 50 (1987), 225–40

Coomans, Thomas, 'From Flanders to Scotland: The Choir Stalls of Melrose Abbey in the Fifteenth Century', in *Perspectives for an Architecture of Solitude: Essays on Cistercians, Art and Architecture in Honour of Peter Fergusson*, ed. by Terryl Nancy Kinder, Medieval Church Studies, 11 (Turnhout: Brepols, 2004), pp. 235–52

Corbin, Alain, *The Lure of the Sea: The Discovery of the Seaside in The Western World, 1750–1840*, trans. by Joceyln Phelps (Berkeley: University of California Press, 1994)

Coy, Jenny, and Sheila Hamilton-Dyer, 'The Bird and Fish Bone', *Excavations at Iona, 1988*, ed. by Finbar McCormick, *Ulster Journal of Archaeology*, 56 (1993), 78–108

Crawford, Iain A., 'Structural Discontinuity and Associable Evidence for Settlement Disruption: Five Crucial Episodes in a Continuous Occupation (the Udal Evidence)', in *Settlement and Society in Scotland: Migration, Colonisation and Integration*, ed. by Roger Alexander Mason, Association of Scottish Historical Studies Symposium, 2 (Glasgow: Association for Scottish Historical Studies, 1988), pp. 1–34

Creighton, Charles, *History of Epidemics in Britain from AD 664 to the Extinction of Plague*, 2 vols (Cambridge: Cambridge University Press, 1891)

Crone, Anne, and Fiona Watson, 'Sufficiency to Scarcity: Medieval Scotland, 500–1600', in *People and Woods in Scotland: A History*, ed. by T. Christopher Smout (Edinburgh. Edinburgh University Press, 2005), pp. 60–81

——, and Coralie M. Mills, 'Seeing the Wood and the Trees: Dendrochronological Studies in Scotland', *Antiquity*, 76 (2002), 788–94

Crowley, Thomas J., and Thomas S. Lowery, 'How Warm was the Medieval Warm Period?' *AMBIO*, 29.1 (2000), 51–54

Cunningham, Andrew, 'Identifying Disease in the Past: Cutting the Gordon's Knot', *Asclepio*, 54 (2002), 13–34

Currie, Christopher K., 'The Role of Fishponds in the Monastic Economy', in *The Archaeology of Rural Monasteries*, ed. by Roberta Gilchrist and Harold Mytum, B.A.R. British series, 203 (Oxford: B.A.R., 1989), pp. 147–51

D'Arrigo, Rosanne, Patrick Klinger, Timothy Newfield, Miloš Rydval, and Roy Wilson, 'Complexity in Crisis: The Volcanic Cold Pulse of the 1690s and the Consequences of Scotland's Failure to Cope', *Journal of Volcanology and Geothermal Research*, 389 (2020), 106746

Dale, Corinne, *The Natural World in the Exeter Book Riddles*, Nature and Environment in the Middle Ages (Cambridge: Brewer, 2017)

Danker-Carstensen, Peter, 'Wieviel Störfleisch verträgt die Dienstmagd in der Woche? Regional Legendenbildung in einem kurzen Kapitel Fischereigeschichte', in *Som en rejselysten flåde: tilegnet Ole Lisberg Jensen (Festkrift Ole Lisberg Jensen på 60-årsdagen 21. juni 1999)*, ed. by Frank Allan Rasmussen and Flemming Rieck, Orlogsmuseets Skriftenrække, 2. (Christianshavn: Orlogsmuseet, 1999), pp. 30–46

——, 'Stör oder Lachs — aber auf keinen Fall mehr als zweimal in der Woche? Legendenbildung und Erzähltradition in einem Kapitel deutscher Fischereigeschichte', in *Mythen der Vergangenheit: Realität und Fiktion in der Geschichte. Jörgen Bracker zum 75. Geburtstag*, ed. by Ortwin Pelc (Göttingen: Vandenhoeck, 2012), pp. 265–85

Davies, David, *The Cases of Labourers in Husbandry Stated and Considered* (Bath: Cruttwell, 1795)

Dawson, Alastair, *So Fair and Foul a Day: A History of Scotland's Weather and Climate* (Edinburgh: Birlinn, 2009)

——, Kieran R. Hickey, Paul Andrew Mayewski, and Atle Nesje, 'Greenland (GISP2) Ice Core and Historical Indicators of Complex North Atlantic Climate Changes during the Fourteenth Century', *The Holocene*, 17.4 (2007), 427–34

Dawson, Susan, Alasdair G. Dawson, and Jason T. Jordan, 'North Atlantic Climate Change and Late Holocene Windstorm Activity in the Outer Hebrides, Scotland', *Scottish Archaeological Internet Reports*, 48: *Aeolian Archaeology: the Archaeology of Sand Landscapes in Scotland*, ed. by David Griffiths and Patrick Ashmore (2011), 25–36 online at: <http://soas.is.ed.ac.uk/index.php/sair/article/view/3065> [accessed 7 July 2022]

De Raymond, Jean-François, *Querelle de l'inoculation, ou, préhistoire de la vaccination*, Problèmes et controverses (Paris: Vrin, 1982)

De Ricci, Seymour, with William Jerome Wilson, *Census of Medieval and Renaissance Manuscripts in the United States and Canada*, 2 vols (Washington, DC: The Library of Congress, 1937)

Degroot, Dagomar, *The Frigid Golden Age: Climate Change, the Little Ice Age, and the Dutch Republic, 1560–1720*, Studies in Environment and History (Cambridge: Cambridge University Press, 2018)

Devroey, Jean-Pierre, *Économie rurale et société dans l'Europe franque (VIe–IXe siècles)* (Paris: Belin, 2003)

——, 'Catastrophe, crise et changement social: à propos des paradigmes d'interprétation du développement médiéval (500–1100)', in *Vers une anthropologie des catastrophes:*

Actes des 9e journées anthropologiques de Valbonne, ed. by Luc Buchet, Catherine Rigeade, Isabelle Séguy, and Michel Signoli (Paris: Antibes, 2009), pp. 139–61

Ditchburn David, *Scotland and Europe: The Medieval Kingdom and its Contacts with Christendom, c. 1215–1545, I: Religion, Culture and Commerce* (East Linton: Tuckwell, 2000)

——, and Alastair John MacDonald, 'Medieval Scotland, 1100–1560', in *The New Penguin History of Scotland*, ed. by Robert Allan Houston and William Knox (London: Penguin, 2001), pp. 96–181

Dixon, Cyril William, *Smallpox* (London: Churchill, 1962)

Dixon, Piers, 'Crops and Livestock in the Pre-Improvement Era', in *Scottish Life and Society. A Compendium of Scottish Ethnology, II: Farming and the Land*, ed. by Alexander Fenton and Kenneth Veitch (Edinburgh: Donald, 2011), pp. 229–43

——, 'Hunting, Summer Grazing and Settlement: Competing Land Use in the Uplands of Scotland', *Ruralia*, 7 (2009), 27–46

Dobson, Andrew P., and Robin E. Carper, 'Infectious Diseases and Human Population History', *BioScience*, 46 (1996), 115–26

Dols, Michael W., 'Plague in Early Islamic History', *Journal of the American Oriental Society*, 94 (1974), 371–83

Donkin, Robin Arthur, 'Cistercian Sheep-Farming and Wool-Sales in the Thirteenth Century', *The Agricultural History Review*, 6 (1958), 2–8

Donnelly, Joseph, 'Skinned to the Bone: Durham Evidence for Taxations of the Church in Scotland, 1254–1366', *The Innes Review*, 50 (1999), 1–24

Dougall, Martin, and Jim Dickson, 'Old Managed Oaks in the Glasgow Area', in *Scottish Woodland History*, ed. by T. Christopher Smout (Edinburgh: Scottish Cultural Press, 1997), pp. 75–84

Downie, Allan W., and Keith Dumbell, 'Pox Viruses', *Annual Review of Microbiology*, 10 (1956), 237–52

Duggan, Ana T., Maria F Perdomo, Dario Piombino-Mascali, Stephanie Marciniak, Debi Poinar, Matthew V. Emery, Jan P. Buchmann, Sebastian Duchêne, Rimantas Jankauskas, Margaret Humphreys, G. Brian Golding, John Southon, Alison Devault, Jean-Marie Rouillard, Jason W. Sahl, Olivier Dutour, Klaus Hedman, Antti Sajantila, Geoffrey L. Smith, Edward C. Holmes, and Hendrik N. Poinar, '17th Century Variola Virus Reveals the Recent History of Smallpox', *Current Biology*, 26 (2016), 3407–12

Dumayne-Peaty, Lisa, 'Late Holocene Human Impact on the Vegetation of Southeastern Scotland: a Pollen Diagram From Dogden Moss, Berwickshire', *Review of Palaeobotany and Palynology*, 105 (1999), 121–41

Duncan, Archibald Alexander McBeth, *Scotland: The Making of the Kingdom*, Edinburgh History of Scotland, 1 (Edinburgh: Oliver & Boyd, 1978)

Duncan-Jones, Richard Phare, 'The Impact of the Antonine Plague', *Journal of Roman Archaeology*, 9 (1996), 108–36

Düx, Ariane, Sebastian Lequime, Livia Victoria Patrono, Bram Vrancken, Sengül Boral, Jan F. Gogarten, Antonia Hilbig, David Horst, Kevin Merkel, Baptiste Prepoint,

Sabine Santibanez, Jasmin Schlotterbeck, Marc A. Suchard, Markus Ulrich, Navena Widulin, Annette Mankertz, Fabian H. Leendertz, Kyle Harper, Thomas Schnalke, Philippe Lemey, and Sébastien Calvignac-Spencer, 'The History of Measles: From a 1912 Genome to an Antique Origin', *bioRxiv: The Preprint Server for Biology* (2019), online at <https://doi.org/10.1101/2019.12.29.889667>

——, Sebastian Lequime, Livia Victoria Patrono, Bram Vrancken, Sengül Boral, Jan F. Gogarten, Antonia Hilbig, David Horst, Kevin Merkel, Baptiste Prepoint, Sabine Santibanez, Jasmin Schlotterbeck, Marc A. Suchard, Markus Ulrich, Navena Widulin, Annette Mankertz, Fabian H. Leendertz, Kyle Harper, Thomas Schnalke, Philippe Lemey, and Sébastien Calvignac-Spencer, 'Measles Virus and Rinderpest Virus Divergence Dated to the Sixth Century BCE', *Science*, 368 (2020), 1367–70

Dyer, Christopher, 'The Consumption of Fresh-Water Fish in Medieval England', in *Medieval Fish, Fisheries and Fishponds in England*, ed. by Michael Aston, B.A.R. British Series, 182 (Oxford: B.A.R., 1988), pp. 27–35

Eckenode, Thomas R., 'The English Cistercians and Their Sheep During the Middle Ages', *Cîteaux: commentarii cistercienses*, 24 (1973), 250–66

Eden, Frederick Morton, *The State of the Poor, or, an History of the Labouring Classes in England from the Conquest to the Present Period*, 3 vols (London: Davis, 1797)

Edwardes, Edward J., *A Concise History of Small-Pox and Vaccination in Europe* (London: Lewis, 1902)

Edwards, Anthony Stockwell Garfield, 'The McGill Fragment of Lydgate's "Fall of Princes"', *Scriptorium*, 28 (1974), 75–77

Enghoff, Inge Bodker, 'Fishing in the Southern North Sea Region from the 1st to the 16th Century AD: Evidence from Fish Bones', *Archaeofauna*, 9 (2000), 59–132

Evans, Nicholas, *The Present and the Past in Medieval Irish Chronicles*, Studies in Celtic History, 27 (Woodbridge: Boydell, 2010)

——, 'The Irish Chronicles and the British to Anglo-Saxon Transition in Seventh-Century Northumbria', in *The Medieval Chronicle VII*, ed. by Juliana Dresvina and Nicholas Sparks (Amsterdam: Rodopi, 2011), pp. 15–38

Evelyn, John, *Silva, or, a Discourse of Forest-Trees, and the Propagation of Timber in his Majesty's Dominions*, 5th edn, repr. (London: Stobart, 1979)

Eversley, George Shaw-Lefevre, *Commons, Forest and Footpaths: The Story of the Battle During the Last Forty-Five Years for Public Rights over the Commons, Forests and Footpaths of England and Wales*, rev. edn (London: Cassell, 1910)

Face, R. D., 'Techniques of Business in Trade between the Fairs of Champagne and the South of Europe in the 12th and 13th Centuries', *The Economic History Review*, 10 (1957–1958), 427–38

Faure, Éric, and Natacha Jacquemard, 'L'émergence du paludisme en Gaule: Analyse comparée des écrits de Sidoine Apollinaire et Grégoire de Tours', in *Présence de Sidoine Apollinaire*, ed. by Rémy Poignault and Annick Stoehr-Monjou, Caesarodunum, 44–45 (Clermont-Ferrand: Centre des Recherches A. Piganiol, 2014), pp. 55–70

Fawcett, Richard, and Richard Oram, *Melrose Abbey* (Stroud: Tempus, 2004)

Feeny, David, Fikret Berkes, Bonnie J. McCay, and James M. Acheson, 'The Tragedy of the Commons: Twenty-Two Years Later', *Human Ecology*, 18.1 (1990), 1–19

Fenner, Frank John, 'The Effects of Changing Social Organisation on the Infectious Diseases of Man', in *The Impact of Civilisation on the Biology of Man*, ed. by Stephen Vickers Boyden (Canberra: Australian National University Press, 1970), pp. 48–76

——, Donald A. Henderson, Isao Arita, Zdenek Jezek, and Ivan Danilovich Ladnyi, *Smallpox and its Eradication*, History of International Public Health, 6 (Geneva: World Health Organization, 1988)

Fenton, Alexander, 'Cultivating Tools and Tillage: The Old Scotch Plough, the Improved Plough and Spades', in *Scottish Life and Society. A Compendium of Scottish Ethnology, II: Farming and the Land*, ed. by Alexander Fenton and Kenneth Veitch (Edinburgh: Donald, 2011), pp. 655–78

——, *Scottish Life and Society. A Compendium of Scottish Ethnology, V: The Food of the Scots* (Edinburgh: Donald. 2007)

Fenton, James H. C., 'A Postulated Natural Origin for the Open Landscape of Upland Scotland', *Plant Ecology and Diversity*, 1 (2008), 115–27

Ferrari, Giada, Judith Neukamm, Helle T. Baalsrud, Abagail M. Breidenstein, Mark Ravinet, Carina Phillips, Frank Rühli, Abigail Bouwman, and Verena J. Schuenemann, 'Variola Virus Genome Sequenced from an Eighteenth-Century Museum Specimen Supports the Recent Origin of Smallpox', *Philosophical Transactions of the Royal Society B: Biological Sciences*, 375 (2020), 20,190,572

Fiennes, Richard Nathaniel T.-W., *Zoonoses and the Origins and Ecology of Human Diseases* (London: Academic Press, 1978)

Fisher, Greg, *Rome, Persia and Arabia: Shaping the Middle East from Pompey to Muhammad* (Abingdon: Routledge, 2019)

Flanagan, Marie Therese, 'Irish Royal Charters and the Cistercian Order', in *Charters and Charter Scholarship in Britain and Ireland*, ed. by Marie Therese Flanagan and Judith A. Green (Basingstoke: Palgrave, 2005), pp. 122–25

Fletcher, John, *Gardens of Earthly Delight: The History of Deer Parks* (Oxford: Windgather, 2011)

de Fonseca, Rodrigo, *Consultationes medicae singularibus remediis refertae* (Venice: Guerilius, 1619)

Foot, Sarah, 'Plenty, Portents and Plague: Ecclesiastical Readings of the Natural World in Early Medieval Europe', *Studies in Church History*, 46 (2010), 15–41

Fraser-Mackintosh, Charles, *The Last Macdonalds of Isla* (Glasgow: Celtic Monthly Office, 1895)

Freind, John, *The History of Physick from the Time of Galen to the Beginning of the Sixteenth Century*, 2 vols (London: [n. pub.], 1727)

——, *Histoire de la Medecine depuis Galien jusqu'au comencement du seizieme siècle*, trans. by Étienne Coulet, 2 vols (Leiden: Langerak, 1727)

——, *Histoire de la Medecine depuis Galien jusqu'au XVI siecle*, trans. by Jean-Baptiste Senac (Paris: Vincent, 1728)

——, *Historia Medicinae a Galeni Tempore usque ad initium saeculi decimi sexti*, trans. by John Wigan (Venice: Coleti, 1735)

Furuse, Yuki, Akira Suzuki, and Hitoshi Oshitani, 'Origin of Measles Virus: Divergence from Rinderpest Virus Between the 11th and 12th Centuries', *Virology Journal*, 7 (2010), 52, available online at <https://doi.org/10.1186/1743-422X-7-52>

Geist, Valerius, *Deer of the World: Their Evolution, Behavior, and Ecology* (Mechanicsburg: Stackpole Books, 1998)

Gemmill, Elizabeth, and Nicholas Mayhew, *Changing Values in Medieval Scotland: A Study of Prices, Money, and Weights and Measures* (Cambridge: Cambridge University Press, 1995)

Gilbert, John M., *Hunting and Hunting Reserves in Medieval Scotland* (Edinburgh: Donald, 1979)

——, 'Falkland Park to 1603', *Tayside and Fife Archaeological Journal*, 19–20 (2014), 69–77

Gilbertson, David Dennis, Jean-Luc Schwenninger, Robert A. Kemp, and Edward J. Rhodes, 'Sand-drift and Soil Formation Along an Exposed North Atlantic Coastline: 14,000 Years of Diverse Geomorphological, Climatic and Human Impacts', *Journal of Archaeological Science*, 26.4 (1999), 439–69

Gilliam, James Frank, 'The Plague Under Marcus Aurelius', *American Journal of Philology*, 82 (1961), 225–51

Glaser, Rüdiger, *Klimageschichte Mitteleuropas. 1000 Jahre Wetter, Klima, Katastrophen*, 2nd edn (Darmstadt: Wissenschaftliche Buchgesellschaft, 2008)

——, and Dirk Riemann, 'A Thousand Year Record of Temperature Variations for Germany and Central Europe Based on Documentary Data', *Journal of Quaternary Science*, 24.5 (2009), 437–49

Graham, Rose, 'The Taxation of Pope Nicholas IV', *English Historical Review*, 23 (1908), 434–54

Grant, Alexander, *Independence and Nationhood: Scotland, 1306–1469* (Edinburgh: University Press, 1984)

Green, Monica H., 'Climate and Disease in Medieval Eurasia', in *The Oxford Research Encyclopedia of Asian History*, ed. by David Ludden (Oxford: Oxford University Press, 2018), online at: <https://doi.org/10.1093/acrefore/9780190277727.013.6> [accessed 3 August 2022]

Greenberg, James, 'Plagued by Doubt: Reconsidering the Impact of a Mortality Crisis in the 2nd c. A.D.', *Journal of Roman Archaeology*, 16 (2003), 413–25

Guillet, Sébastien, Christophe Corona, Markus Stoffel, Myriam Khodri, Franck Lavigne, Pablo Ortega, Nicolas Eckert, Pascal Dkengne Sielenou, Valérie Daux, Olga V. Churakova (Sidorova), Nicole Davi, Jean-Louis Edouard, Yong Zhang, Brian H. Luckman, Vladimir S. Myglan, Joël Guiot, Martin Beniston, Valérie Masson-Delmotte, and Clive Oppenheimer, 'Climate Response to the Samalas Volcanic Eruption in 1257 Revealed by Proxy Records', *Nature Geoscience*, 10 (2017), 123–29

Gunderson, Lance H., and Crawford Stanley Holling, eds, *Panarchy: Understanding Transformations in Human and Natural Systems* (Washington: Island Press, 2002)

Haeser, Heinrich, *Lehrbuch der Geschichte der Medicin und der Volkskrankheiten* (Jena: Mauke, 1845)

Hahn, Johann Gottfried von, *Variolarum antiquitates, nunc primum e Graecis erutae* (Brieg: Trampius, 1733)

Haidvogl, Gertrud, Dmitry Lajus, Didier Pont, Martin Schmid, Mathias Jungwirth, and Julia Lajus, 'Typology of Historical Sources and the Reconstruction of Long-Term Historical Changes of Riverine Fish: A Case Study of the Austrian Danube and Northern Russian Rivers', *Ecology of Freshwater Fish*, 23.4 (2014), 498–515

Hardin, Garrett, 'The Tragedy of the Commons', *Science*, 162 (1968), 1243–48

Hartley, Marie, and Joan Ingilby, *Life and Tradition in the Moorlands of North-East Yorkshire* (Otley: Smith Settle, 1972)

Hatcher, John, *The History of the British Coal Industry, I: Before 1700, Towards the Age of Coal* (Oxford: Clarendon, 1993)

Hayman, Richard, *Trees, Woodlands and Western Civilization* (London: Hambledon, 2003)

Heath, John, *The Illustrated History of Derbyshire* (Buckingham: Barracuda, 1982)

Hecker, Justus Friedrich Carl, *De peste antoniniana commentario* (Berlin: Schade, 1835)

Hindmarch, Erlend, and Oram, Richard, 'Eldbotle; the Archaeology and Environmental History of a Medieval Rural Settlement in East Lothian', *Proceedings of the Society of Antiquaries of Scotland*, 142 (2012), 245–300

Hirsch, August, 'Blattern', in *Handbuch der Historisch-Geographischen Pathologie*, ed. by August Hirsch, 3 vols (Stuttgart: Enke, 1860), I, pp. 214–16

History of the Feuds and Conflicts Among the Clans in the Northern Parts of Scotland and in the Western Isles (Glasgow: Foulis Press, 1764)

Hitzbleck, Herbert, 'Die Bedeutung des Fisches für die Ernährungswirtschaft Mitteleuropas in vorindustrieller Zeit unter besonderer Berücksichtigung Niedersachsens' (unpublished doctoral thesis, Georg-August University, 1971)

Hodgson, Victoria, 'The Cistercian Abbey of Coupar Angus, *c.* 1164–*c.* 1560' (unpublished doctoral thesis, University of Stirling, 2016)

——, 'The Landholding and Landscape Exploitation of Coupar Angus Abbey: Granges and Glenisla', in *Monastic Europe: Medieval Communities, Landscapes and Settlements*, ed. by Edel Bhreathnach, Malgorzata Krasnodębska-D'Aughton, and Keith Smith, Medieval Monastic Studies, 4 (Turnhout: Brepols, 2020), pp. 431–50

Hoffmann, Richard C., 'Economic Development and Aquatic Ecosystems in Medieval Europe', *American Historical Review*, 101 (1996), 631–69

——, '*Salmo salar* in Late Medieval Scotland: Competition and Conservation for a Riverine Resource', *Aquatic Sciences*, 77 (2015), 355–66, online at <https://doi.org/10.1007/s00027-015-0397-4>

——, and Alasdair Ross, *Inland Fishings in Medieval Scotland*, online at <https://www.stir.ac.uk/about/faculties/arts-humanities/our-research/centre-for-environment-heritageand-policy/projects/past-projects/inland-fishings-in-medieval-scotland/> [accessed 15 August 2022]

——, and Alasdair Ross, 'This Belongs to Us! Competition between the Royal Burgh of Stirling and the Augustinian Abbey of Cambuskenneth over Salmon Fishing Rights on the River Forth', in *Monastic Europe: Medieval Communities, Landscapes and Settlements*, ed. by Edel Bhreathnach, Malgorzata Krasnodębska-D'Aughton, and Keith Smith, Medieval Monastic Studies, 4 (Turnhout: Brepols, 2020), pp. 451–76

Holwell, John Zephaniah, *An Account of the Manner of Inoculating for the Small Pox in the East Indies* (London: De Hondt, 1767)

Hopkins, Donald R., 'Ramses V: Earliest Known Smallpox Victim?', *World Health*, May (1980), 22–26

——, *Princes and Peasants: Smallpox in History* (Chicago: University of Chicago Press, 1983); and rev. edn, *The Greatest Killer: Smallpox in History* (Chicago: University of Chicago Press, 2002)

Horden, Peregrine, 'Disease, Dragons and Saints: The Management of Epidemics in the Dark Ages', in *Epidemics and Ideas: Essays on the Historical Perception of Pestilence*, ed. by Terence Ranger and Paul Slack, Past and Present Publications (Cambridge: Cambridge University Press, 1992), pp. 45–76

——, 'Mediterranean Plague in the Age of Justinian', in *The Cambridge Companion to the Age of Justinian*, ed. by Michael Maas (Cambridge: Cambridge University Press, 2005), pp. 134–60

Hough, Carole, '"Find the Lady": The Term Lady in English and Scottish Place-Names', in *Names in Multi-Lingual, Multi-Cultural and Multi-Ethnic Contact: Proceedings of the 23rd International Congress of Onomastic Sciences, August 17–22, 2008, York University, Toronto, Canada*, ed. by Wolfgang Ahrens, Sheila Embleton, and André Lapierre, with Grant Smith and Maria Figueredo (Toronto: York University, 2009), pp. 511–18

Hughes, Kathleen, *Early Christian Ireland: Introduction to the Sources* (London: Hodder, 1972)

Humphries, Christopher John, and Elaine Shaughnessy, *Gorse*, Shire Natural History, 9 (Aylesbury: Shire, 1982)

Hunt, Edwin S., and James M. Murray, *A History of Business in Medieval Europe, 1200–1550* (Cambridge: Cambridge University Press, 1999)

Innes, Cosmo, *Lectures on Scottish Legal Antiquities* (Edinburgh: Edmonston, 1872)

James, Tom Beaumont, and Christopher Gerrard, *Clarendon: Landscape of Kings* (Bollington: Windgather, 2007)

Jane, Stephen F., Keith H. Nislow, and Andrew R. Whiteley, 'The Use (and Misuse) of Archaeological Salmon Data to Infer Historical Abundance in North America with a Focus on New England', *Reviews in Fish Biology and Fisheries*, 24.3 (2014), 943–54

Jenner, Edward, *An Inquiry into the Causes and Effects of Variolae Vaccinae* (London: Low, 1798)

——, and William Woodville, *A Comparative Statement of the Facts and Observations Relative to the Cow Pox* (London: Low, 1800)

Jones, Andrew K. G., 'The Fish Bone', in *Perth High Street Archaeological Excavations 1975–1977, Fascicule 4: Living and Working in a Medieval Scottish Burgh, Environ-*

mental Remains and Miscellaneous Finds, ed. by George W. I. Hodgson and Arthur MacGregor (Perth: Tayside and Fife Archeological Committe, 2011), pp. 53–57

Jones, Melvyn, 'Woods, Trees, and Animals: A Perspective from South Yorkshire, England', in *Trees, Forested Landscapes and Grazing Animals: A European Perspective on Woodlands and Grazed Treescapes*, ed. by Ian D. Rotherham (London: Routledge, 2013), pp. 24–34

Jordan, William Chester, *The Great Famine: Northern Europe in the Early Fourteenth Century* (Princeton: Princeton University Press, 1996)

Keith, Robert, *An Historical Catalogue of the Scottish Bishops Down to the Year 1688* (Edinburgh: Bell and Bradfute, 1824)

Keller, Marcel, Maria A. Spyrou, Christiana L. Scheib, Gunnar U. Neumann, Andreas Kröpelin, Brigitte Haas-Gebhard, Bernd Päffgen, Jochen Haberstroh, Albert Ribera I. Lacomba, Claude Raynaud, Craig Cessford, Raphaël Durand, Peter Stadler, Kathrin Nägele, Jessica S. Bates, Bernd Trautmann, Sarah A. Inskip, Joris Peters, John E. Robb, Toomas Kivisild, Dominique Castex, Michael McCormick, Kirsten I. Bos, Michaela Harbeck, Alexander Herbig, and Johannes Krause, 'Ancient Yersinia pestis Genomes from across Western Europe Reveal Early Diversification During the First Pandemic (541–750)', *Proceedings of the National Academy of Sciences*, 116 (2019), 12,363–73

Kelly, Robert L., *The Lifeways of Hunter-Gatherers: The Foraging Spectrum* (Cambridge: Cambridge University Press, 2013)

Kennedy, Hugh N., *The Armies of the Caliphs: Military and Society in the Early Islamic State* (London: Routledge, 2001)

Kerr, William Alexander, *Peat and its Products* (Glasgow: Begg, Kennedy & Elder, 1905)

Kershaw, Ian, 'The Great Famine and Agrarian Crisis in England 1315–1322', *Past and Present*, 59 (1973), 3–50

Khalakdina, Asheena, Alejandro Costa, and Sylvie Briand, 'Smallpox in the Post-Eradication Era', *Weekly Epidemiological Record*, 91.20 (2016), 257

King, Peter, 'Coupar Angus and Cîteaux', *Innes Review*, 27 (1976), 49–69

——, *The Finances of the Cistercian Order in the Fourteenth Century*, Cistercian Studies Series, 85 (Kalamazoo: Cistercian Publications, 1985)

Kirby, Tony, 'WHO Celebrates 40 Years Since Eradication of Smallpox', *Lancet Infectious Diseases*, 20 (2020), 174

Kirk, William, 'Prehistoric Sites at the Sands of Forvie, Aberdeenshire: A Preliminary Examination', *Aberdeen University Review*, 35 (1953), 150–71

Kister, Meir Jacob, 'The Campaign of Hulubān: A New Light on the Expedition of Abraha', *Le Muséon*, 78 (1965), 425–36

Kotar, S. L., and J. E. Gessler, *Smallpox: A History* (Jefferson: McFarland, 2013)

Lajus, Dmitry L., Zoya V. Dmitrieva, Alexei V. Kraikovski, Julia A. Lajus, and Daniel A. Alexandrov, 'Atlantic Salmon Fisheries in the White and Barents Sea Basins: Dynamic of Catches in the 17–18th Century and Comparison with 19–20th Century Data', *Fisheries Research*, 87.2/3 (2007), 240–54

——, Zoya V. Dmitrieva, Alexei V. Kraikovski, Julia A. Lajus, and Daniel A. Alexandrov, 'The Use of Historical Catch Data to Trace the Influence of Climate on Fish

Populations: Examples from the White and Barents Sea Fisheries in the 17th–18th Centuries', *ICES Journal of Marine Science*, 62.7 (2005), 1426–35

Lajus, Julia A., Yaroslava Alekseeva, Ruslan Davydov, Zoya V. Dmitrieva, Alexei V. Kraikovski, Dmitry L. Lajus, Vladimir Lapin, Vadim O. Mokievsky, Alexei Yurchenko, and Daniel A. Alexandrov, 'Status and Potential of Historical and Ecological Studies on Russian Fisheries in the White and Barents Seas: The Case of the Atlantic Salmon (*Salmo salar*)', in *The Exploited Seas: New Directions for Marine Environmental History*, ed. by Paul Holm, Tim Denis Smith, and David John Starkey, Research in Maritime History, 21 (St John's: International Maritime Economic History Association, 2001), pp. 67–96

Lambert, Joyce Mildred, Joseph Newell Jennings, Charles T. Smith, Charles Green, and J. N. Hutchinson, *The Making of the Broads: A Reconsideration of their Origin in the Light of New Evidence* (London: Murray, 1961)

Lavigne, Franck, Jean-Philippe Degeai, Indyo Pratomo, Patrick Wassmer, Irka Hajdas, Danang Sri Hadmoko, Edouard de Belizal, Jean-Christophe Komorowski, Sébastien Guillet, Vincent Robert, Pierre Lahitte, Clive Oppenheimer, Markus Stoffel, Céline M. Vidal, and Surono, 'Source of the Great A.D. 1257 Mystery Eruption Unveiled, Samalas Volcano, Rinjani Volcanic Complex, Indonesia', *Proceedings of the National Academy of Sciences*, 110 (2013), 16742–47

Le Clerc, Daniel, 'Essai d'un plan pour servir à la continuation de l'histoire de la medecine', in his *Histoire de la medecine*, 3 vols (Amsterdam: [n. pub.], 1723), I, 765–820

Lemey, Philippe, and David Posada, 'Molecular Clock Analysis', in *The Phylogenetic Handbook: A Practical Approach to Phylogenetic Analysis and Hypothesis Testing*, ed. by Philippe Lemey, Marco Salemi, and Anne-Mieke Vandamme (Cambridge: Cambridge University Press, 2009), pp. 362–72

Leven, Karl-Heinz, 'Zur Kenntnis der Pocken in der arabischen medizin, im lateinischen Mittelalter und in Byzanz', in *Die Begegnung des Ostens mit dem Westen: Kongreßakten des 4. Symposions des Mediävistenverbandes in Köln 1991 aus Anlaß des 1000. Todesjahresder Kaiserin Theophanu*, ed. by Odilo Engels and Peter Schreiner (Sigmaringen: Thorbecke, 1993), pp. 341–54

Lewin, Peter K., '"Mummy" Riddles Unravelled', *Bulletin of the Microscopy Society of Canada*, 12 (1984), 3–8

Li, Yu, Darin S. Carroll, Shea N. Gardner, Matthew C. Walsh, Elizabeth A. Vitalis, and Inger K. Damon, 'On the Origin of Smallpox: Correlating Variola Phylogenetics with Historical Smallpox Records', *Proceedings of the National Academy of Sciences*, 104 (2007), 15,787–92

Liddiard, Robert, 'Introduction', in *The Medieval Park: New Perspectives*, ed. by Robert Liddiard (Oxford: Windgather, 2007), pp. 1–8

——, ed., *The Medieval Park: New Perspectives* (Oxford: Windgather, 2007)

Little, Andrew George, 'The Authorship of the Lanercost Chronicle', *English Historical Review*, 31 (1916), 269–79

Little, Lester K., ed., *Plague and the End of Antiquity: The Pandemic of 541–750* (Cambridge: Cambridge University Press, 2007)

Littman, Robert J., 'The Athenian Plague: Smallpox', *Transactions and Proceedings of the American Philological Association*, 100 (1969), 261–75
——, 'The Plague of Syracuse: 396 BC', *Mnemosyne*, 37 (1984), 110–16
——, 'The Plague of Athens: Epidemiology and Paleopathology', *Mount Sinai Journal of Medicine*, 76 (2009), 456–67
——, and Michael L. Littman, 'Galen and the Antonine Plague', *The American Journal of Philology*, 94 (1973), 243–55
Litzenburger, Laurent, *Une ville face au climat: Metz à la fin du moyen âge, 1400–1530* (Nancy: PUN Éditions Universitaires de Lorraine, 2015)
Lloyd, Terence Henry, *The English Wool Trade in the Middle Ages* (Cambridge: Cambridge University Press, 1977)
——, *The Movement of Wool Prices in Medieval England* (Cambridge: Cambridge University Press, 1973)
Ludlow, Francis, 'The Utility of the Irish Annals as a Source for the Reconstruction of Climate', 2 vols (unpublished doctoral thesis, Trinity College Dublin, 2010)
Lunt, William Edward, 'The Financial System of the Medieval Papacy in the Light of Recent Literature', *The Quarterly Journal of Economics*, 23.2 (1909), 251–95
——, 'Papal Taxation in England in the Reign of Edward I', *English Historical Review*, 30 (1915), 398–417
——, 'A Papal Tenth Levied in the British Isles from 1274 to 1280', *English Historical Review*, 32 (1917), 49–89
——, *Papal Revenues in the Middle Ages*, Records of Civilization, Sources and Studies, 19, 2 vols (New York: Columbia University Press, 1934)
MacArthur, William P., 'The Identification of Some Pestilences Recorded in the Irish Annals', *Irish Historical Studies*, 6 (1949), 179–81
——, comments on Shrewsbury's 'The Yellow Plague', *Journal of the History of Medicine and Allied Sciences*, 5 (1950), 214–15
MacDonald, Angus, *The Place-Names of West Lothian* (Edinburgh: Oliver and Boyd, 1941)
Macdonald, Angus, and Archibald Macdonald, *Clan Donald*, 3 vols (Inverness: Northern Counties, 1896–1904)
Mackintosh, Alexander, *Historical Memoirs of the House and Clan of Mackintosh and of the Clan Chattan* (London: Clay, 1880)
Mackie, John Duncan, ed., *Thomas Thomson's Memorial on the Old Extent*, Stair Society, 10.2 (Edinburgh: Stair Society, 1946)
Macklin, Mark G., Clive Bonsall, Fay M. Davies, and Mark R. Robinson, 'Human-Environment Interactions during the Holocene: New Data and Interpretations from the Oban Area, Argyll, Scotland', *The Holocene*, 10.1 (2000), 109–21
Maddicott, John Robert, 'Plague in Seventh-Century England', *Past and Present*, 156 (1997), 7–54
Maggs Bros, *The Art of Writing, 2800 BC to 1930 AD* (London: Maggs Bros, 1930)
Malloy, Kevin, and Derek Hall, 'Medieval Hunting and Wood Management in the Buzzart Dykes Landscape', *Environment and History*, 25.3 (2019), 365–90

——, and Derek Hall, 'Archaeological Excavations of the Medieval Royal Kincardine Landscape, Aberdeenshire, Scotland', *Medieval Archaeology*, 61.1 (2018), 157–76

Mann, Michael E., Zhihua Zhang, Scott Rutherford, Raymond S. Bradley, Malcolm K. Hughes, Drew Shindell, Caspar Ammann, Greg Faluvegi, and Fenbiao Ni, 'Global Signatures and Dynamical Origins of the Little Ice Age and Medieval Climate Anomaly', *Science*, 326 (2009), 1256–60

Mann, S., *Cattle, Cotton and Commerce. Life around Gatehouse of Fleet in 1792. A Reprint of the First Statistical Account of Scotland for the Parishes of Anworth and Girthon 1792* (Cambridge: Beechwood Publishing, 1993)

Mantovani, Adriano, 'Notes on the Development of the Concept of Zoonoses', *Historia Medicinae Veterinariae*, 26 (2001), 41–52

Mather, Anne, 'Geology, Soils, Climate and Vegetation', in *Scottish Life and Society. A Compendium of Scottish Ethnology, II: Farming and the Land*, ed. by Alexander Fenton and Kenneth Veitch (Edinburgh: Donald, 2011), pp. 63–84

Mayewski, Paul Andrew, Loren David Meeker, Sallie I. Whitlow, Mark Stephen Twickler, Margaret C. Morrison, Peter Bloomfield, Gerard Clark Bond, Richard Blane Alley, Anthony J. Gow, Debra A. Meese, Pieter Meiert Grootes, Michael Ram, Kendrick C. Taylor, and Mark A. Wumkes, 'Changes in Atmospheric Circulation and Ocean Ice Cover over the North Atlantic during the Last 41,000 years', *Science*, 263 (1994) 1747–51

——, Eelco E. Rohling, J. Curt Stager, Wibjörn Karlén, Kirk A. Maasch, David L. Meeker, Eric A. Meyerson, Francoise Gasse, Shirley van Kreveld, Karin Holmgren, Julia Lee-Thorp, Gunhild Rosqvist, Frank Rack, Michael Staubwasser, Ralph R. Schneider, and Eric G. Steig, 'Holocene Climate Variability', *Quaternary Research*, 62 (2004), 243–55

Mayhew, Nicholas J., 'Alexander III — A Silver Age? An Essay in Scottish Medieval History', in *Scotland in the Reign of Alexander III 1249–1286*, ed. by Norman A. Reid (Edinburgh: Donald, 1990), pp. 55–73

McCarthy, Dan, 'Chronological Synchronisation of the Irish Annals (last modified 14 December 2011)' <www.irish-annals.cs.tcd.ie> [accessed 7 July 2022]

McCollum, Andrea M., Yu Li, Kimberly Wilkins, Kevin L. Karem, Whitni B. Davidson, Christopher D. Paddock, Mary G. Reynolds, and Inger K. Damon, 'Poxvirus Viability and Signatures in Historical Relics', *Emerging Infectious Diseases*, 20 (2014), 177–84

McCormick, Michael, 'Gregory of Tours on Sixth-Century Plague and Other Epidemics', *Speculum*, 96 (2021), 38–96

McDonnell, J., *Inland Fisheries in Medieval Yorkshire 1066–1300*, Borthwick Papers, 60 (York: University of York, 1981)

McGill Library, 'Archival Collections and Manuscripts – Medieval European Manuscripts', <https://www.mcgill.ca/library/branches/rarebooks/special-collections/archival-collections-and-manuscripts> [accessed 21 September 2022]

McGladdery, Christine A., *James II*, 2nd edn (Edinburgh: Donald, 2015)

McKechnie, Hector, 'Early Land Valuations', *Juridical Review*, 70 (1930), 70–77

McKeown, Thomas, *The Origins of Human Disease* (Oxford: Blackwell, 1988)

McNeill, Peter G. B., and Hector L. MacQueen, with Anona May Lyons, *Atlas of Scottish History to 1707* (Edinburgh: Edinburgh University Press, 1996)

McNeill, William H., *Plagues and Peoples* (New York: Anchor, 1976)

Mead, Richard, *De Variolis et Morbillis Liber* (London: [n. pub.], 1747)

——, *A Discourse on the Smallpox and Measles* (London: [n. pub.], 1748)

Mileson, Stephen Anthony, *Parks in Medieval England*, Medieval History and Archaeology (Oxford: Oxford University Press, 2009)

——, 'The Sociology of Park Creation in Medieval England', in *The Medieval Park: New Perspectives*, ed. by Robert Liddiard (Oxford: Windgather, 2007), pp. 11–26

Mills, Coralie M., 'Historic Pine and Dendrochronology in Scotland', *Scottish Woodland History Discussion Group: Notes*, 13 (2008), 9–14

——, and Anne Crone, 'Dendrochronological Evidence of Scotland's Native Timber Resources Over the Last 1000 Years', *Scottish Forestry*, 66.1 (2012), 18–34

Monro, Alexander, *Observations on the Different Kinds of Smallpox* (Edinburgh: Constable, 1818)

Moore, James, *The History of the Smallpox* (London: Longman, 1815)

Mordechai, Lee, Merle Eisenberg, Timothy P. Newfield, Adam Izdebsky, Janet E. Kay, and Hendrik Poinar, 'The Justinianic Plague: An Inconsequential Pandemic?', *Proceedings of the National Academy of Sciences*, 116 (2019), 25, 546–54

Moreno Chamarro, Eduardo, Davide Zanchettin, Katja Lohmann, Jürg Luterbacher, and Johann H. Jungclaus, 'Winter Amplification of the European Little Ice Age Cooling by the Subpolar Gyre', *Scientific Reports*, 7.1 (2017), 9981–88, online at: <https://doi.org10.1038/s41598-017-07969-0>

Morgan, J. L., 'Economic Administration of Coupar Angus Abbey, 1440–1560', 3 vols (unpublished doctoral thesis, University of Glasgow, 1929)

Morris, Christopher D., Colleen E. Batey, and D. James Rackham, *Freswick Links, Caithness. Excavation and Survey of a Norse Settlement*, North Atlantic Biocultural Organisation and Highland Archaeology Monograph, 1 (Inverness: Highland Libraries, 1995)

Mühlemann, Barbara, Lasse Vinner, Ashot Margaryan, Helene Wilhelmson, Constanza de la Fuente Castro, Morten E. Allentoft, Peter de Barros Damgaard, Anders Johannes Hansen, Sofie Holtsmark Nielsen, Lisa Mariann Strand, Jan Bill, Alexandra Buzhilova, Tamara Pushkina, Ceri Falys, Valeri Khartanovich, Vyacheslav Moiseyev, Marie Louise Schjellerup Jørkov, Palle Østergaard Sørensen, Yvonne Magnusson, Ingrid Gustin, Hannes Schroeder, Gerd Sutte, Geoffrey L. Smith, Christian Drosten, Ron A. M. Fouchier, Derek J. Smith, Eske Willerslev, Terry C. Jones, and Martin Sikora, 'Diverse Variola Virus (Smallpox) Strains were Widespread in Northern Europe in the Viking Age', *Science*, 369 (2020), 391, online at: <https://doi.org/10.1126/science.aaw8977>

Munro, John H., 'Medieval Woollens: The Western European Woollen Industries and Their Struggles for International Markets, c. 1000–1500', in *The Cambridge History of Western Textiles*, ed. by David T. Jenkins, 2 vols (Cambridge: Cambridge University Press, 2003), I, pp. 228–324

——, 'Wool-Price Schedules and the Qualities of English Wools in the Later Middle Ages, c. 1270–1499', *Textile History*, 9 (1978), 118–69

Nash, Roderick, *Wilderness and the American Mind* (Newhaven: Yale University Press, 2001)

Nauwerck, Arnold, 'Der Lachsfang in der Kinzig: In biologischer Beleuchtung. Nach Akten des Fürstlich Fürstenbergischen Rentamtes Wolfach aus dem 18. Jahrhundert', *Die Ortenau: Zeitschrift des Historischen Vereins für Mittelbaden*, 66 (1986), 499–525, online at: <http://dl.ub.uni-freiburg.de/diglit/ortenau1986/0499>

Newfield, Timothy P., 'A Cattle Panzootic in Early Fourteenth-Century Europe', *Agricultural History Review*, 57 (2009), 155–90

——, 'Early Medieval Epizootics and Landscapes of Disease: The Origins and Triggers of European Livestock Pestilences, 400–1000 CE', in *Landscapes and Societies in Medieval Europe East of the Elbe: Interactions Between Environmental Settings and Cultural Transformations*, ed. by Sunhild Kleingärtner, Timothy P. Newfield, Sébastien Rossignol, and Donat Wehner, Papers in Medieval Studies, 23 (Toronto: Pontifical Institute of Medieval Studies, 2013), pp. 73–113

——, 'Human-Bovine Plagues in the Early Middle Ages', *Journal of Interdisciplinary History*, 46 (2015), 1–38

——, 'Malaria and Malaria-Like Disease in the Early Middle Ages', *Early Medieval Europe*, 25 (2017), 251–300

——, Ana T. Duggan, and Hendrik Poinar, eLetter response to Mühlemann and others, 'Diverse Variola Virus', *Science*, 369 (2020) online at: <https://www.academia.edu/44264499/_RE_Diverse_variola_virus_smallpox_strains_were_widespread_in_northern_Europe_in_the_Viking_Age_Science_369_2020_eLetter_https_science_sciencemag_org_content_369_6502_eaaw8977_tab_e_letters> [accessed 3 August 2022]

Neville, Cynthia J., *Land, Law and People in Medieval Scotland* (Edinburgh: Edinburgh University Press, 2012)

Nicholas, Ralph W., 'The Goddess Śītalā and Epidemic Smallpox in Bengal', *Journal of Asian Studies*, 41 (1981), 21–44

Nicholson, Lachlan, 'From the River Farrar to the Loire Valley. The MacDonald Lord of the Isles, the Scottish Crown, and International Diplomacy, 1428–1438', in *The Lordship of the Isles*, ed. by Richard D. Oram, Northern World, 68 (Leiden: Brill, 2014), pp. 88–100

Nie, Giselle de, 'The Spring, the Seed and the Tree: Gregory of Tours on the Wonders of Nature', *Journal of Medieval History*, 11.2 (1985), 89–135

Noble, Gordon, and Nicholas Evans, *The King in the North: The Pictish Kingdoms of Fortriu and Ce* (Edinburgh: Donald, 2019)

Ogurtsov, Maxim G., 'The Spörer Minimum Was Deep', *Advances in Space Research*, 64 (2019), 1112–16

Oldland, John, 'Cistercian Clothing and Its Production at Beaulieu Abbey, 1269–70', in *Medieval Clothing and Textiles IX*, ed. by Robin Netherton and Gale Owen-Crocker (Woodbridge: Boydell, 2013), pp. 73–96

Oppenheimer, Clive, 'Ice Core and Palaeoclimatic Evidence for the Timing and Nature of the Great Mid-13th Century Volcanic Eruption', *International Journal of Climatology*, 23 (2003), 417–26

Oram, Richard D., 'A Fit and Ample Endowment? The Balmerino Estate, 1228–1603', in *Life on the Edge: the Cistercian Abbey of Balmerino, Fife (Scotland)*, ed. by Terryl L. Kinder (Pontigny: Cîteaux, 2008), pp. 61–79

——, 'Royal and Lordly Residence in Scotland c. 1050 to c. 1250', *The Antiquaries Journal*, 88 (2008), 165–89

——, 'Innse Gall: Culture and Environment on a Norse Frontier in the Scottish Western Isles', in *The Norgesveldet in the Middle Ages*, ed. by Steinar Imsen, 'Norgesveldet' Occasional Papers, 2 (Trondheim: Tapir, 2010), pp. 125–48

——, 'Disease, Death and the Hereafter in Medieval Scotland', in *A History of Everyday Life in: Medieval Scotland 1000 to 1600*, ed. by Edward J. Cowan and Lizanne Henderson, A History of Everyday Life in Scotland, 1 (Edinburgh: Edinburgh University Press, 2011), pp. 196–225

——, *Domination and Lordship Scotland 1070–1230* (Edinburgh: Edinburgh University Press, 2011)

——, 'Social Inequality in the Supply and Use of Fuel in Scottish Towns c. 1750–1850', in *Environmental and Social Justice in the City: Historical Perspectives*, ed. by Geneviève Massard-Guilbaud and Richard Rodger (Banbury: White Horse, 2011), pp. 211–31

——, *Alexander II 1214–1249: King of Scots* (Edinburgh: Donald, 2012)

——, 'The Salt Industry in Medieval Scotland', *Studies in Medieval and Renaissance History*, 3rd ser., 9 (2012), 209–32

——, 'Estuarine Environments and Resource Exploitation in Eastern Scotland c. 1125–c. 1400: A Comparative Study of the Forth and Tay Estuaries', in *Landscapes or Seascapes?: The History of the Coastal Environment in the North Sea Area Reconsidered*, ed. by Erik Thoen, Guus J. Borger, Tim Soens, Adriaan M. J. de Kraker, Dries Tys, Lies Vervaert, and Henk J. T. Weerts, Comparative Rural History of the North Sea Area, 13 (Brepols: Turnhout, 2013), pp. 353–77

——, 'Arrested Development? Energy Crises, Fuel Supplies, and the Slow March to Modernity in Scotland, 1450–1850', in *Energy Transitions in History Global Cases of Continuity and Change*, ed. by Richard W. Unger, Rachel Carson Centre Perspectives, 2 (Munich: Rachel Carson Centre, 2013), 17–24

——, ed., *The Lordship of the Isles* (Leiden: Brill, 2014)

——, 'Introduction: A Celtic Dirk at Scotland's Back? The Lordship of the Isles in Mainstream Scottish Historiography since 1828', in *The Lordship of the Isles*, ed. by Richard D. Oram (Leiden: Brill, 2014), pp. 1–39

——, 'Between a Rock and a Hard Place: Climate, Weather and the Rise of the Lordship of the Isles', in *The Lordship of the Isles*, ed. by Richard D. Oram (Leiden: Brill, 2014), pp. 40–61

——, 'From "Golden Age" to Depression: Land Use and Environmental Change in the Medieval Earldom of Orkney', in *Northern Worlds: Landscapes, Interactions and*

Dynamics, ed. by Hans Christian Golløv, Studies in Archaeology & History, 22 (Odense: University Press of Southern Denmark, 2014), pp. 203–14

——, '"The Worst Disaster Suffered by the People of Scotland in Recorded History": Climate Change, Dearth and Pathogens in the Long Fourteenth Century', *Proceedings of the Society of Antiquaries of Scotland*, 144 (2015), 223–44

——, 'Fuel Transitions, Supply Crises and Climate Change in Lowland Scotland c. 1200–c. 1550', in *Sous le soleil: Systèmes et transitions énergétiques du Moyen Âge à nos jours. Homme et société*, ed. by Charles-François Mathis and Geneviève Massard-Guilbaud (Paris: Éditions de la Sorbonne, 2019), pp. 211–24

——, and Paul W. Adderley, 'Lordship and Environmental Change in Central Highland Scotland, c. 1300–c. 1400', *Journal of the North Atlantic*, 1 (2008), 74–84

——, and Paul W. Adderley, 'Lordship, Land and Environmental Change in West Highland and Hebridean Scotland c. 1300–c. 1450', in *Le Interazioni Fra Economia E Ambiente Biologico Nell'Europa Preindustriale Secc.XIII–XVIII = Economic and biological interactions in pre-industrial Europe, from the 13th to the 18th century: atti della Quarantunesima Settimana di studi, 26–30 aprile 2009*, ed. by Simonetta Cavaciocchi, Pubblicazioni, 41 (Florence: University of Florence Press, 2010), pp. 257–68

——, and Paul W. Adderley, 'Re Innse Gall: A Norse Colony in the Irish Sea and Hebrides?', in *The Norwegian Domination and the Norse World c. 1100–c. 1400*, ed. by Steinar Imsen (Trondheim: Rostra, 2010), pp. 125–48

Pajer, Petr, Jiri Dresler, Hana Kabíckova, Libor Písa, Pavel Aganov, Karel Fucik, Daniel Elleder, Tomas Hron, Vitezslav Kuzelka, Petr Velemínsky, Jana Klimentova, Alena Fucikova, Jaroslav Pejchal, Rita Hrabakova, Vladimir Benes, Tobias Rausch, Pavel Dundr, Alexander Pilin, Radomir Cabala, Martin Hubalek, Jan Stríbrny, Markus H. Antwerpen, and Hermann Meyer, 'Characterization of Two Historic Smallpox Specimens from a Czech Museum', *Viruses*, 9 (2017), 200, online at: <https://doi.org/10.3390/v9080200>

Parker Pearson, Mike, Niall Sharples, Jacqui Mulville, and Helen Smith, *Between Land and Sea: Excavations at Dun Vulan, South Uist*, Sheffield Environmental and Archaeological Research Campaign in the Hebrides, 3 (Sheffield: Sheffield Academic Press, 1999)

——, Helen Smith, Jacqui Mulville, and Mark Brennand, 'Cille Pheadair: the Life and Times of a Norse-period Farmstead c. 1000–1300', in *Land, Sea and Home: Proceedings of a Conference on Viking Period Settlement*, ed. by John Hines, Alan Lane, and Mark Redknap, Society for Medieval Archaeology Monograph Series, 20 (2004), pp. 235–54

——, Mark Brennand, Jacqui Mulville, and Helen Smith, eds, *Cille Pheadair: A Norse Farmstead and Pictish Burial Cairn in South Uist* (Oxford: Oxbow, 2018)

Parry, Martin Lewis, 'Secular Climatic Change and Marginal Agriculture', *Transactions of the Institute of British Geographers*, 64 (1975), 1–13

Parsons, David N., *The Vocabulary of English Place-Names (CEAFOR–COCK-PIT)* (Nottingham: English Place-Name Society, 2004)

Paton, Henry, *The Mackintosh Muniments, 1442–1820* (Edinburgh: Privately printed, 1903)

Paulet, Jean-Jacques, *Histoire de la petite vérole*, 2 vols (Paris: Ganeau, 1768)

——, *Recherches historiques et physiques sur les maladies épizootiques* (Paris: Ruault, 1775)

Paulus, Carolus, *Unpublished Pages Relating to The Manor and Parish of Ecclesall Including the Enclosure of the Common and Waste Lands There* (Sheffield: Northend, 1927)

Peck, Kayla M., Adam S. Lauring, and Christopher S. Sullivan, 'Complexities of Viral Mutation Rates', *Journal of Virology*, 92 (2018), e01021–17

Penman, Michael A., *David II, 1329–71* (Edinburgh: Donald, 2005)

——, *Robert the Bruce, King of the Scots* (Newhaven: Yale University Press, 2014)

Perlin, John, *A Forest Journey: The Role of Wood in the Development of Civilization* (Cambridge: Harvard University Press, 1989)

Perry, Robert T., Walter A. Orenstein, and Neal A. Halsey, 'The Clinical Significance of Measles: A Review', *Journal of Infectious Diseases*, 189 (2004), S4–S16

Pfister, Christian, Rudolf Brázdil, and Rüdiger Glaser, *Climatic Variability in Sixteenth-Century Europe and its Social Dimension* (Boston: Kluwer, 1999)

Pinches, Sylvia, 'From Common Rights to Cold Charity: Enclosure and Poor Allotments in the Eighteenth and Nineteenth Centuries', in *Medicine, Charity and Mutual Aid: The Consumption of Health and Welfare in Britain, c. 1550–1950*, ed. by Anne Borsay and Peter Shapely, Historical Urban Studies (Aldershot: Ashgate, 2007), pp. 35–53

Pirenne, Henri, *Economic and Social History of Medieval Europe* (London: Paul, Trench, Trubner & Company, 1936)

Pluskowski, Aleks, 'The Social Construction of Medieval Park Ecosystems: An Interdisciplinary Perspective', in *The Medieval Park: New Perspectives*, ed. by Robert Liddiard (Macclesfield: Windgather, 2007), pp. 63–78

Porchon, Antoine, *Nouveau traitte du pourpre, de la rougeole et petite verole* (Paris: Villery, 1688)

Porter, Ashleigh, Ana T. Duggan, Hendrik N. Poinar, and Edward C. Holmes, 'Comment: Characterization of Two Historic Smallpox Specimens from a Czech Museum', *Viruses*, 9 (2017), 276, online at: <https://doi.org/10.3390/v9100276>

Porter, Valerie, *Yesterday's Countryside* (Newton Abbot: David & Charles, 2000)

Powicke, F. Maurice, *The Loss of Normandy, 1189–1204: Studies in the Angevin Empire*, 2nd edn (Manchester: Manchester University Press, 1961)

Prestwich, Michael, *Edward I* (London: Yale University Press, 1997)

Proctor, Christopher A., Andy Baker, William L. Barnes, and Mabs Ann Gilmour, 'A Thousand Year Speleothem Proxy Record of North Atlantic Climate from Scotland', *Climate Dynamics*, 16 (2000), 815–20

——, Andy Baker, and William L. Barnes, 'A Three Thousand Year Record of North Atlantic Climate', *Climate Dynamics*, 19 (2002), 449–54

——, Andy Baker, and William L. Barnes, 'Northwest Scotland Stalagmite Data to 3600 BP' <https://www.ncdc.noaa.gov/paleo/study/5418> [accessed 22 August 2022]

Pummer, Reinhard, 'A Samaritan Manuscript in McGill Library', *Fontanus*, 5 (1992), 161–72

Pyne, Stephen J., *Fire — A Brief History* (London: British Museum, 2001)
Quelch, Peter R., 'Ancient Trees in Scotland', in *Scottish Woodland History*, ed. by T. Christopher Smout (Dalkeith: Scottish Cultural Press, 2002), pp. 23–38
Quinton, Eleanor Jane Powys, 'The Drapers and the Drapery Trade of Late Medieval London, *c.* 1300–*c.* 1500' (unpublished doctoral thesis, University of London, 2001)
Rackham, Oliver, *The History of the Countryside* (Oxford: Dent, 1986)
——, *Woodlands* (London: Collins, 2012)
——, 'Woodland and Wood-Pasture', in *Trees, Forested Landscapes and Grazing Animals: A European Perspective on Woodlands and Grazed Treescapes*, ed. by Ian D. Rotherham (London: Routledge, 2013), pp. 11–22
Rajerison, Minoarisoa, Marie Melocco, Voahangy Andrianaivoarimanana, Soloandry Rahajandraibe, Feno Rakotoarimanana, André Spiegel, Maherisoa Ratsitorahina, and Laurence Baril, 'Performance of Plague Rapid Diagnostic Test Compared to Bacteriology: A Retrospective Analysis of the Data Collected in Madagascar', *BMC Infectious Diseases*, 20 (2020), 90
Reid, Norman H., *Alexander III, 1249–1286: First Among Equals* (Edinburgh: Donald, 2019)
——, ed., *Scotland and the Reign of Alexander III, 1249–1286* (Edinburgh: Donald, 1990)
——, and Geoffrey Wallis Steuart Barrow, *Sheriffs of Scotland: An Interim List to c. 1306* (St Andrews: University of St Andrews Library, 2002)
Reid, Robert Corsane, 'De Veteripont', *Transactions of the Dumfriesshire and Galloway Natural History and Antiquarian Society (3rd Series)*, 33 (1955), 91–106
Reiske, Johann Jacob, *Miscellaneas aliquot observationes medicas ex Arabum monumentis* (Leiden: [n. pub.], 1746)
Robin, Christian Julien, 'The Peoples Beyond the Arabian Frontier in Late Antiquity: Recent Epigraphic Discoveries and Latest Advances', in *Inside and Out: Interactions between Rome and the People's on the Arabian and Egyptian Frontiers in Late Antiquity*, ed. by Jitse Dijkstra and Greg Fisher, Late Antique History and Religion, 8 (Leuven: Peeters, 2014), pp. 65–71
——, 'Himyar, Aksūm, and Arabia Deserta in Late Antiquity: The Epigraphic Evidence', in *Arabs and Empires before Islam*, ed. by Greg T. Fisher (Oxford: Oxford University Press, 2015), pp. 127–71
Rodriguez-Trelles, Francisco, Rosa Tarrío, and Francisco J. Ayala, 'Molecular Clocks: Whence and Whither?', in *Telling the Evolutionary Time: Molecular Clocks and the Fossil Record*, ed. by Phillip C. J. Donoghue and M. Paul Smith, Systematics Association Special Volume (London: Taylor & Francis, 2005), pp. 5–26
Rogers, James Edwin Thorold, *A History of Agriculture and Prices in England*, 8 vols (Oxford: Clarendon, 1902)
Rorke, Martin, 'Scottish Overseas Trade, 1275/86–1597' (unpublished doctoral thesis, University of Edinburgh, 2001)
——, 'English and Scottish Overseas Trade, 1300–1600', *The Economic History Review*, 59 (2006), 265–88

Ross, Alasdair, '"Harps of their Owne Sorte"? A Reassessment of Pictish Chordophone Depictions', *Cambrian Medieval Celtic Studies*, 36 (1998), 37–60

——, 'Men for All Seasons? The Strathbogie Earls of Atholl and the Wars of Independence, *c.* 1290–*c.* 1335: Part 1', *Northern Scotland*, 20 (2000), 1–30

——, 'Men for All Seasons? The Strathbogie Earls of Atholl and the Wars of Independence, *c.* 1290–*c.* 1335: Part 2', *Northern Scotland*, 21 (2001), 1–15

——, 'The Lords and Lordship of Glencarnie', in *The Exercise of Power in Medieval Scotland c. 1200–1500*, ed. by Steve Boardman and Alasdair Ross (Dublin: Four Courts, 2003), pp. 158–74

——, 'The Province of Moray c. 1000–1230', 2 vols (unpublished doctoral thesis, University of Aberdeen, 2003)

——, *Assessing the Impact of Past Grazing Regimes: Transhumance in the Forest of Strathaan, Banffshire* (Stirling: University of Stirling, 2004)

——, 'The Dabhach in Moray: A New Look at an Old Tub', in *Landscape and Environment in Dark Age Scotland*, ed. by Alex Woolf, St John's House Papers, 11 (St Andrews: University of St Andrews, 2006), pp. 57–74

——, 'Scottish Environmental History and the (Mis)use of Soums', *Agricultural History Review*, 54 (2006), 213–28

——, 'The Bannatyne Club and the Publication of Scottish Ecclesiastical Cartularies', *Scottish Historical Review*, 85.2 (2006), 202–33

——, 'The Identity of the "Prisoner of Roxburgh": Malcolm Son of Alexander or Malcolm Macheth', in *Fil súil nglais = A Grey Eye Looks Back: A Festschrift in Honour of Colm O Baoill*, ed. by Sharon Arbuthnot and Kaarina Hollo (Ceann Drochaid: Clann Tuirc, 2007), pp. 269–83

——, 'Moray, Ulster and the MacWilliams', in *The World of the Galloglass: Kings, Warlords and Warriors in Ireland and Scotland, 1200–1600*, ed. by Seán Duffy (Dublin: Four Courts, 2007), pp. 24–44

——, *Literature Review of the History of Grassland Management in Scotland*, Scottish National Heritage Commissioned Report, 313 (Edinburgh: Scottish Natural Heritage, 2008)

——, *The Kings of Alba: c. 1000–c. 1130* (Edinburgh: Donald, 2011)

——, 'Improvement on the Grant Estates in Strathspey in the Eighteenth Century: Theory, Practice and Failure?', in *Custom, Improvement and the Landscape in Early Modern Britain*, ed. by Richard W. Hoyle (Farnham: Ashgate, 2011), pp. 289–311

——, 'Two 1585 × 1612 Surveys of Vernacular Buildings and Tree Usage in the Lordship of Strathavon, Banffshire', *Miscellany of the Scottish History Society*, 14 (2012), 1–52

——, 'Ghille Chattan Mhor and Clann Mhic an Tòisich Lands in the MacDonald Lordship of Lochaber', in *The Lordship of the Isles*, ed. by Richard D. Oram, Northern Worlds, 68 (Leiden: Brill, 2014), pp. 101–22

Rotherham, Ian D., 'Peat Cutters and Their Landscapes: Fundamental Change in a Fragile Environment', in *Peatland Ecology and Archaeology: Management of a Cultural Landscape: The Proceedings and Invited Papers from the 1997 Sheffield Conference of the Landscape Conservation Forum*, ed. by Ian D. Rotherham, Landscape Archaeology and Ecology, 4 (Sheffield: Wildtrack, 1999), pp. 28–51

——, 'Fuel and Landscape — Exploitation, Environment, Crisis and Continuum', *Landscape Archaeology and Ecology*, 5 (2005), 65–81

——, 'The Historical Ecology of Medieval Parks and the Implications for Conservation', in *The Medieval Park: New Perspectives*, ed. by Robert Liddiard (Oxford: Windgather, 2007), pp. 79–96

——, 'The Importance of Cultural Severance in Landscape Ecology Research', in *Landscape Ecology Research Trends*, ed. by Arthur Dupont and Hugo Jacobs (Hauppauge: Nova Science, 2008), pp. 71–87

——, *Peat and Peat Cutting* (Oxford: Shire, 2009)

——, 'War & Peat: Exploring Interactions Between People, Human Conflict, Peatlands, and Ecology', in *War & Peat: The Remarkable Impacts of Conflicts on Peatlands and of Peatlands on Conflicts. A Military Heritage of Moors, Heaths, Bogs and Fens*, ed. by Ian D. Rotherham and Christine Handley, Landscape Archaeology and Ecology, 10 (Sheffield: Wildtrack, 2013), pp. 7–44

——, *The Lost Fens* (Stroud: History, 2013)

——, 'Reinterpreting Wooded Landscapes, Shadow Woods and the Impacts of Grazing', in *Trees, Forested Landscapes and Grazing Animals: A European Perspective on Woodlands and Grazed Treescapes*, ed. by Ian D. Rotherham (London: Routledge, 2013), pp. 72–86

——, 'History and Heritage in the Bog — Examples from Cumbria and the Surrounding Areas', in *History and Heritage of the Bogs and Peatlands of Cumbria and Surrounding Areas*, ed. by Ian D. Rotherham, Landscape Archaeology and Ecology, 14 (Sheffield: Wildtrack, 2021), pp. 3–42

——, 'The History of Domestic Peat Fuel Exploitation in Relation to Carbon and Climate Change', in *In The Bog*, ed. by Ian D. Rotherham and Christine Handley (Sheffield: Wildtrack, 2022), in press

——, and Dave Egan, 'The Economics of Fuel Wood, Charcoal and Coal: An Interpretation of Coppice Management of British Woodlands', in *History and Sustainability*, ed. by Mauro Agnoletti, Marco Armiero, Stefania Barca, and Gabriela Corona (European Society for Environmental History, 2005), pp. 100–04

——, Dave Egan, and Paul A. Ardron, 'Fuel Economy and the Uplands: The Effects of Peat and Turf Utilisation on Upland Landscapes', in *Society, Landscape and Environment in Upland Britain*, ed. by Ian D. Whyte and Angus J. L. Winchester, Society for Landscape Studies Supplementary Series, 2 (Liverpool: Society for Landscape Studies, 2004), pp. 99–109

Ruffer, M. Armand, 'An Eruption Resembling that of Variola in the Skin of a Mummy of the Twentieth Dynasty (1200–1100 BC)', *Journal of Pathology and Bacteriology*, 15 (1911), 32–34

——, 'Pathological Notes on the Royal Mummies of the Cairo Museum', *Mitteilungen zur Geschichte der Medizin und der Naturwissenschaften*, 13 (1914), 239–48

Russell, Josiah C., 'That Earlier Plague', *Demography*, 5 (1968), 174–84

Russell, Paul, Peter McClure, and David Rollason, 'Celtic Names', in *The Durham Liber Vitae*, ed. by David W. Rollason and Lynda Rollason, with Elizabeth Briggs, 3 vols (London: British Library, 2007), II, pp. 35–43

Ryder, Michael Lawson, 'Medieval Sheep and Wool Types', *Agricultural History Review*, 32 (1984), 14–28

Rydval, Miloš, Neil J. Loader, Björn E. Gunnarson, Daniel L. Druckenbrod, Hans W. Linderholm, Steven G. Moreton, Cheryl V. Wood, and Rob Wilson, 'Reconstructing 800 Years of Summer Temperatures in Scotland from Tree Rings', *Climate Dynamics*, 49 (2017), 2951–74

Saldiva, Paulo Hilário Nascimento, and Simone Georges El Khouri Miraglia, 'Health Effects of Cookstove Emissions', *Energy for Sustainable Development*, 8.3 (2004), 13–19

Sarris, Peter, 'The Justinianic Plague: Origins and Effects', *Continuity and Change*, 17 (2002), 169–82

Saumaise, Claude, *De annis climactericis et antiqua astrologia diatribae* (Leiden: Elzevier, 1648)

Schnurrer, Friedrich, *Chronik der seuchen, I: Vom Anfang der Geschichte bis in die Mitte des fünfzehnten Jahrhunderts* (Tübingen: Osiander, 1823)

Schreg, Rainer, 'Ecological Approaches in Medieval Rural Archaeology', *European Journal of Archaeology*, 17.1 (2014), 83–119

Schwarz, Klaus, 'Der Weserlachs und die bremischen Dienstboten: Zur Geschichte des Fischverbrauchs in Norddeutschland', *Bremisches Jahrbuch*, 74/75 (1995/96), 134–73, online at: <https://brema.suub.uni-bremen.de/periodical/pageview/53037> [accessed 29 June 2022]

——, 'Nochmals: Der Lachs und die Dienstboten', *Bremisches Jahrbuch*, 77 (1998), 277–83, online at: <https://brema.suub.uni-bremen.de/periodical/pageview/53867> [accessed 29 June 2022]

Sennert, Daniel, *De Febribus Libri IV* (Paris: [n. pub.], 1633)

Serjeantson, Dale, *Farming and Fishing in the Outer Hebrides* AD *600 to 1700: The Udal, North Uist* (Southampton: Highfield, 2013)

Sharples, Niall, with Julie Bond, J. Cartledge, A. Clarke, S. College, I. Dennis, Rowena Gale, M. Hamilton, Claire Ingrem, Alan Lane, J. Light, P. MacDonald, P. Marshall, Karen Milek, Julie Mulville, A. Smith, H. Smith, and T. Young, *A Norse Farmstead in the Outer Hebrides: Excavations at Mound 3, Bornais, South Uist*, Cardiff Studies in Archaeology (Oxford: Oxbow, 2005)

Shaw, Alexander Mackintosh, *Historical Memoirs of the House and Clan of Mackintosh* (London: Clay, 1880)

Shaw-Lefevre, George, *English Commons and Forests* (London: Cassell, 1894)

Shepherd, Nan, *The Living Mountain: A Celebration of the Cairngorm Mountains of Scotland* (Edinburgh: Canongate, 2011)

——, *In the Cairngorms* (Plymouth: Galileo, 2018)

Shrewsbury, John F. D., 'The Yellow Plague', *Journal of the History of Medicine and Allied Sciences*, 4 (1949), 5–47

Sigl, Michael, Mai Winstrup, Joseph R. McConnell, Kees C. Welten, Gill Plunkett, Francis Ludlow, Ulf Büntgen, Marc William Caffee, Nathan J. Chellman, Dorthe Dahl-Jensen, H. Fischer, Sepp Kipfstuhl, Conor Kostick, Olivia J. Maselli, Florian Mekhaldi, Robert Mulvaney, Raimund Muscheler, Daniel R. Pasteris, Jonathan R. Pilcher, Matthew W. Salzer, Simon Schüpbach, Jørgen Peder Steffensen, Bo Møllesøe Vinther, and Thomas E. Woodruff, 'Timing and Climate Forcing of Volcanic Eruptions for the Past 2,500 Years', *Nature*, 523 (2015), 543–49

Slavin, Phillip, 'The Great Bovine Pestilence and its Economic and Environmental Consequences in England and Wales, 1318–50', *Economic History Review*, 65.4 (2012), 1239–66

——, 'Epizootic Landscapes: Sheep Scab and Regional Environment in England in 1279–1280', *Landscapes*, 17 (2016), 156–70

——, *Experiencing Famine in Fourteenth-Century Britain*, Environmental Histories of the North Atlandic World, 4 (Turnhout: Brepols, 2019)

——, 'Mites and Merchants: The Crisis of English Wool and Textile Trade Revisited, *c.* 1275–1330', *Economic History Review*, 73.4 (2020), 885–913 online at: <https://doi.org/10.1111/ehr.12969>

Smith, Albert Hugh, *English Place-Name Elements*, English Place-Name Society, 25–26, 2 vols (Cambridge: Cambridge University Press, 1956)

Smith, Andrew, 'The Kelso Abbey Cartulary: Context, Production and Forgery' (unpublished doctoral thesis, University of Glasgow, 2011)

Smith, Ashby, 'Preface', in *The Miscellaneous Works of the Late Robert Willan M.D.* (London: Cadell, 1821), pp. v–xxvii

Smith, Catherine, 'Conclusions: The Environment of Medieval Perth', in *Perth High Street Archaeological Excavations 1975–1977, Fascicule 4: Living and Working in a Medieval Scottish Burgh, Environmental Remains and Miscellaneous Finds*, ed. by George W. I. Hodgson and Arthur MacGregor (Perth: Tayside and Fife Archaeological Committee, 2011), pp. 81–94

Smith, Dennis, ed., *The Third Statistical Account of Scotland, XXIX: The County of Kincardine* (Edinburgh: Scottish Academic Press, 1951)

Smith, Grafton Elliot, and Mathaf al-Miṣrī, *The Royal Mummies*, Catalogue général des antiquités égyptiennes du Musée du Caire (Cairo: L'institut français d'archéologie orientale, 1912)

Smithson, Chad, Jacob Imbery, and Chris Upton, 'Re-Assembly and Analysis of an Ancient Variola Virus Genome', *Viruses*, 9 (2017), 253

Smout, T. Christopher, and Mairi Stewart, *The Firth of Forth: An Environmental History* (Edinburgh: Birlinn, 2012)

Smyth, Alfred P., 'The Earliest Irish Annals: Their First Contemporary Entries, and the Earliest Centres of Recording', *Proceedings of the Royal Irish Academy*, 72 (1972), 1–48

——, *Warlords and Holy Men: Scotland AD 80–1000*, New History of Scotland, 1 (London: Edward Arnold, 1984)

Spyrou, Maria A., Kirsten I. Bos, Alexander Herbig, and Johannes Krause, 'Ancient Pathogen Genomics as an Emerging Tool for Infectious Disease Research', *Nature Reviews Genetics*, 20 (2019), 323–40

Stamp, Laurence Dudley, and William George Hoskins, *The Common Lands of England & Wales* (London: Collins, 1963)

Stathakopoulos, Dionysios Ch., *Famine and Pestilence in the Late Roman and Early Byzantine Empire: A Systematic Survey of Subsistence Crises and Epidemics*, Birmingham Byzantine and Ottomon Monographs, 9 (Farnham: Ashgate, 2004)

Stein, Claudia, '"Getting" the Pox: Reflections by a Historian on How to Write the History of Early Modern Disease', *Nordic Journal of Science and Technology Studies*, 2 (2014), 53–60

Stenseth, Nils Christian, Bakyt B. Atshabar, Mike Begon, Steven R. Belmain, Eric Bertherat, Elisabeth Cariel, Kenneth L. Gage, Herwig Leirs, and Lila Rahalison, 'Plague: Past, Present, and Future', *PLoS Medicine*, 5 (2008), e3

Stevenson, Alexander, 'Taxation in Medieval Scotland', in *Atlas of Scottish History to 1707*, ed. by Peter G. B. McNeill and Hector L. MacQueen, with Anona May Lyons (Edinburgh: Edinburgh University Press, 1996), pp. 298–305

——, 'Trade between Scotland and the Low Countries in the Later Middle Ages' (unpublished doctoral thesis, University of Aberdeen, 1982)

Stevenson, Wendy B., 'The Monastic Presence: Berwick in the Twelfth and Thirteenth Centuries', in *The Scottish Medieval Town*, ed. by Michael Lynch, Michael Spearman, and Geoffrey Stell (Edinburgh: Donald, 1988), pp. 99–115

The Stockland Turbaries. Information Leaflet (Stockland Parish Council: Devon, [n. d.])

Stone, A., 'Heaths, Commons and Wastes: An Investigation into the Character, Management and Perceptions of Heathland Landscapes in the Medieval and Post-Medieval Periods, with Particular References to the Counties of Norfolk, Suffolk, Essex, and Hertfordshire' (unpublished doctoral thesis, University of East Anglia, 2017)

Stone, David, 'The Impact of Drought in Early Fourteenth-Century England', *Economic History Review*, 67.2 (2014), 435–62

Stothers, Richard B., 'Climatic and Demographic Consequences of the Massive Volcanic Eruption of 1258', *Climatic Change*, 45 (2000), 361–74

Stringer, Keith J., *Earl David of Huntingdon: A Study in Anglo-Scottish History* (Edinburgh: Edinburgh University Press, 1985)

Stuart, John, 'The Erroll Papers', in *The Miscellany of the Spalding Club*, ed. by John Stuart, 5 vols (Aberdeen: Spalding Club, 1842), II, 209–350

Sydenham, Thomas, *Observationes medicae circa acutorum historiam et curationem* (London: Kettilby, 1685)

Sykes, Naomi, 'Animal Bones and Animal Parks', in *The Medieval Park: New Perspectives*, ed. by Robert Liddiard (Oxford: Windgather, 2007), pp. 49–62

Szabo, Vicki Ellen, *Monstrous Fishes and the Mead-Dark Sea: Whaling in the Medieval North Atlantic*, Northern World, 35 (Leiden: Brill, 2008)

Taylor, Christopher C., 'Problems and Possibilities', in *Medieval Fish, Fisheries and Fishponds in England*, ed. by Michael Aston, B.A.R. British Series, 182 (Oxford: B.A.R., 1988), pp. 465–73

Thibault, Max, and Anne-Françoise. Garçon, 'Un problème d'écohistoire: le saumon dans les contrats de louage, une origine médiévale?', in *Pêche et pisciculture en eau douce: la rivière et l'étang au Moyen Âge, Actes des 1res Rencontres Internationales de Liessies, 27–29 avril 1998*, ed. by Paul Benoît, Frédéric Loridant and Olivier Mattéoni, Actes des Rencontres internationales de Liessies, 1 [unpaginated booklet and CD-Rom] (Lille: Conseil Général du Nord, 2004), 23 pp.

Thomas, Sarah, 'Bishops, Priests, Monks and Their Patrons. The Lords of the Isles and the Church', in *The Lordship of the Isles*, ed. by Richard D. Oram, Northern World, 68 (Leiden: Brill, 2014)

Tipping, Richard, 'Climatic Variability and "Marginal" Settlement in Upland British Landscapes: A Re-Evaluation', in *Landscapes*, 3 (2002), 10–29

——, *Bowmont: An Environmental History of the Bowmont Valley and the Northern Cheviot Hills, 10,000 BC — AD 2000* (Edinburgh: The Society of Antiquaries of Scotland, 2010)

——, and James Adams, 'Structure, Composition and Significance of Medieval Storm Beach Ridges at Caerlaverock, Dumfries and Galloway', *Scottish Journal of Geology*, 43.2 (2007), 115–23

Tucker, Jonathan B., *Scourge: The Once and Future Threat of Smallpox* (New York: Grove, 2001)

Turner, Michael, ed., *Home Office Acreage Returns HO67, List and Analysis, I: Bedfordshire — Isle of Wight, 1801*, List & Index Society, 189 (London: Swift, 1982)

Tytler, Patrick Fraser, *The History of Scotland*, 9 vols (Edinburgh: Tait, 1828–1843)

Van Dam, Petra J. E. M., 'Fish for Feast and Fast: Fish Consumption in the Netherlands in the Late Middle Ages', in *Beyond the Catch: Fisheries of the North Atlantic, the North Sea and the Baltic, 900–1850*, ed. by Louis Sicking and Darlene Abreu-Ferreira, Northern World, 41 (Leiden: Brill, 2009), pp. 321–27

Van Engelen, Aryan F. V., Jan Buisman, and Folkert Ijnsen, 'A Millennium of Weather, Winds and Water in the Low Countries', in *History and Climate: Memories of the Future?*, ed. by Phil D. Jones, Astrid E. J. Ogilvie, Trevor D. Davies, and Keith R. Briffa (New York: Kluwer, 2001), pp. 101–24

Van Ess, Josef, *Der Fehltritt des Gelehrten: die 'Pest von Emmaus' und ihre theologischen Nachspiele*, Supplemente zu den Schriften der Heidelberger Akademie der Wissenschaften, Philosophisch-Historische Klasse, 13 (Heidelberg: Winter, 2001)

Verhulst, Adriaan, 'De inlandse wol in de textielnijverheid van de Nederlanden van de 12e tot de 17e eeuw: produktie, handel en verwerking', *Bijdragen en Mededelingen betreffende de Geschiedenis der Nederlanden*, 85 (1970), 6–18

Vetter, Johannes, *Die Schiffart, Flötzerei und Fischerei auf dem Oberrhein* (Karlsruhe: Braun, 1864)

Vinther, Bo Møllesøe, Stephanie J. Johnsen, Kevin K. Andersen, Henrik B. Clausen, and Aksel Walløe Hansen, 'NAO Signal Recorded in the Stable Isotopes of the Greenland Icecores', *Geophysical Research Letters*, 30 (2003), 1387

Voltaire, *Histoire de l'empire de Russie sous Pierre le Grand* (Lyon: [n. pub.], 1761)

Warde, Paul, *Energy Consumption in England & Wales 1560–2000* (Rome: Consiglio Nazionale delle ricerche, Instituto di Studi sulle Società del Mediterraneo, 2007)

——, and Tom Williamson, 'Fuel Supply and Agriculture in Post-Medieval England', *Agricultural History Review*, 62.1 (2014), 61–82

Watt, Donald Elmslie Robertson, 'Bagimond di Vezza and his "Roll"', *The Scottish Historical Review*, 80.1 (2001), 1–23

——, and Athol Laverick Murray, eds, *Fasti Ecclesiae Scoticanae Medii Aevi Ad Annum 1638*, rev. edn (Edinburgh: Scottish Record Society, 2003)

Webb, Nigel R., *Heathlands* (London: Collins, 1986)

——, 'The Traditional Management of European Heathlands', *Journal of Applied Ecology*, 35 (1998), 987–90

Wellard, Gordon, *200 Years of Frimley's History: The Story of the Frimley Fuel Allotments Charity and Pine Ridge Golf Centre* (Camberley: Pine Ridge Golf Centre and Frimley Fuel Allotments Charity, 1995)

Werlhof, Paul Gottlieb, *Disquisitio medica et philologica de variolis et anthracibus ubi de utriusque affectus antiquitatibus signis differentiis medelis disserit* (Hanover: Förster, 1735)

Wertheim, Joel O., and Sergei L. Kosakovsky Pond, 'Purifying Selection Can Obscure the Ancient Age of Viral Lineages', *Molecular Biology and Evolution*, 28 (2011), 3355–65

White, Lynn, Jr, 'The Historical Roots of Our Ecologic Crisis', *Science*, 155, no. 3767 (1967), 1203–07

White, William, *History, Gazetteer, and Directory of Suffolk, and the Towns Near Its Borders* (Sheffield: Leader, 1844)

Whyte, Ian D., 'Rural Society and Economy', in *Scotland: The Making and Unmaking of the Nation c. 1100–1707*, ed. by Robert Harris and Alan R. MacDonald, 2 vols (Dundee: Dundee University Press, 2006), I, 158–73

Wilks, Mark, *Historical Sketches of the South of India*, 3 vols (London: Longman, 1817)

Willan, Robert, 'An Inquiry into the Antiquity of the Smallpox, Measles, and Scarlet Fever', in *The Miscellaneous Words of the Late Robert Willan M.D.*, ed. by Ashby Smith (London: Cadell, 1821), pp. 1–115

Williams, David Henry, *The Cistercians in the Early Middle Ages* (Leominster: Gracewing, 1998)

Williamson, May G., 'The Non-Celtic Place-Names of the Scottish Border Counties' (unpublished doctoral thesis, University of Edinburgh, 1942), online at: <https://spns.org.uk/resources/the-non-celtic-place-names-of-the-scottish-border-counties-may-g-williamson>

Willis, Thomas, 'De febribus', in his *Opera medica & physica*, 2 vols (Geneva: De Tournes, 1680), I, pp. 65–222

Wilson, James, and Archibald Campbell Lawrie, 'Charter of the Abbot and Convent of Cupar, 1220', *Scottish Historical Review*, 8 (1910–1911), 172–77

——, 'Original Charters of the Abbey of Cupar, 1219–1448', *Scottish Historical Review*, 10 (1912–1913), 272–86

Wilson, Rob, Neil J. Loader, Miloš Rydval, H. Patton, A. Frith, Coralie Mary Mills, Anne Crone, C. Edwards, Lars-Åke Larsson, and B. E. Gunnarson, 'Reconstructing Holocene Climate from Tree Rings: The Potential for a Long Chronology from the Scottish Highlands', *The Holocene*, 22.1 (2011), 3–11

Winchester, Angus J. L., *The Harvest of the Hills — Rural Life in Northern England and the Scottish Borders, 1400–1700* (Edinburgh: Edinburgh University Press, 2000)

——, 'Shielings and Common Pastures', in *Northern England and Southern Scotland in the Central Middle Ages*, ed. by Keith John Stringer and Angus J. L. Winchester (Woodbridge: Boydell, 2017), pp. 273–97

Winstead, Karen A., 'Vulfolaic the Stylite: Orientalism and Performing Holiness in Gregory's Histories', in *East of West: Cross-Cultural Performance and the Staging of Difference*, ed. by Claire Sponsler and Xiaomei Chen (New York: Palgrave, 2000), pp. 63–74

Withington, Lothrop, ed., *Elizabethan England* (London: Scott, 1899)

Wohlsein, Peter, and Jeremiah Saliki, 'Rinderpest and Peste des Petits Ruminants — the Diseases: Clinical Signs and Pathology', in *Rinderpest and Peste des Petits Ruminants: Virus Plagues of Large and Small Ruminants*, ed. by Thomas Barrett, Paul-Pierre Pastoret, and William P. Taylor (London: Academic, 2006), pp. 68–85

Wolter, Christian, 'Historic Catches, Abundance, and Decline of Atlantic Salmon Salmo salar in the River Elbe', *Aquatic Sciences*, 77.3 (2015), 367–80

Woodville, William, *The History of Inoculation of the Smallpox in Great Britain*, 2 vols (London: [n. pub.], 1796)

Woolf, Alex, *From Pictland to Alba, 789–1070*, New Edinburgh History of Scotland, 2 (Edinburgh: Edinburgh University Press, 2007)

Wright, Lawrence, *Home Fires Burning: The History of Domestic Heating and Cooking* (London: Routledge, 1964)

Young, Arthur, *General View of the Agriculture of the County of Norfolk* (London: The Board of Agriculture, 1804)

Zehender, Gianguglielmo, Alessia Lai, Carla Veo, Annalisa Bergna, Massimo Ciccozzi, and Massimo Galli, 'Bayesian Reconstruction of the Evolutionary History and Cross-Species Transition of Variola Virus and Orthopoxviruses', *Journal of Medical Virology*, 90 (2018), 1134–41

Zinsser, Hans, *Rats, Lice, and History* (London: Routledge, 1935)

Zylberberg, David, 'Plants and Fossils: Household Fuel Consumption in Hampshire and the West Riding of Yorkshire 1750–1830' (unpublished doctoral thesis, York University Toronto, 2014)

Index

Environmental Histories of the North Atlantic World

Title in Series

Thomas Finan, *Landscape and History on the Medieval Irish Frontier: The King's Cantreds in the Thirteenth Century* (2016)

Environment, Colonization, and the Baltic Crusader States: Terra Sacra I, ed. by Aleksander Pluskowski (2019)

Ecologies of Crusading, Colonization, and Religious Conversion in the Medieval Baltic: Terra Sacra II, ed. by Aleksander Pluskowski (2019)

Philip Slavin, *Experiencing Famine in Fourteenth-Century Britain* (2019)